T0331195

Materials and Devices for
Smart Systems III

MATERIALS RESEARCH SOCIETY
SYMPOSIUM PROCEEDINGS VOLUME 1129

Materials and Devices for Smart Systems III

Symposium held December 1–4, 2008, Boston, Massachusetts, U.S.A.

EDITORS:

Ji Su (acting Editor)
NASA Langley Research Center
Hampton, Virginia, U.S.A.

Li-Peng Wang
TricornTech Corporation
San Jose BioCenter
San Jose, California, U.S.A.

Yasubumi Furuya
Hirosaki University
Hirosaki, Japan

Susan Trolier-McKinstry
The Pennsylvania State University
University Park, Pennsylvania, U.S.A.

Jinsong Leng
Harbin Institute of Technology
Harbin, China

Materials Research Society
Warrendale, Pennsylvania

CAMBRIDGE
UNIVERSITY PRESS

University Printing House, Cambridge CB2 8BS, United Kingdom

One Liberty Plaza, 20th Floor, New York, NY 10006, USA

477 Williamstown Road, Port Melbourne, VIC 3207, Australia

314-321, 3rd Floor, Plot 3, Splendor Forum, Jasola District Centre, New Delhi - 110025, India

79 Anson Road, #06-04/06, Singapore 079906

Cambridge University Press is part of the University of Cambridge.

It furthers the University's mission by disseminating knowledge in the pursuit of education, learning and research at the highest international levels of excellence.

www.cambridge.org
Information on this title: www.cambridge.org/9781605111018

Materials Research Society
506 Keystone Drive, Warrendale, PA 15086
http://www.mrs.org

© Materials Research Society 2009

First published 2009
First paperback edition 2012

Single article reprints from this publication are available through University Microfilms Inc., 300 North Zeeb Road, Ann Arbor, MI 48106

CODEN: MRSPDH

A catalogue record for this publication is available from the British Library

ISBN 978-1-605-11101-8 Hardback
ISBN 978-1-107-40841-8 Paperback

This work was supported in part by the Army Research Office under Grant Number W911NF-09-1-0053. The views, opinions, and/or findings contained in this report are those of the author(s) and should not be construed as an official Department of the Army position, policy, or decision, unless so designated by other documentation.

CONTENTS

*Invited Paper

NANO-MATERIALS AND NANO-FABRICATION

vi

SHAPE MEMORY ALLOYS AND
MAGNETO-MATERIALS

*Invited Paper

*Invited Paper

*Invited Paper

*Invited Paper

PREFACE

This volume is a record of Symposium V, "Materials, Devices, and Characterization for Smart Systems," held December 1–4 at the 2008 MRS Fall Meeting in Boston, Massachusetts. This symposium is the third MRS symposium on the rapidly developing research field of materials and devices for smart systems; the successful first and second were held by the Materials Research Society at the 2003 Fall Meeting and the 2005 Fall Meeting, respectively. These symposia have functioned as a bridge to connect new development and achievements in the areas of smart materials, sensing and actuating devices, and intelligent/smart systems. They have also provided a public forum for communication and have enhanced the cooperation among the researchers from related disciplines to build an interdisciplinary research field and community.

Smart/intelligent systems have been recognized as one of the primary and critical technologies for the present and the future. They are demanded in every facet of applications: from everyday life to space exploration missions, from civilian products to military needs, from robots to information technology (IT)/ communication technology, etc. Smart devices are fundamental components to realize smart systems. Smart materials are the critical foundation for high-performance smart devices. Materials, devices, and systems cannot be separated. The right way to achieve significant progress in this field is to conduct research as a coherent interdisciplinary field.

The organizers hope this effort will keep inspiring future cooperative and interdisciplinary research and promoting applications of Smart Materials And Related Technologies (SMART).

Ji Su (acting Editor)
Li-Peng Wang
Yasubumi Furuya
Susan Trolier-McKinstry
Jinsong Leng

February 2009

ACKNOWLEDGMENTS

The organizers would like to take this opportunity to thank the U.S. Army Research Office (ARO), and the Hirosaki University, Japan, for funding the symposium. The organizers would also like to thank the following people for their help with the manuscript peer-review process and/or for serving as symposium technical session chairs, and/or for presenting invited presentations. We are grateful to them for volunteering their time and expertise, which contributed significantly to the success of the symposium and to the quality of the proceedings.

Hiroshi Asanuma (Japan)
Jayasimha Atulasimha (USA)
Frederic Dumas-Bouchiat (France)
Shanyi Du (China)
Kwang Kim (USA)
Qing Ma (USA)
Esashi Masayoshi (Japan)
Shuichi Miyazaki (Japan)
Paul Mulvaney (Australia)
Paul Muralt (Switzerland)
Qibing Pei (USA)
Ramamoorthy Ramesh (USA)
Koukou Suu (Japan)
Ichiro Takeuchi (USA)
Koichi Tsuchiya (Japan)
Baoxiang Wang (Norway)

The organizers are grateful to the 2008 MRS Fall Meeting Chairs, especially Dr. Young-Chang Joo for his encouragement and support, and to the MRS staff for their help in preparing this proceedings volume.

MATERIALS RESEARCH SOCIETY SYMPOSIUM PROCEEDINGS

MATERIALS RESEARCH SOCIETY SYMPOSIUM PROCEEDINGS

Prior Materials Research Society Symposium Proceedings available by contacting Materials Research Society

Ferroelectrics and Piezoelectrics

Mater. Res. Soc. Symp. Proc. Vol. 1129 © 2009 Materials Research Society   1129-V10-01-C08-01

Ferroelectric Random Access Memory as a Non-Volatile Cache Solution in a Multimedia Storage System

Dong Jin Jung and Kinam Kim
Memory Business Division, Samsung Electronics Co. Ltd. San #16, Banwol-Dong, Hwasung-City, Gyunggi-Do, S. Korea, 445-701

ABSTRACT

We demonstrate that ferroelectric memory is very eligible to become a non-volatile cache solution, in particular, in a multimedia storage system such as solid-state disk. It could provide benefits both of performance and of reliability. In performance, a FRAM cache allows us to rid overhead of power-off recovery. Random WRITE performance has been improved by 250%. In assertion of endurance, we investigate acceleration factors to evaluate cycle-to-failure of the ferroelectric memory both in device-level and in capacitor-level. What has been found is that ferroelectric memory cells have 6.0×10^{14} of the cycle-to-failure at the operational condition of 85 °C and 2.0V. This cycle-to-failure is well above lifetime READ/WRITE cycles of 9.5×10^{13} in such system. From 2-dimensional stress simulation, it has also been concluded that the number of dummy cells plays a critical role in qualifying the high temperature life tests.

INTRODUCTION

There has been enormous improvement in VLSI (very large-scale integration) technology to implement system performance of computing platform in many ways over the past decades. For instance, data throughput of central processing unit (CPU) has been increased by thousand times faster (e.g., several GHz in Quad-core, 2006) than that of Intel 286 (6 MHz) emerged in the beginning of 1980s. Alongside, other important platform, a latest version of dynamic random access memory (DRAM) reaches a clock speed of 1 GHz. By contrast, state-of-the-art HDD (hard disk drive) transfers data at 600 MB/sec around (see figure 1). Note that data rate of the latest HDD is still orders of magnitude slower than the processor/system-memory clock speed. To achieve the throughput performance in more effective way, it is therefore needed to bridge performance gap in between each component. To compensate the gap between CPU and system memory, a CPU cache* has been required and adopted. In this paper, authors are trying to attempt not only how ferroelectric random access memory (FRAM) provides NV-cache solutions in a multimedia storage system such as solid state disk (SSD) with performance benefits but also what should be satisfied in terms of lifetime endurance in such applications. Also, we demonstrate that what integration technology is critical for qualifying high temperature lifetime tests.

EXPERIMENT

---

*File system cache is an area of physical memory that stores recently used data as long as possible to permit access to the data without having read from the disk.

150-nm technology has been adopted to integrate an 1T1C FRAM in 64 Mb density, organization of which has 16 IOs. Figure 2 shows micrographic views of cross-sectional images

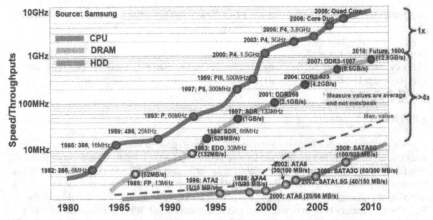

Figure 1. Evolution of electronic components in data throughput performance.

after full integration of the FRAM both (a) in a peripheral circuitry region and (b) in a cell array region, containing $15F^2$ cells. In process features, 80-nm MOCVD PZT serves as a ferroelectric film; $SrRuO_3$ as a top electrode (TE); and Ir as a bottom electrode (BE). The 64 Mb FRAM operates at $V_{DD} = 1.8 \pm 0.2$ V; cycle/read-access time is 120 ns/100 ns, respectively; and CMOS stand-by current is less than 20 μA[1]. Capacitor-level tests have been carried out on the cell arrays of test-element-group (TEG) in fully integrated wafers[1-3]. Package-level reliabilities have also been evaluated[3] both in the high temperature life tests and in endurance. After wafer-level tests, individual prime dies have proceeded to a conventional packaging process for the standard device-level tests. 2-dimensional (2-D) stress simulation in cell arrays has been done by utilizing a commercial tool, ABAQUS CAE ver. 6.7.1[4].

DISCUSSION

Thanks to the bi-stable state of ferroelectrics at near ambient temperature, ferroelectric memory has two important characteristics worth mentioning from the operational point of view. First, since core circuitry for the memory does not require stand-by power during quiescent state and the information remains unchanged even with no power supplied, it is thus *non-volatile*. Second, as the core needs to return the original state after being read, it is called a *destructive* read-out device. This is because the original information is destroyed after READ. As a consequence, it is essential to return the information back to its original state, which is so-called RESTORE, necessarily following the READ. This operation is so inevitable in the destructive read-out memory. In particular, when the ferroelectric memory are used as one of the storage devices in computing system, such as a byte-addressable non-volatile (NV)-cache device, the memory has to ensure lifetime endurance, which is regarded as the number of READ/WRITE (or ERASE if such operation is required) cycles that memory can withstand before loss of any of

4

entire bit information. Here, we begin to discuss FRAM as a NV-cache solution in SSD. Next, since such a NV-cache memory has to meet the requirement of lifetime READ/WRITE cycles,

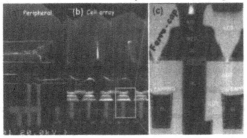

Figure 2. Cross-sectional micrographs both (a) in a peripheral circuitry region and (b) in a cell region, (c) in which one of the cell capacitors is pictured.[1]

we evaluate device endurance necessary for ensuring 10-year lifetime through acceleration factors (AFs) in terms of temperature and voltage. Also, we compare these AFs obtained from device-level tests with those from capacitor-level. Finally, we investigate what integration technology was critical for qualifying the device life tests by means of 2-D stress simulation.

Performance benefits as a NV-cache solution

SSD, one of the multimedia storage systems, in general, consists of 4 important devices. First is a micro-controller having a few hundreds of clock speed in MHz, with real-time operating system (firmware). Second is solid-state storage device such as HDD or NAND-Flash memory, which has several hundreds of memory size in gigabyte. Third is host interface that has the primary function of transferring data between the motherboard and the mass storage device. In particular, SATA (serial advanced technology attachment) 6G offers sustainable 100 MB/s of

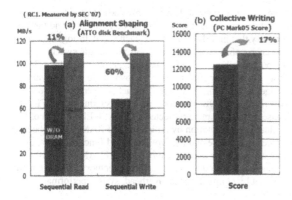

Figure 3. Impact of DRAM utilization in SSD on system performance. (a) Increase in sequential READ/WRITE by IO shaping. (b) Performance improvement by collective WRITE.

data disk rate in HDD. In addition, bandwidth required in DRAM is dominated by the serial I/O ports whose maximum speed can reach 600MB/s. SATA adapters can communicate over a high-speed serial cable. Last but not least is a buffer memory playing a considerable role in system performance. As such, DRAM utilization in SSD brings us many advantages as a buffer memory. For example, in DRAM-employed SSD, not only does IO shaping in DRAM allow us to align WRITE-data unit fitted into NAND flash page/block size but collective WRITE could also be possible. As a result of sequential WRITE, the former brings a performance benefit improved by 60% at maximum, and also the latter gives us another performance benefit improved by 17% due to increase in cache function, as shown in figure 3.

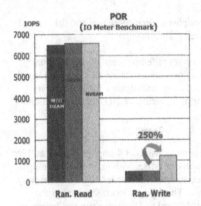

Figure 4. Additional performance benefit for DRAM plus FRAM in SSD.

As an attempt to implement system performance further, not only does DRAM have been considered but FRAM has also been taken into account because of its non-volatility and random accessibility. Before that, it is noteworthy that in SSD with no NV-cache, system-log manager is needed to record and maintain log of each transaction[†] in order to ensure that file system maintains consistency even during a power-failure. A log file that contains all the changes in metadata, generally serves as a history list of transactions performed by the file system over a certain period of time. Once the changes are recorded to this log, the actual operation is now executed. This is so-called power-off recovery (POR). By contrast, POR is redundant in FRAM-employed-SSD as a NV-cache because metadata can be protected by FRAM. Elimination of POR overhead is the single most critical implementation by utilization of FRAM. This is because FRAM provides such system with byte-addressable and non-volatile RAM function. Thus, in spite of sudden power failure, system can safely be protected by adopting FRAM even without POR overhead, ensuring integrity of metadata stored in the ferroelectric memory. Through many benchmark tools, we have confirmed that by eliminating this overhead, system performance has

---

[†] Each set of operations for performing a specific task.

been increased by 250% in random WRITE (see figure 4). This also brings the system to no need of FLUSH operation in file system. As a consequence, additional 9.4% increase in performance, maximizing cache hitting ratio. Since metadata frequently updated do not necessarily go to NAND flash medium, endurance of the flash memories can be increased by 8% at maximum as well. Besides, failure rate of operations can be reduced by 20% due to firmware robustness increased mostly by elimination of the POR overhead.

Endurance

In FRAM, it is not readily achieved to assure whether or not a memory device can endure virtually infinite READ/WRITE cycles. This is because of memory size that is several tens or hundreds megabits typically. For instance, a HTOL (high temperature operational life) test during 2 weeks at 125 °C, is merely a few millions of endurance cycles for each memory cell in 64-Mb memory size. Even taking into account minimum number of cells (in this case 128 bits because of 16 IOs), time to take evaluation of 10^{13} cycles is at least more than 20 days. Therefore, it is essential to find acceleration factors to estimate device endurance through measurable quantities such as voltage and temperature. However, direct extraction of acceleration factors from memory chips is not as easy in practice as it seems to be in theory. This is because VLSI circuit consists of many discrete CMOS components that have a temperature and voltage range to work. Generally, more than 125 °C is supposed to be a limit to operate properly. A voltage range of a memory device is also specified in given technology node (±10% of V_{DD}=1.8 V in this case). Despite those difficulties, we have attempted to figure out acceleration factors in terms of temperature and voltage, together with information obtained from capacitor-level tests.

First, in regard to package-level endurance, figure 5 represents changes in (a) peak-to-peak

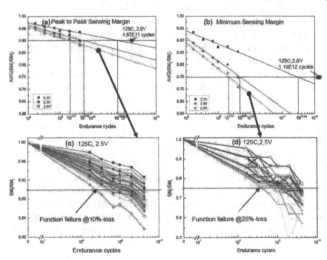

7

Figure 5. Changes in (a) peak-to-peak sensing margin (SMpp) and (b) tail-to-tail sensing margin (SMtt) as a function of endurance cycles at 125 °C. (c) SMpp vs. endurance cycles at 125 °C, 2.5V. (d) SMtt vs. endurance cycles at 125 °C, 2.5V. SMt and SMi of the ordinate in figure 5(a) and (b) is sensing margin at time t and initial time, respectively.

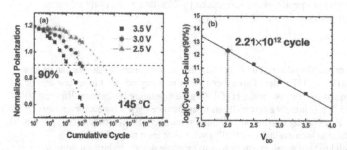

Figure 6. (a) A normalized polarization plot against cumulative fatigue cycles at 145 °C in a variable voltage range. (b) Logarithm of CTF vs. stress voltage, V_{DD} at 145 °C.

sensing margin (SMpp) and (b) tail-to-tail sensing margin (SMtt) as READ/WRITE cycles continues to stress devices cumulatively at 125 °C. Both SMpp and SMtt have been obtained by averaging out 30 package samples for each stress voltage. Function-failed packages have been observed when SMpp and SMtt reach 10% and 25% loss of each initial value, respectively. As seen in figure 5, voltage acceleration factors (AF_V) between 2.0V and 2.5V has been calculated by these criteria ($AF_V = 81$ at SMpp and $AF_V = 665$ at SMtt). In other words, the test FRAMs can endure 1×10^{12} of READ/WRITE cycles at the condition of 125 °C and 2.0V.

Second, in capacitor-level endurance, figure 6 shows a normalized polarization plot against cumulative fatigue cycles at 145 °C in a variable voltage range. Here, we introduce a term of CTF (cycle-to-failure) which is referred to as an endurance cycle at which remanent polarization (or sensing margin) has a reasonable value for cell capacitors (or memory) to operate. Polarization drops gradually as fatigue cycles increase and the collapsing rate is accelerated as stress voltage goes higher. Likewise, provided 10% loss of polarization is criteria of CTF, the CTF at 145 °C and 2.0V approximates 2.2×10^{12}. (NB. This is reasonable because samples of 10% loss in SMpp turned out to be defective functionally.) Figure 7 is a logarithm plot of CTF as a function of stress voltage in a various range of temperature. Considering temperature- and voltage-acceleration factors from figure 7, acceleration condition of 145 °C, 3.5V is more stressful in 5 orders of magnitude than that of 85 °C, 2.0V. In other words, 1.0×10^9 of CTF at 145 °C, 3.5V is equivalent to 6.0×10^{14} at 85 °C, 2.0V.

Results of the acceleration factors obtained from device-level tests differ from those in capacitor-level. For example, while $AF_V(2.5V/2.0V)$ [‡]~81 in device-level tests, that ~16 in capacitor-level. We have yet to find a reasonable clue of what makes this difference. But it could be thought that the difference might arise from the fact that a memory device contains many different functional circuitries such as voltage-latch sense amplifier, word-line/plate-line drivers,

---

[‡] It is thought that AF_V in capacitor-level tests follows AF_V of SMpp in device-level rather than that of SMtt because of nature of capacitor-level tests that average out all the cell capacitor connected in parallel.

all of which make tiny amount of voltage difference magnify each effect on cell capacitors. This tendency can also be observed in the big gap of AF_V obtained from two different definitions between SMtt (AF_V=665) and SMpp (AF_V=81). Tail-bit behaviors of memory cells could include a certain amount of extrinsic imperfection, in general. Thus, we believe that results tested in capacitor-level seem to be close to a fundamental nature of CTF than those in device-level

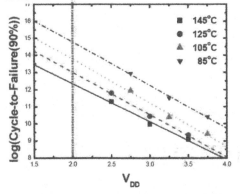

Figure 7. A logarithm plot of CTF as stress voltage increases in a various range of temperature.

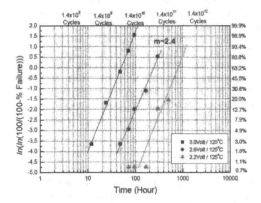

Figure 8. Weibull distributions of endurance life in device-level tests at 125 °C.

tests due to lack of extrinsic components. Figure 8 is Weibull distribution of endurance life in package samples tested at 125 °C in a various voltage range. The distributions in a 2.2-3.0V range of voltage have a similar shape parameter, m~2.4. This suggests that evaluation of endurance tests in device-level makes sense in physical term. As seen in figure 8, voltage-endurance stress at less than 2.0V does not allow us to obtain any sign of degradation in sensing margins within a measurable time span. Nor does temperature-endurance stress above 125 °C due to off-limits of operational specifications of the device.

9

Meanwhile, how many endurance cycles are necessary for use in applications of NV-cache solutions such as data memory and code memory? To answer this question, we need to understand access patterns of NV-cache devices in multimedia system. Now, we take into account the followings: First is the ratio of READ and WRITE per cycle in data memory (likewise, number of data fetching per cycle in code memory). Generally, the ratio for data

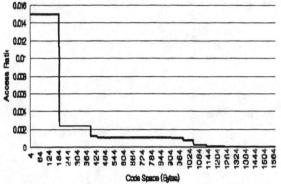

Figure 9. Data locality of FRAM as a code memory.

memory and code memory is 0.75 and 1.00, respectively. Second is data locality[§]. Figure 9 is a simulation result showing strong locality of 1.5% when FRAM has been considered a code memory. As shown in figure 9, less than 200 bytes of code space is more frequently accessed. Provided wear-leveling in READ/WRITE against the strong locality and taking an example of 20 MHz clock frequency of main memory (CPU clock ~ 200 MHz), what has been found is that the endurance cycles for 10-year lifetime becomes less than 9.5×10^{13}. This number of cycles is far less than the cycles we presumably assumed, which is ~ more than 10^{15} cycles. Thus, authors believe that the CTF of 6.0×10^{14} at 85 °C, 2.0V is big enough to ensure that the developed ferroelectric memory as a NV-cache is so endurance-free as to be adopted to a multimedia storage system.

Integration technology

Defective cells in function-failed packages during high-temperature life tests are localized at the corner of the memory cell arrays. Figure 10(a) represents a plan view of the failed cell capacitor by FIB (focused ion beam). Figure 10(b) is a cross-sectional micrograph of the defective spot observed in figure 10(a). The IrO_2 layer disappeared and formed a gap between ATE Ir and TE Ir. The gap has turned out to be an empty, confirmed by rastering of EDX (Energy Dispersive X-ray)[4]. This empty space could come from the two kinds of origins. One is the reduction of IrO_2 with involvement of hydrogen-related products contained in the inter-layer dielectric (ILD) and inter-metal dielectric (IMD)

[§] The locality of reference is the phenomenon that the collection of data locations often consists of relatively well predictable clusters of code space in bytes.

10

$<IrO_2> \rightarrow <Ir> + (O_2) : Kp = P_{O2}$ (1),

↑

Hydrogen-related products

where angular bracket denotes solid state and parenthesis bracket represents the gaseous state. From this reaction formula, IrO_2 is readily reduced to Ir. Also, atomic hydrogen or hydrogen-related products may accelerate this dissociation process, combine with oxygen, and then finally make gaseous phase of water. The other is possible IrO_2 de-lamination by the opposite state of stress distribution between ATE Ir and TE Ir. From 2-D stress simulation at the very edge of the suspected cell arrays, we can obtain a S22 value that is defined as stress projected on the surface normal, the IrO_2 layer. Figure 11 shows stress projected on the IrO_2 layer as a function of the number of memory cells located from the very edge boundary. In case of having a couple of dummy cells, the stress S22 is relatively uniform. By contrast, S22 distribution of the case with no dummy cell shows not only an abrupt change in the stress at the boundary but fluctuates so much that the value of the stress has many ups and downs even on the middle of memory cells in location. Combined with the reduction of IrO_2 layer by hydrogen and its related products, we think that the dummy-cell absence could cause a de-lamination between the two. Accordingly, it

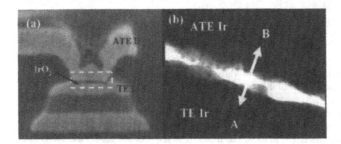

Figure 10. (a) Micrographs of the failed cell capacitor after a HTS test at 150 °C during 500 hours in package-level and (b) a STEM image of the dashed rectangle in (a).

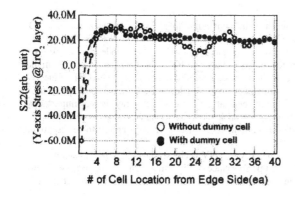

Figure 11. Stress projected on the IrO_2 layer as a function of the number of memory cells located from the edge boundary.

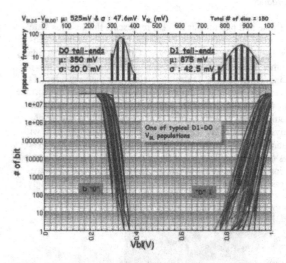

Figure 12. Tail-bit populations of V_{BLD1} and V_{BLD0} for an integration scheme with adequate dummy cells applied. The number of dies is 150 in total.

is desirable to make the stress distribution in the IrO_2 layer as uniform as possible. Figure 12 shows tail-bit populations of V_{BLD1} and V_{BLD0} for an integration scheme with adequate dummy cells. The number of samples is 150 in total. The mean value μ of tail-bits for D1 and D0 is 350 and 875 mV, respectively. The standard deviation σ for D1 is 42.5 mV, which is a bit dispersed, compared with that of 20 mV for D0.

CONCLUSIONS

Utilization of FRAM as a NV-cache solution in a multimedia storage system such as SSD, gives users critical advantages. By elimination of POR overhead, random WRITE throughput can be enhanced by 250%. In spite of strong data locality of FRAM as a code memory, 10-year lifetime endurance has been estimated to be less than 1.0×10^{14} cycles in such system. From the investigation of acceleration factors both in device-level and in capacitor-level, CTF of the developed 64 Mb FRAM has been estimated to approximately 6.0×10^{14} at 85 °C, 2.0 V. It has also been found that the number of dummy cells at the corner of cell arrays is very important to mitigate stress distribution between the ATE stack and TE, playing a subsequent role in protecting the ATE stack from a de-lamination. At least 2 dummy cells appear to be essential to qualify the high temperature life tests.

To be in a nutshell, ferroelectric memory as a NV-cache seems to be a very plausible scenario for increase in data throughput performance of SSD. In assertion of endurance, lifetime endurance is no longer problematic even in the FRAM based on a destructive read-out scheme.

Also, it is in mass productiveness that the FRAM developed has been confirmed by means of applying dummy cells in the edge location of cell arrays.

REFERENCES

1. D. J. Jung, W. S. Ahn, Y. K. Hong, H. H. Kim, Y. M. Kang, J. Y. Kang, E. S. Lee, H. K. Ko, S. Y. Kim, W. W. Jung, J. H. Kim, S. K. Kang, J. Y. Jung, H. S. Kim, D. Y. Choi, S. Y. Lee, K. H. A, C. Wei, and H. S. Jeong, *Symp. on VLSI Tech. Dig.*, pp.102-103 (2008).
2. Y. M. Kang, H. J. Joo, J. H. Park, S. K. Kang, J-H. Kim, S. G. Oh, H. S. Kim, J. Y. Kang, J. Y. Jung, D. Y. Choi, E. S. Lee, S. Y. Lee, H. S. Jeong, and Kinam Kim, *Symp. on VLSI Tech. Dig.*, pp.152-153, (2006).
3. J.-H. Kim, Y.M. Kang, J. H. Park, H. J. Joo, S. K. Kang, D. Y. Choi, H. S. Rhie, B. J. Koo, S. Y. Lee, H. S. Jeong, and Kinam Kim., *IEDM Tech. Dig.*, pp. 889-892 (2005).
4. J.-H. Kim, D. J. Jung, H. H. Kim, Y. K. Hong, E. S. Lee, S. Y. Kim, J. Y. Jung, H. K. Koh, D. Y. Choi, S. K. Kang, H. Kim, W. W. Jung, J. Y. Kang, Y. M. Kang, S. Y. Lee and H. S. Jeong, *Extended Abstract in 40th International Conference on Solid-State Devices and Materials*, Tsukuba, Japan, pp. 1156-1157 (2008).

Mater. Res. Soc. Symp. Proc. Vol. 1129 © 2009 Materials Research Society 1129-V11-08

Harmonic Analysis of AlN Piezoelectric Mesa Structures

M. Mazumdar[1], Adam Kabulski[2], Richard Farrell[2], Sridhar Kuchibhatla[2], V. Pagàn[2], and

D. Korakakis[2,3]

Department of Computer Science, Davis & Elkins College[1]

Lane Department of Computer Science and Electrical Engineering, West Virginia University[2]

National Energy Technology Laboratory, 3610 Collins Ferry Road, Morgantown, WV-26507-0880, USA[3]

ABSTRACT

In this work we report the displacement response of piezoelectric Aluminum Nitride (AlN) thin film MESA in an electrical circuit consisting of a circular MESA in series with a resistance subjected to time varying electrical loads. ANSYS was utilized for the simulation of 3D piezoelectric structures; using coupled field analysis to understand the electro-mechanical behavior of AlN thin film mesas. ANSYS applies finite element analysis (FEM) method to simulate the transient piezoelectric trends. Ringing and overshoot effects were observed in the thin simulation results on applying pulse voltages of varying frequencies to the circuit. The fast rise time of the voltage pulse could be exciting these effects. The effect of fast rising pulse voltages on the RC time constant of the circuit is still unclear at this point and needs to be further investigated.

INTRODUCTION

Some of the widely used applications of piezoelectric circular composite plates include but are not limited to synthetic jets [1, 2], micropumps [3], and medical applications [8], energy harvesting devices [4-7]. When compared to the typically used piezoelectric material such as PZT and other metal oxides, group III nitrides offer several advantages like high temperature stability and high biocompatibility in addition they have the ability to be integrated into NEMS devices [9]. Of all the group III nitrides, AlN has good chemical and mechanical stability under harsh conditions. AlN has the ability to sustain its piezoelectric properties at high temperatures. Therefore AlN is suitable for fabricating MEMS device as it can be deposited at low temperatures and the thickness of AlN films can be controlled down to a few nanometers.

Some of these clamped circular piezoelectric composites applications such as micropumps are driven by pulse oscillating waveform [10].Overshoot and ringing effects were not observed. However it has been demonstrated prior that pulse voltages can excite overshoot and ringing in piezoelectric films. Controlling the rise time of the pulse waveform [11] and implementing RC networks [12], were used to reduce the overshoot and ringing effects in the piezoelectric actuators. But damping techniques such as an RC network have not been thoroughly investigated for thin films having thicknesses in the submicron range to damp these overshooting effects.

In order to fully understand the behavior of the displacement of a piezoelectric material, both the electrical and mechanical equations must be considered. These constitutive equations include time dependent variables, which can be heavily influenced by the thin film nature of these MEMS devices. When analyzing a time dependent equation, time constants play a significant role. Typical piezoelectric films are relatively thick (several micrometers) so they exhibit low capacitances (<1pF) which relates to an insignificant time constant. However, the films being studied in this work are thin and have a much larger capacitance (20pF) so the time constant becomes more apparent. For these films, a theoretical model relating the behavior of capacitance to displacement is desired. The phase lag between the displacement graph and the voltage graph is then compared with respect to time. Under static response the expansion of a piezoelectric film is linear with respect to the applied voltage. In dynamic electrical loading conditions the piezoelectric response cannot be described by simple equations. The dynamic forces generated within the film create charges on the surface which add to the drive voltage. This study deals with investigating the dynamic response of AlN thin film piezoelectric circular MESA electrically connected in series to a resistor and clamped at the bottom surface. ANSYS finite element analysis tool has been utilized to simulate the piezoelectric response of the clamped MESA.

FINITE ELEMENT 3D MODELLING (FEM)

The interaction between the electric and structural properties of a piezoelectric device subjected to boundary conditions can be modeled using Finite Element Analysis (FEM). FEM techniques were implemented using ANSYS software which is a general-purpose finite element software package. In the FEM technique the complete domain is discretized by establishing nodal solution variables and element shape functions. The MESA was meshed into nodes and elements using 3D coupled-field element SOLID5. In the FEM mesh each node point is connected to a limited number of nodes and each node point consisting of four field variables. With KEYOPT (1) set to 3 for the SOLID 5 element, the field variables for this case are the electric potential and three components of the displacement. The variables are calculated from the FEM equations. A resistor and an independent pulse voltage source were connected in series to the meshed MESA. To model the series resistor and independent pulse voltage source the CIRCU94 element with KEYOPT (1) set to 0 and 4 respectively were used as shown in Figure 1. ANSYS FEM software solves the field equations using Transient Analysis set up taking into consideration the initial conditions of the system and the transient response of the system was computed for pulse voltages of varying amplitudes and oscillating frequencies.

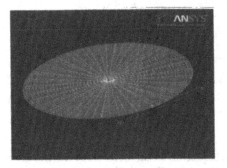

Figure 1. Piezoelectric circular MESA connected in series with a pulse voltage source and series resistance.

Prior to simulation boundary conditions were applied as loads by clamping lower surface of meshed MESA in X, Z,Y directions. This circular mesa was connected in series to a resistor having the following values: (1e3 Ω, 4e3 Ω, 1e4 Ω, and 4e5 Ω). A 10V amplitude pulse was applied across the thickness of this structure and displacements on the top electrode were computed, while keeping the lower surface of the piezoelectric film clamped down in X, Y, and Z directions. The rise time of the pulse voltage source was set at 1e-8 sec. Then, transient analysis was setup to determine the dependence of displacement of the mesa as a function of time for different low frequencies ranging from 1 KHz-10 KHz. The value of the piezoelectric coefficient assumed for d_{33} was equal to 3.9pm/V.

RESULTS AND DISCUSSION

The simulation results for the first few cycles for some of the circuit configurations show a ringing and overshooting effects which could be attributed to the fast rising pulse voltage as shown in. Shown in Figure 2 is an example of ringing effect observed for a frequency of 1KHz and a series resistance of 4e5 Ω.

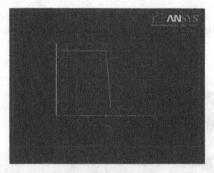

Figure 2. Ringing and overshooting effects in AlN circular thin film MESA subjected to a frequency of 1KHz and having a series resistance of 4e5Ω.

The overshooting of the voltage across the MESA should be reflected in the observed displacement of the top surface of the MESA. A film without overshoot must have a peak displacement corresponding to 3.9pm/V. A plot of the peak displacement of the thin film for varying series resistances and frequencies is shown in Figure 3 shows that true displacement per volt i.e. 3.9pm/V is observed at low series resistances implying that ringing effects are small for these conditions. Also with increasing series resistance the displacement increases which implies an increased overshoot voltage across the MESA. The peak displacement values in Figure 4 indicate that at higher frequencies the behavior of the MESA does not follow a trend with increasing series resistance which could be due to increased RC effects which dampen the ringing effect.

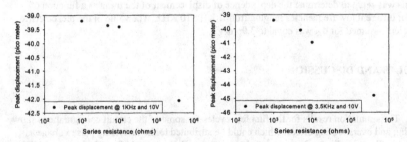

Figure 3. Left: Peak displacement of the AlN circular MESA for 1 KHz frequency as a function of series resistance, Right: Peak displacement of the AlN circular MESA for 3.5 KHz frequency as a function of series resistance.

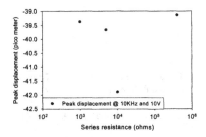

Figure 4. Peak displacement of the AlN circular MESA for 10 KHz frequency as a function of series resistance.

CONCLUSIONS

The sudden rise in the pulse voltage sets in ringing and overshooting effects which influence the observed displacement and the effects of RC time constant of the circuit. At lower frequencies the overshooting is less pronounced when a low series resistance is used. At higher frequencies the displacement does not follow a trend which could be due to the prominent effect of RC time constant. The effect of RC time constant on the reduction of ringing and overshoot needs be simulated for several time cycles to understand the response of the circuit for dynamic electrical loads.

ACKNOWLEDGEMENTS

This technical effort was performed in support of the National Energy Technology Laboratory's on-going research in high temperature flow control hardware for advanced power systems under the RDS contract DE-AC26-04NT41817. This work was also supported in part by, AIXTRON and NSF RII contract EPS 0554328 for which WV EPSCoR and WVU Research Corp matched funds. The authors would like to thank Dr.Kolin Brown and other material growth and characterization lab members for their technical support.

19

REFERENCES

1. Q. Gallas, G.Wang, M. Papila, M. Sheplak, L. Cattafesta,Optimization of synthetic jet actuators, 41st AIAAAerospace SciencesMeeting and Exhibit, Reno, NV, USA, AIAA-2003-0635, (Jan. 2003).
2. Q. Gallas, R. Holman, T. Nishida, B. Carroll, M. Sheplak, L. Cattafesta, AIAA J. 41 (2), 240–247 (2003).
3. C.J. Morris, F.K. Forster, Optimization of a circular piezoelectric bimorph for a micropump driver, J. Micromech. Microeng. 10, 459–465 (2000).
4. S. Kim,W.W. Clark, Q.M.Wang, Piezoelectric energy harvesting with a clamped circular plate: analysis, J. Intel. Mater. Syst. Str. 16, 847–854 (2005).
5. S. Kim,W.W. Clark, Q.M.Wang, Piezoelectric energy harvesting with a clamped circular plate: experimental study, J. Intel. Mater. Syst. Str. 16, 855–863 (2005).
6. S. Horowitz, M. Sheplak, L. Cattafesta, T. Nishida, A MEMS acoustic energy harvester, J. Micromech. Microeng. 16 (9), 174–181 (2006).
7. F. Liu, A. Phipps, S. Horowitz, K. Ngo, L. Cattafesta, T. Nishida, M. Sheplak, Acoustic energy harvesting using an electromechanical Helmholtz resonator, J. Acoust. Soc. Am. 123 (4), 1983–1990 (2008).
8. T. Liu, M. Veidt, S. Kitipornchai, Modelling the input–output behaviour of piezoelectric structural health monitoring systems for composite plates, Smart Mater. Struct. 12, 836–844 (2003).
9. V Cimalla, J Pezoldt and O Ambacher, J. Phys. D: Appl. Phys. 40 6386–6434 (2007).
10. Y.H.Mu, N.P.Hung, K.A.Ngoi, Int.J.Adv.Manu.tech. 15, 573-576 (1999).
11. S.Sugiyama, K.Uchino, Applications of Ferroelectrics,Sixth IEEE Int symp,637-640 (1986).
12. J.M.Thompson , www.ansys.com/events/proceedings/2006./papers/286.pdf (2006).

Mater. Res. Soc. Symp. Proc. Vol. 1129 © 2009 Materials Research Society   1129-V09-02

Erbium Alloyed Aluminum Nitride Films for Piezoelectric Applications

A. Kabulski[1], V. Pagán[1], and D. Korakakis[1,2]
[1]Lane Department of Computer Science and Electrical Engineering, West Virginia University, Morgantown, WV, U.S.A.
[2]National Energy Technology Laboratory, 3610 Collins Ferry Road, Morgantown, West Virginia 26507-0880

ABSTRACT

Aluminum nitride (AlN) films have been explored for sensor and actuator applications, but the resultant piezoelectric coefficient is still too low to make the films more competitive with more commonly used piezoelectric materials such as lead zirconate titanate (PZT). While AlN does have the disadvantage of a lower piezoelectric response, it does have the ability to maintain its piezoelectric properties above 400°C, something that is not possible with other piezoelectric materials. It is desirable to achieve a larger piezoelectric response for AlN in order to facilitate the integration of nitride based devices into existing technologies but conventional methods of improving the response by growing higher quality film only result in slight improvements in the piezoelectric response. A method of improving the d_{33} piezoelectric coefficient beyond any values found in literature may be possible by exploring methods of improving PZT films.

Rare earth doping has been reported to improve the piezoelectric properties of PZT resulting in significant increases in the piezoelectric coefficient. Research has been conducted using rare earth dopants to improve upon the optical properties of AlN, but the impact on piezoelectric effect has never been considered.

Thin, 250-1000 nm, AlN:Er films have been reactively sputtered using erbium (Er)/aluminum alloyed targets to explore any improvement in piezoelectric properties of the AlN:Er films as compared to AlN films. AlN films with 0.5 and 1.5% Er concentrations have been found to have piezoelectric coefficients that are larger than comparable 'Er-free' AlN films. AlN films with only 0.5% Er quantities were found to increase the d_{33} coefficient compared to a similar AlN film depending on the thickness of the film. This increase results in $d_{33,f}$ values greater than 7pm/V which is larger than most values found in literature. By increasing the Er content to 1.5%, values of $d_{33,f}$ were found to be as large as 15 pm/V. This enhanced piezoelectric response is still lower than that of PZT, but can be used to create superior actuator devices than that of typical AlN films.

INTRODUCTION

Much research has been performed on methods of improving the material properties of lead zirconate titanate (PZT) using rare earth dopants [1-6]. This research has led to PZT-based films that exhibit superior piezoelectric, hydrostatic piezoelectric and pyroelectric coefficients as well as improvements in the dielectric constant, fracture toughness, and strength of the material [1,2]. Similar research has also been conducted on other piezoelectric materials such as BNT6BT. The addition of a rare earth dopant was found to create large interfacial polarization due to the creation of defects which may account for the large piezoelectric coefficients observed [7]. Rare earth dopants in PZT have also been found to increase the polarization and the coexistence of randomly quenched defects and nonrandomly distributed defects. The addition of the rare earth material is believed to lead to the presence of combinatory hard and soft piezoelectrics, explaining the increases in d_{33} [4]. While AlN is not believed to be ferroelectric and rare earth doping will not influence AlN as it has PZT, it is still possible that the addition of Er can lead to AlN films with an improved piezoelectric response.

EXPERIMENT

AlN films and AlN:Er films ranging in thickness from ~150-1000 nm were reactively sputtered on (100) p-type, double-side polished, 1-20 Ω-cm resistivity silicon wafers using a CVC 610 DC magnetron sputter deposition system. The target used for AlN deposition was a 50 mm-diameter, 99.999% Al disk while the targets used for AlN:Er deposition were 50 mm-diam, 99.9% Al disks that have been alloyed with 1 and 3% erbium concentrations to create AlN:Er(1%) and AlN:Er(3%), respectively. Before deposition, the sputter system was pumped to a base pressure of less than $5x10^{-6}$ mbar. A pre-sputter cleaning process was performed to remove oxygen and other contamination from the surface of the sputter targets. Targets were cleaned using Ar gas and a target power of 200 W for 10 min. After the cleaning process was completed, N_2 and Ar gases were introduced into the chamber at flow rates of 27.0 and 3.0 sccm respectively. Depositions were performed at a chamber pressure of ~45 mTorr at room temperature. A target power of 200 W was used for all depositions. After the AlN or AlN:Er films were deposited, a Pt backside contact was deposited on the silicon wafer.

After deposition, the thickness and surface roughness of each sample were measured using a J.A. Woollam spectroscopic ellipsometer to determine the effects of erbium alloying on deposition rates and thin film properties. Standard photolithography techniques were used to deposit 800µm-diameter solid, circular Pt topside contacts on top of the samples for piezoelectric characterization. The $d_{33,f}$ piezoelectric coefficient of each film was measured by applying a low frequency AC signal ranging from a few volts to a maximum of 40 volts peak to peak across the top and bottom contacts of the sample and measuring the displacement of the film using a Polytec MSV-100 scanning head laser Doppler vibrometer (LDV). All measurements were performed at room temperature in ambient environment.

DISCUSSION

Ellipsometry confirmed that the refractive index and roughness of the AlN films agreed with similar films found in literature. Despite the addition of larger Er atoms, the surface roughness of each AlN:Er film was found to be within error of the surface roughness of similar AlN films. The surface roughness of each film was found to increase with thickness, but the percentage roughness decreased with thickness, agreeing with literature [8]. Due to the frequency dependence typical of sputtered aluminum nitride films, a range of low frequency measurements were performed in order to choose an optimal measurement frequency. The frequency response of AlN and both AlN:Er samples of two thicknesses, ~500 and ~1000 nm, was measured over a range of frequencies from 3-10 kHz. The samples were found to exhibit definite frequency dependence with large changes resulting when the frequency is greater than 7 kHz. Frequencies below 3 kHz were not considered because high background vibrational noise was present at these frequencies. Therefore, the measurement frequency of 3.5 kHz was chosen to assure that the changes in displacement observed were not artifacts of the measurement setup.

Piezoelectric measurements of the AlN films show that the values of the piezoelectric coefficient $d_{33,f}$ increase with thickness, but the increase becomes less apparent for thicknesses above 1 μm. This trend agrees with the most extensive piezoelectric displacement versus thickness study in literature with the only difference being that the films in this work do not exhibit a piezoelectric effect for thicknesses below 200 nm [8]. For the purposes of this paper, the maximum measured values for the $d_{33,f}$ coefficient of the AlN films are referenced when examining the behavior of the AlN:Er films. Fig. 1 shows the behavior of AlN:Er(1%) as compared to the AlN films.

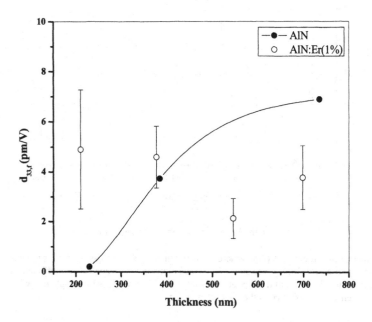

Fig. 1. $d_{33,f}$ vs. thickness for AlN and AlN:Er(1%) films

The piezoelectric displacement of the AlN:Er(1%) films is larger than the AlN reference for the films less than 400 nm thickness but begins to drop off as the thickness is increased. The values of $d_{33,f}$ were also found to vary across the sample as shown by the relatively large error bars. This may be attributed to the more apparent polycrystalline nature of the sputtered films with the addition of erbium. An enhanced piezoelectric effect for the thinner films was also observed in the AlN:Er(3%) films.

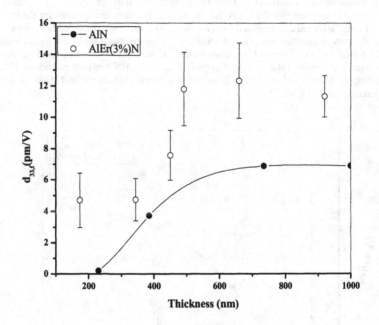

Fig. 2. $d_{33,f}$ vs. thickness for AlN and AlN:Er(3%) films

In these films, the values of $d_{33,f}$ increase as the thickness is increased, similar to what was observed in the AlN films. Once again, large variations in the piezoelectric coefficient were measured, but with all values of $d_{33,f}$ larger than that of the comparable AlN sample and with some values being twice as large as the AlN counterpart.

CONCLUSIONS

AlN:Er films exhibit a significantly higher piezoelectric response than AlN films. While this is only true for thicknesses less than 400nm in the AlN:Er(1%) samples, the AlN:Er(3%) showed an improvement in $d_{33,f}$ for all thicknesses measured. In the AlN:Er(3%) samples, values of $d_{33,f}$ were measured to be between 10-15 pm/V, an increase of 3-7 pm/V over the comparable AlN films and also larger than any values reported in literature for both sputtered and epitaxially grown AlN. It should be noted that it is possible that the results included in this work may be overestimating the piezoelectric effect due to the use of LDV as the measurement method, though the trends are still apparent. Since AlN is not ferroelectric, it is unlikely that the enhancement in the piezoelectric effect is due to an improvement in polarization as is observed in rare earth doped PZT. Instead, it is possible that the larger Er atoms are causing the increase in observed piezoelectric response though this behavior is currently being further explored. The enhanced $d_{33,f}$ films were found to exhibit excellent piezoelectric and optical properties allowing for the fabrication of superior piezoelectric devices based on AlN.

ACKNOWLEDGMENTS

This technical effort was performed in support of the National Energy Technology Laboratory's on-going research in high temperature flow control hardware for advanced power systems under the RDS contract DE-AC26-04NT41817. This work was also supported in part by NSF RII contract EPS 0554328 for which WV EPSCoR and WVU Research Corp matched funds.

REFERENCES

1. T.C. Goel, P.K.C. Pillai, H.D. Sharma, A.K. Tripathi, A. Tripathi, C. Pramila, and A. Govindan, International Symposium on Electrets (1994) 720-724
2. A. Garg and D.C. Agrawal, Materials Science and Engineering B, 86 (2001) 134-143
3. S.R. Shannigrahi, F.E.H. Tay, K. Yao, and R.N.P Choudhary, Journal of the European Ceramic Society, 24 (2004) 163-170
4. Y. Gao, K. Uchino, and D. Viehland, J. Appl. Phys., 92, 4 (2002)
5. S.B. Majumder, A. Dixit, B. Roy, W. Jia, and R.S. Katiyar, IEEE International Symposium on Applications of Ferroelectrics (2002) 111-114
6. S.R. Shannigrahi and S. Tripathy, Ceramics International 33 (2007) 595-600
7. H. Fan and L. Liu, J. Electroceram. (2007)
8. F. Martin, P. Muralt, M.-A. Dubois, and A. Pezous, J. Vac. Sci. Technol. A, v 22, n 2 (2004) 361-365

Mater. Res. Soc. Symp. Proc. Vol. 1129 © 2009 Materials Research Society 1129-V11-06

Piezoelectric Response of Lanthanum Doped Lead Zirconate Titanate Films for Micro Actuators Application

Takashi Iijima, Bong-yeon Lee, Seiji Fukuyama
National Institute of Advanced Industrial Science and Technology (AIST),
AIST Tsukuba central 5, Tsukuba 305-8565, Japan

ABSTRACT

La doped lead zirconate titanate (PLZT) films are prepared using a chemical solution deposition process. The effect of La substitution on the piezoelectric response was investigated to clear the possibility of the micro actuator application for the PLZT films. Nominal compositions of the 10% Pb excess PLZT precursor solutions were controlled like La/Zr/Ti= 0/65/35, 3/65/35, 6/65/35, and 9/65/35. These precursor solutions were deposited on the Ir/Ti/SiO$_2$/Si substrates, and the thickness of the PLZT films was 2µm. 10 to 20- µm- diameter Pt top electrodes are formed with a sputtering and a photolithography process. The polarization-field (P-E) hysteresis curves and the longitudinal displacement curves were measured with a twin beam laser interferometer connected with a ferroelectric test system. With increasing La substitution amount, the P-E hysteresis curves became slim shape, and remnant polarization (Pr) decreased. The hysteresis of the piezoelectric longitudinal displacement curves also decreased with increasing La substitution amount. The amount of the displacement under unipolar electric field showed a peak at La/Zr/Ti= 3/65/35. The calculated effective longitudinal piezoelectric constant ($d_{33}eff$) is 129.2 pm/V at 3/65/35. This amount was relatively higher than that of PZT films at morphotropic phase boundary (MPB: 0/53/47) composition prepared the same film preparation process.

INTRODUCTION

In the field of the micro-scale smart systems, high performance piezoelectric film micro actuators are demanded. Therefore, static actuators are well developed as the micro actuator, while the structure of the micro static actuator is very complicated and the driving voltage is relatively high [1]. To overcome these subjects of the micro static actuators, piezoelectric film micro actuators such as lead zirconate titanate (PZT) are investigating [2]. For the PZT film micro devices, large displacement is required and the PZT films are driven with higher electric field than coercive field (Ec) of the film. To develop the piezoelectric micro actuator devices, therefore, low leakage current, better linearity of displacement, and high piezoelectric constant are required for the PZT films. It is well known that a hysteresis of the piezoelectric longitudinal displacement of the lanthanum (La) substitute lead zirconate titanate (PLZT) decrease with increase of lanthanum substitution amount for bulks and thin films [3, 4]. In this study, to clear the possibility of the micro actuator application of the PLZT films, the effect of La substitution for the PZT films on the piezoelectric response were investigated.

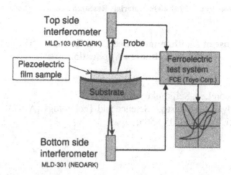

Figure 1. Schematic illustration of the twin beam laser interferometer (TBI) system for ferroelectric and piezoelectric response measurement

EXPERIMENTAL PROCEDURES

La doped lead zirconate titanate (PLZT) films are prepared using a chemical solution deposition (CSD) process. To prepare the precursor solutions, lead acetate trihydrate (99.9%, Nacalai tesque), titanium iso-propoxide (99.999%, Aldrich), zirconium n-propoxide (70% in propanol, Azmax) and lanthanum acetate hydrate (99.99%, Wako chemical) were used, and 2-methoxyethanol was used as a solvent. The method of Blum and Gurkovitch [5] was modified to prepare the precursor solutions. The composition of Zr/Ti rate was kept constant as 65/35, and 10 mol% excess Pb was added to the PLZT precursor solutions. The amount of La substitution for Pb was changed from 0 to 9 at % like $(Pb_{1.1-x}La_x)(Zr_{0.65}Ti_{0.35})O_3$, x= 0, 0.03, 0.06, 0.09 that were denoted in La/Zr/Ti= 0/65/35, 3/65/35, 6/65/35, and 9/65/35. The solutions were deposited on 2-inch Ir/Ti/SiO$_2$/Si substrates with a spin coater. A sequence of spin coating, drying at 120 °C for 1 min and pyrolysis treatment at 500 °C for 2 min was performed three times, and the samples were fired at 700 °C for 3 min. The thickness of one deposition layer was about 80 nm. This process was repeated several times to fabricate dense and crack free 2-μm-thick PLZT films. Details of the PZT thick films' preparation process have been described elsewhere [6]. To evaluate electrical properties, a Pt top electrode was sputtered onto the PLZT films. The electrode was then patterned into a diameter of 15 to 20 μm using photolithography.

Leakage current (I-V) characteristic was measured with pico-ampere meter (4140B, Agilent technology). Two individual laser Doppler interferometers were connected to the ferroelectric test system (FCE; Toyo Corp.) to measure the longitudinal displacement of the PLZT films. The top side interferometer (MLD-102; Neoark Corp.) measures the PLZT film surface longitudinal displacement; the bottom side interferometer (MLD-301A; Neoark Corp.) measures the Si substrate surface longitudinal displacement. These two laser beams are aligned precisely to examine the same position; both sides' displacements are measured with the same timing with the polarization hysteresis (P-E) curve at 100 Hz. Figure 1 shows schematic illustration of the twin beam laser interferometer (TBI). A subtracted displacement of the bottom side from the top side corresponds to the differential displacement measured using double-beam laser interferometry (DBI) [7].

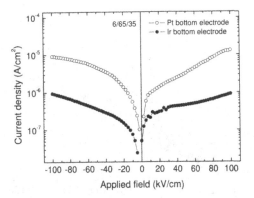

Figure 2. Leakage current (*I-V*) property of 6 at% La doped PLZT (6/65/35) films for Pt and Ir bottom electrode.

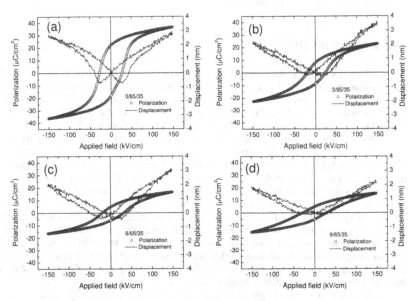

Figure 3. Polarization hysteresis (*P-E*) curves and longitudinal piezoelectric displacements of PLZT films deposited on Ir bottom electrode for (a) 0/65/35, (b) 3/65/35, (c) 6/65/35, and (d) 9/65/35.

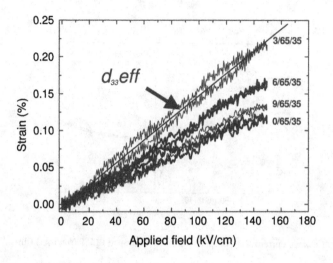

Figure 4. Unipolar driven longitudinal strain-field curve for 0/65/35, 3/65/35, 6/65/35, and 9/65/35 films deposited on Ir bottom electrode.

RESULTS AND DISCUSSION

Figure 2 shows leakage current (I-V) property of 6 at% La doped PLZT (6/65/35) for Pt and Ir bottom electrode. The leakage current density of Ir bottom electrode is about order of magnitude lower than that of Pt bottom electrode. The current density at 80 kV/cm is 8.3×10^{-6} A/cm^2 for Pt and 6.5×10^{-7} A/cm^2 for Ir, respectively. Therefore, Ir bottom electrode is concluded to be effective in decreasing the leakage current of PZT type films such as SrRuO$_3$ [8].

Figure 3 shows polarization hysteresis (P-E) curves and longitudinal piezoelectric displacements of PLZT films deposited on Ir bottom electrode for (a) 0/65/35, (b) 3/65/35, (c) 6/65/35, and (d) 9/65/35. It can be seen that all of the hysteresis curves and butterfly shaped displacements are well saturated. In the case ferroelectric properties, P-E hysteresis curves become slim shape and maximum polarization and remanent polarization (Pr) decrease with increasing the La concentration. This tendency is consistent with bulk PLZT and epitaxial thin films [4, 9]. On the other hand, electric field induced longitudinal piezoelectric displacement curves shows slim butterfly shape whereas the total displacement (difference between maximum and minimum displacement) did not show remarkable difference. Moreover, amount of negative displacement, which means shrinkage of the film thickness, around coercive field (Ec) decrease with increasing the La dope amount. These results indicate that hysteresis of the displacement decrease because domain switching happens easily with increasing La dope amount.

To estimate effective longitudinal piezoelectric constant ($d_{33}eff$), longitudinal displacements under unipolar field were measured. Figure 4 shows unipolar driven longitudinal strain-field curve for 0/65/35, 3/65/35, 6/65/35, and 9/65/35 films deposited on Ir bottom

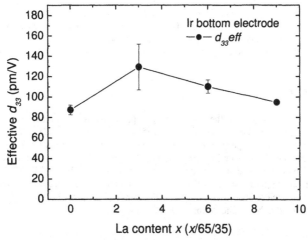

Figure 5. Relationship between the effective longitudinal piezoelectric constant and La dope amount.

electrode. The longitudinal strain of La doped films clearly increase compared with that of without La doping (0/65/35). The effective longitudinal piezoelectric constant is estimated from slope of the strain line.

Figure 5 shows relationship between the effective longitudinal piezoelectric constant and La dope amount. Error bars indicate mean error. It can be seen that the effective longitudinal piezoelectric constant shows a peak at 3/65/35, and the amount is $d_{33}eff$ = 129.2 pm/V. This amount was relatively higher than that of PZT films at morphotropic phase boundary (MPB: 0/53/47) composition prepared the same film preparation process. Table I summarize of the leakage current (J) at 80kV/cm, 2Pr, 2Ec, and $d_{33}eff$. From these result, it turns out that 3/65/35

Table I. leakage current (J) at 80kV/cm, ferroelectric properties (2Pr, 2Ec), and piezoelectric property ($d_{33}eff$) for 0/65/35, 3/65/35, 6/65/35, and 9/65/35 films deposited on Ir bottom electrode.

	J at 80kVcm (A/cm^2)	2Pr (μC/cm^2)	2Ec (kV/cm)	$d_{33}eff$ (pm/V)
0/65/35	5.9x10^{-8}	41.0	55.8	87.2
3/65/35	1.9x10^{-8}	18.0	46.2	129.2
6/65/35	1.1x10^{-7}	9.5	48.6	110.1
9/65/35	2.6x10^{-7}	9.8	58.0	94.9

31

film show the lowest leakage current, relatively slim P-E hysteresis curve, and the highest piezoelectric constant. In the case of 9/65/35, meanwhile, the leakage current and 2Ec showed relatively large amount. One of the reasons for the degraded electric properties seems to be insufficient heat treatment conditions in the film preparation process. Therefore, further investigation of the film preparation process for higher La amount doped PZT are required.

CONCLUSIONS

To clear the possibility of the micro actuator application for the piezoelectric films, La doped PZT films are prepared for La/Zr/Ti= 0/65/35, 3/65/35, 6/65/35, and 9/65/35. The effect of La amount on the electric properties such as leakage current, ferroelectricity, and piezoelectricity was investigated. With increasing La dope amount, P-E hysteresis curve and butterfly shape piezoelectric displacement curve become slim shape. In the case of 3/65/35, PLZT film show the lowest leakage current, relatively slim P-E hysteresis and butterfly shaped displacement curve, and the highest piezoelectric constant, d_{33eff} = 129.2 pm/V. Therefore PLZT film composed of 3/65/35 is considered to be candidate for micro actuator devices.

REFERENCES

1. W. C. Tang, T. Cuong, H. Nguyen and R. T. Howe, *Sensore and Actuators* **20**, 25 (1989).
2. P. Muralt, *J Micromech Microeng.* **10**, 136 (2000).
3. K. Furuta and K. Uchino, *Adv. Ceram. Mater.* **1**, 61 (1986).
4. H. Shima, H. Naganuma, T. Iijima, H. Funakubo and S. Okamura, *Trans. Mater. Res. Soc. Japan* **32**, 79 (2007)
5. J.B.Blum and S.R.Gurkovich, *J. Materials Science*, **20**, 4479 (1985).
6. T. Iijima, S. Osone, Y. Shimojo and H. Nagai, *Int. J Appl. Ceram. Technol.* **3**, 442 (2006).
7. A. L. Kholkin, Ch. Wuetchrich, D. V. Taylor and N. Setter, *Rev. Sci. Instrum.* **67**, 1935 (1996).
8. K. S. Liu and T. F. Tseng, *Appl. Phycs. Lett.* **72**, 1182 (1998).
9. K. Nishida, Y. Honda, S. Yokoyama, H. Funakubo, T. Yamamoto, K. Saito and T. Katoda, *Appl. Phys. Lett.* **90**, 262902 (2007).

Mater. Res. Soc. Symp. Proc. Vol. 1129 © 2009 Materials Research Society   1129-V04-17

Wireless Remote 2-D Strain Sensor Using SAW Delay Line

T. Nomura, and A. Saitoh
Faculty of Engineering, Shibaura Institute of Technology
3-7-5 Toyosu, Koto-ku, Tokyo 135-8548, Japan

ABSTRACT

Two-dimensional (2-D) strain sensor utilizing surface acoustic wave (SAW) devices is demonstrated. SAW devices offer many attractive features for applications as chemical and physical sensors. In this paper, a novel SAW strain sensor that employs SAW delay lines has been designed. Two crossed delay lines were used to measure the two-dimensional strain. A wireless sensing system is also proposed for effective operation of the strain sensor.

INTRODUCTION

In many fields, a measurement of strain and its distribution is required. It is the need for health and safety monitoring for the sake of safety and comfort, which requires constant monitoring of strains within complex structures and buildings as well as the strain distributions in mobile objects, typically the wings of an aircraft [1], [5], [8], [9].

The traditional method to measure strain is using a resistance strain gauge, which is mainly structured to measure unidirectional strains. However, as is the case when measuring stress, rosette analysis is required when measuring strains on a plane. Rosette analysis requires multiple elements to be positioned together at a single location and used simultaneously. Consequently there a large number of elements and accordingly a large number of electrical supply lines, making the measurement process very complex.

A surface acoustic wave (SAW) can be easily excited on a piezoelectric substrate using an interdigital transducer (IDT). Its propagation characteristics are easily affected by external physical and electrical changes. This is why applications of SAW's include sensors for measuring physical parameters. Recently, there have been many reports on SAW sensors that take advantage of SAW's propagation characteristics, which change easily in response to external factors [2], [3], [4], [5],.

In this paper, we first clarify the relationship between changes in the propagation path length of a SAW and the phase of the wave. We then discuss a strain sensor that utilizes changes in the velocity of a SAW caused by strain in the SAW device.

Although some reports on sensors using SAWs have been published, they mostly discussed one-dimensional strains [6], [7], [9]. We herein propose a two-dimensional (2-D) strain sensor that simultaneously measures strains in two directions using two delay lines that cross each other at right angles. These two delay lines exist on an anisotropic piezoelectric substrate (128° Y-cut, X-propagating LiNbO$_3$). The delay lines are used as a 2-D strain sensor. The 2-D strain sensor is passive sensor that can be attached to a structure and the remotely interrogated though a wireless interfaces. As its working principle, this strain sensor monitors changes in the propagation path length caused by strains applied to the piezoelectric substrate.

MEASUREMENT OF STRAIN USNG SAW

Principle of measurement

Figure 1 shows the two-dimensional strain sensor and the principle of measurement by using an SAW. The SAW strain sensor was constructed with two delay lines crossed at right angles (Fig.1 (a)). In each delay line, when a strain is applied to a substrate containing a SAW delay line, the propagation path length changes Fig.(b)). Let this change in propagation path length be denoted by Δl, and we then obtain the following expression for $\Delta\theta$, the phase difference between the input and output waves:

$$\theta + \Delta\theta = \omega\frac{(l \pm \Delta l)}{v}$$

$$\Delta\theta = \omega\frac{(l \pm \Delta l)}{v} - \omega\frac{l}{v} = \pm\omega\frac{\Delta l}{v} \quad\cdots\cdots\cdots\cdots(1)$$

l: Propagation distance (m),　v: SAW velocity (m/s),
θ: Phase difference (rad.),　ω : Frequency (rad. /s),

Therefore, by measuring the amount of change between the input and output of a delay line caused by a strain, we can determine the magnitude of the strain.

This magnitude of the strain can be calculated using Eqn. (1), using the amount of change caused by the strain, with reference to the phase difference when no strain is applied.

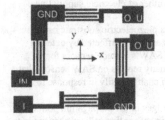

(a) 2-D strain sensor constructed by two SAW delay lines.

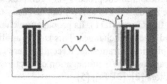

(b) 1-D SAW strain sensor and the change of the propagation length due to the applied strain.

Fig.1. Schematic diagram of 2-D and 1-D strain sensor using SAW delay lines.

CHARACTERISTICS OF A SAW STRAIN SENSOR

Before discussing two-dimensional measurements, we first clarify the characteristics of a one-dimensional strain sensor with a delay line, in order to understand the basic characteristics of a SAW strain sensor.

We first constructed a 1-D strain sensor on a 128° Y-cut X-propagating LiNbO₃ substrate (see Fig.1 (b)), and examined its characteristics. The sensor was composed of a delay line with a propagation path length of 100 λ and a frequency of 40 MHz.

ONE-DIMENSIONAL SAW SENSOR

Figure 2 shows, for the case when a strain in the x direction is applied to the delay line of Fig. 1, the strain in the substrate in the x direction as well as the phase difference caused by the change of the SAW propagating length in the x direction.

A positive (+) strain (strain is plotted on the horizontal axis) indicates that the propagation path is extended, while a negative strain shows that the path is shortened. We also measured the strain on the substrate using a resistance strain gauge attached close to the SAW propagation path. In these experiments, we varied the strain by 20 (μ strain) each time.

From the figure, we can see that the strain and the phase difference are proportional, which implies that we can measure the strain using a SAW delay line.

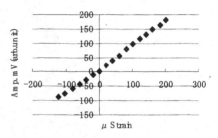

Fig.2. Response of SAW sensor versus static strain

Next, we investigated how the SAW phase difference is related to the strain direction. In our experiment, we used a cantilever to rotate the fixed bronze plate by 90°, setting the direction of the strain to the y direction.

Figure 3 shows the results of this particular experiment together with the results shown in Fig. 2 in the same plot. Thus, Fig. 3 shows the relationship between the strain in the x direction of the substrate and the phase difference caused by a change in the length of the SAW propagating in the x direction for the case when strains are applied in two directions (x and y). In other words, the horizontal axis in the figure denotes the strain in the x direction when strains are applied in both the x and y directions. In addition, the vertical axis shows the phase difference (in degrees), which we obtained by calculating the output of the measurement circuit and then converting it into a phase difference.

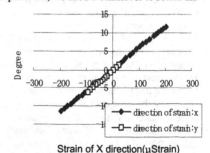

Fig.3. Response of the SAW sensor versus static strain

This plot shows that, regardless of whether the strain is in the x or y direction, the same relationship holds between the strain in the delay line and the phase difference of the propagating SAW. The experimental errors are 0.2 % at the most. Thus, we can conclude that the phase difference of the SAW depends solely on the strain in the delay line, regardless of the direction of the strain.

TWO-DIMENSIONAL SAW STRAIN SENSOR

Figure 4 shows the configuration of the SAW sensors designed to enable the simultaneous measurement of strains in two directions. In each of the two pairs of delay lines, the lines are orientated perpendicularly to each other to enable the simultaneous measurement of the strains in both the x and y directions.

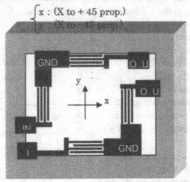

Fig.4. Configuration of two-dimensional strain SAW sensor.

We excited SAWs in two directions, x and y (see Fig. 4), and measured the phase changes in each of these two directions. In this way, we used a single SAW device to measure the strains in the two directions.

However, generally the excitation efficiency and velocity of SAWs in a piezoelectric substrate vary depending on their propagation direction as a result substrate's anisotropy. For this reason, there are cases when the output characteristics of the delay lines in the two directions cannot be made the same resulting in an error in the phase measurements.

To resolve this problem in two-dimensional strain measurement, it is desirable for the SAWs used to have the same excitation efficiency, velocity, and loss.

Since we used 128° Y-cut, X-propagating LiNbO$_3$ in our experiment, the propagation velocity and excitation efficiency differed considerably between the x direction and the perpendicular direction (90°) to x. In our substrate, however, the propagation characteristics were symmetric about the x-axis. We therefore installed the delay lines so that they were symmetric about the x-axis and made them intersect other at right angles (X +/- 45°), when we created our two-dimensional strain sensor.

In the experiment using the 2-D sensor, the two delay lines have been designed to operate at the 50MHz of central frequency.

Figure 5 shows the results of the experiments described above. The two lines in the plot indicate changes in the SAW phases produced by the strain in the delay lines in the two respective directions. The horizontal axis denotes the strain in each propagation path. At larger strains, the slopes of the lines differ significantly in the two directions.

We conjecture that one cause for this was that the adhesion of the cantilever to the sensor was uneven (the piezoelectric substrate, part of the sensor, was rectangular and it was hard to apply the strains evenly over the substrate). However, in both of the two propagation directions, the phase difference of the SAW was proportional to the strain in the propagation path. This implies that it is possible to create a two-dimensional strain sensor device in which two sensors having the same characteristics intersect each other.

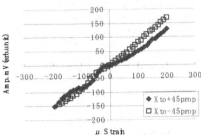

Fig.5. Response of two-dimensional SAW sensor versus static strain

WIRELESS SAW STRAIN SENSOR

In this section, we tried remote sensing of strains, using a two-dimensional SAW strain sensor [5], [8], [9].

Figure 6 outlines the equipment we used for the remote sensing experiment. In this wireless strain measurement, we connected an antenna directly to the IDT of the delay-line SAW sensor shown in Fig. 7. We used a tone burst wave as the signal. The tone burst wave output from the interrogation unit is emitted from the sending antenna and received by the antenna attached to the SAW sensor. The IDT connected to the antenna directly excites a SAW and, after detecting strain, the signal is emitted from the same antenna.

Figure 7 shows how the IDTs were connected to the antenna. These two IDTs were connected directly to each other and a signal from the antenna excited a SAW in both directions.

In our experiment, we set the frequency to 50 MHz and the distance between the two antennas to 55 cm. Also, we used the same strain sensor in the X ±45° direction described in the previous section.

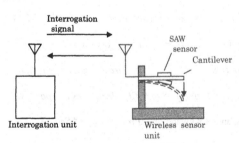

Fig.6. Schematic diagram of the strain monitoring system

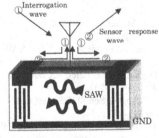

Fig.7. Schematic diagram of the wireless SAW sensor.

37

Figure 8 shows the relationship between the strain and the output in each of the directions, in the wireless measurement. The response was proportional to the strain. The result is similar to the results of the wire measurements we conducted in Section 4. This implies that a SAW strain sensor is capable of measuring two-dimensional strains in a wireless system.

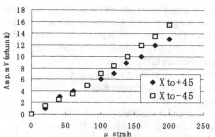

Fig.8. Response of two-dimensional wireless SAW sensor versus applied strain.

CONCLUSION

The two-dimensional strain sensor that utilizes a SAW delay line has been designed and investigated its characteristics and we obtained the following results:

(1) Changes in the phase of a SAW delay line are proportional to the strain in the propagation path. This enables strains to be measured using SAWs.
(2) We also demonstrated that, by placing two delay lines perpendicular to each other, we are able to measure strains in two different directions simultaneously. In such a case, by selecting the right propagation directions of the SAW, we can obtain similar characteristics in both directions.
(3) Finally, we connected the sensor to an antenna and conducted an experiment that showed that wireless and passive measurement of strains is possible.

REFERENCES

1. V.K.Varadan, *Smart Materials and Structural Systems,* no.1-1, pp.135-148 (2001).
2. H.Wohltjen and R.Dessy, *Analytical Chem.*, vol.**51**, no.9, pp.1485-1475(1979).
3. H.Wohltjen, in *Proc. of 4th International Conference on Solid sensors and actuators*, Tokyo, p.471 (1987).
4. M.E.Motamedi and R,M.White, *IEEE Trans. Ultrson. Ferroelec. Freq. Conr.*, vol.**UFFC-34**, no.2, pp.122, (1987).
5. A.Pohl, "A review of wireless SAW sensors", *IEEE Trans. Ultrson. Ferroelec. Freq. Conr.*, vol.**47**, no.2, pp.317-332, March, (2000).
6. D.Hauden, F.Bindler, R.Coquerel, in *Proceedings of IEEE 1985 Ultrasonics Symposium*, pp 486-489 (1985).
7. H.Scherr, G.Scholl, F.Seifert and R.Weigel, in *Proc. of 1996 IEEE Ultrasonics Symposium*, pp.347-350 (1996).
8. J.Hayasaka, K.Tanaka, T.Miura, Y.Ikeda, and H.Kuwano, in *Proc.of the 23rd Sensor Symposium*, pp.455-458 (2006).
9. T.Nomura, A.Saitoh and H.Tokuyama, in *Proc. of the 20th Sensor Symposium*, pp.191-194 (2003).

Mater. Res. Soc. Symp. Proc. Vol. 1129 © 2009 Materials Research Society   1129-V11-02

Growth of Epitaxial Pb(Zr,Ti)O3 Thick Films on (100)CaF2 Substrates with Perfect Polar-axis-orientation and Their Electrical and Mechanical Property Characterization

Takashi Fujisawa[1], Hiroshi Nakaki[1], Rikyu Ikariyama[1], Mitsumasa Nakajima[1], Tomoaki Yamada[1], Mutsuo Ishikawa[1], Hitoshi Morioka[1,2], Takashi Iijima[3], and Hiroshi Funakubo[1]

[1] Department of Innovative and Engineered Materials, Tokyo Institute of Technology, Yokohama, 226-8502, Japan

[2] Application Laboratory, Bruker AXS, Yokohama, 221-0022, Japan

[3] Research Center for Hydrogen Industrial Use and Storage, National Institute of Advanced Industrial Science and Technology, 1-1-1 Higashi, Tsukuba, Ibaraki 305-8568, Japan

ABSTRACT

Crystal structure of the (001)-oriented $Pb(Zr_{0.35}Ti_{0.65})O_3$ films with thickness range from 30 – 3000 nm grown on (100)CaF2 substrates was investigated. Lattice parameter was almost independent of the film thickness and almost agreed with reported ones for the powder of the same composition. Obvious strain gradient along the film thickness direction was not detected by the cross sectional TEM and the Raman spectroscopy analysis. High temperature XRD data suggests that the temperature dependence of the lattice parameter of in-plane, a-axis, of $Pb(Zr_{0.35}Ti_{0.65})O_3$ below T_c was almost the same with that of CaF_2. This similarity is the origin of the small residual strain in the films. The saturated polarization and the field induced strain values were almost the same with the estimated value for the polar-axis-oriented film with the same $Zr/(Zr+Ti)$ ratio.

INTRODUCTION

$Pb(Zr, Ti)O_3$ [PZT] films have been widely investigated for FeRAMs and MEMS applications due to their superior ferroelectric and piezoelectric properties. Growth of polar-axis-oriented films have been also widely investigated not only for fundamental understanding of PZT materials characteristics, but also for their applications using large ferroelectricity. Polar-axis, (001), -oriented tetragonal films were reported on $SrTiO_3$ substrates, but its film thickness was limited to the thin film region below 100 nm [1 - 3]. This is owing to the fact that the polar-axis orientation is achieved mainly by the misfit strain due to the lattice mismatch between the PZT and the underlying layers, such as $SrRuO_3// SrTiO_3$ and $SrTiO_3$. However, this misfit strain is known to strongly depend on the film thickness and is dramatically diminished with increasing film thickness [4]. This is the reason why the polar-axis-oriented films above 1 μm in thickness is hardly reported. In such thick region, the volume fraction of the (001) orientation in tetragonal (100)/(001) -oriented films is found to be changed by the thermal strain above the Curie temperature (T_c) under the cooling process. This shows that the epitaxial films grown on the substrates with large thermal expansion coefficient (α_{sub}) expect to have a large volume fraction of the (001) orientation.   In fact, perfect polar-axis-oriented PZT films were obtained even above 3 μm in thickness by grown on (100)CaF$_2$ substrates with large α_{sub} [5, 6].

However, the large α_{sub} of CaF$_2$ also imagines us the possibility of the large residual strain in the films originated to the large thermal strain.   In this study, the strain state of the polar-axis-oriented films was systematically investigated by changing the film thickness.

EXPERIMENT

Epitaxial Pb(Zr$_{0.35}$Ti$_{0.65}$)O$_3$ films with 30 – 3000 nm in thickness were grown on (100)$_c$SrRuO$_3$// (100)LaNiO$_3$// (100)CaF$_2$ substrates at 600 °C by pulsed-metal organic chemical vapor deposition (MOCVD) from Pb(C$_{11}$H$_{19}$O$_2$)$_2$ - Zr(O-t-C$_4$H$_9$)$_4$ - Ti(O-i-C$_3$H$_7$)$_4$ - O$_2$ system [5].   The Zr/(Zr+Ti) ratio and the film thickness of the films were controlled by the input gas concentration of the source gasses and the deposition time, respectively. In this study, the Zr/(Zr+Ti) ratio was set to be 0.35 under Pb/(Pb+Zr+Ti) = 0.50 ascertained by an X-ray fluorescence spectrometer.   Epitaxial (100)$_c$SrRuO$_3$ and LaNiO$_3$ films were grown by rf magnetron sputtering method on (100)CaF$_2$ substrates.

The orientation and the crystal structure of the deposited films was analyzed by the high-resolution X-ray diffraction (XRD) using a four-axis diffractometer.   The XRD reciprocal space mapping (*XRD-RSM*) was employed for more detail analysis of the crystal structure (orientation, in-plane and out-of-plane lattice parameters, and the temperature dependency of these lattice parameters).   The cross-sectional microstructure of the film was evaluated using transmission electron microscopy (TEM) as well as the Raman spectroscopy using the focused beam around 1 μm in diameter.

RESULTS AND DISCUSSION

Figure 1 shows the XRD θ-2θ scan profiles with different film thickness in the 2θ angle region from 42 to 48°.   The in-plane epitaxial relationship was ascertained for all films by the X-ray pole figures measurement because fourfold symmetry was ascertained, suggesting that the cube-on-cube epitaxial relationship between SrRuO$_3$ and PZT films.   Only *002* diffraction peaks of PZT were observed as shown in Fig.1.   All films were found to be polar-axis-orientation, (001) orientation, irrespective of the film thickness taking account of the *XRD-RSM* results.   A noticeable point in Fig.1 was that peak position was hardly independent of the film thickness as summarized in Figure 2(a).

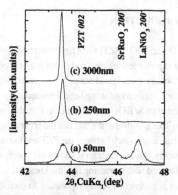

Figure 1   XRD θ-2θ patterns of (a) 50 nm (b) 250 nm and (c) 3000nm-thick PZT films.

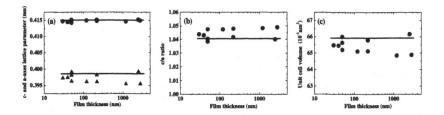

Figure 2 Film thickness dependence of (a) lattice parameter; a-axis (▲) (in-plane) and c-axis (●) (out-of-plane), (b) c/a ratio and (c) unit cell volume for the PZT films grown on CaF$_2$ substrates together with the reported ones for the powder by Kagegawa *et al* shown as solid lines [7].

Figure 2 summarize the film thickness dependence of the lattice parameter of in-plane a-axis, out-of plane c-axis as well as the c/a ratio and the unit cell volume calculated from the lattice parameters of a- and c-axis, which were obtained by the asymmetry spots observations in *XRD-RSM* for the PZT films grown on CaF$_2$ substrates. Reported data for PZT powder by Kagegawa *et al.* [7] was also shown in Figure 2. The lattice parameters of a- and c-axes were almost constant from 0.396 to 0.399 nm and 0.414 to 0.415 nm, respectively, irrespective of the film thickness from 30nm to 3000nm. These value were basically agreed with the reported ones for the powder value with the same Zr/(Zr+Ti) ratio taking account of the accuracy of the film composition analysis. The c/a ratio and unit cell volume were also constant with increase of the film thickness. These results suggest that there is no detectable residual strain in the film at room temperature (*R.T.*) in spite of the fact that the large thermal strain were induced in the PZT film above Curie temperature (*T$_c$*) for realizing the polar-axis orientation. Nevertheless, the XRD analysis only gives us the macroscopic information of the strain in the film.

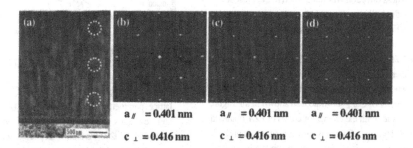

	a$_{//}$ = 0.401 nm	a$_{//}$ = 0.401 nm	a$_{//}$ = 0.401 nm
	c$_\perp$ = 0.416 nm	c$_\perp$ = 0.416 nm	c$_\perp$ = 0.416 nm

Figure 3 (a) Cross-sectional TEM images of 3000 nm-thick PZT film and selected-area electron diffraction patterns (SADP) measured at positions around (b) bottom part, (c) middle part and (d) upper part of the PZT film marked in figure (a).

41

Figures 3 show the cross-sectional TEM images and the selected-area electron diffraction patterns (SADP) of 3000 nm -thick PZT film. Figure 3(a) shows the growth of dense and crack-free film without the existence of the $90°$ domains. Film was found to have perfectly polar-axis-oriented epitaxial growth from top to bottom parts of the film from the results shown in Figure 3(b), (c) and (d). In addition, a- and c-axes lattice parameters calculated from the SADP shown in Figure 3 were almost constant along the film thickness direction and these values almost agreed with the macroscopic ones measured by *XRD-RSM*.

The residual strain graduation along film thickness was also measured by Raman spectroscopy. Figure 4 shows the change of the Raman spectra around $A_1(1TO)$-mode along the film thickness direction for the same 3000nm-thick PZT film shown in Figure 3. The frequency of $A_1(1TO)$-mode is associated with strain and the ferroelectricity. Indeed, the peak shift of $A_1(1TO)$-mode was reported to be sensitively shift by the residual strain [8]. However, the obvious peak shift was not detected in Figure 4, suggesting that the large gradient in residual strain along film thickness direction was not existed in the present PZT thick film grown on CaF_2 substrate.

Figure 5 shows the lattice parameter change with temperature measured by *XRD-RSM* around PZT *204* for 3000 nm-thick PZT film from room temperature to $600°C$. The temperature dependency of the inverse square root of the lattice parameter of CaF_2 substrate also shown in the Figure 5 together with the change of c-axis of PZT obtained by θ-2θ scan. Lattice parameters of a-axis and c-axis increased and decreased, respectively with the increase of the temperature almost in good agreement with the estimated ones by M. J. Haun *et al* for $Pb(Zr, Ti)O_3$ with the $Zr/(Zr+Ti)$ ratio of 0.32 [9]. This shows that the film was under the small

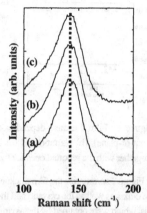

Figure 4 Raman spectra around the $A_1(1TO)$-mode along film thickness direction for the 3000 nm-thick film. (a) bottom part (b) middle part and (c)

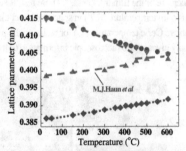

Figure 5 Temperature dependency of the lattice parameters of PZT films [a- (▲) and c- axes (▼) obtained from XRD-RSMs *204*, c-axis (●) obtained from θ-2θ], and the inverse square root of the lattice parameter of CaF_2 substrate (a/2 $^{1/2}$) (◆) . The estimated one for $Pb(Zr,Ti)O_3$ with the $Zr/(Zr+Ti)$ ratio of 0.32 is also shown as dashed line [9].

strain for wide temperature range.   Noticeable point in Figure 5 was that the temperature dependency of the a-axis lattice parameter of PZT was almost the same with that of CaF_2 substrate.

When the film is under cooling process after the deposition, the stress applied to the film, mainly composed of the thermal stress in thick film case, was almost released at T_c by changing the volume fraction of (001) orientation in (100)/(001)-oriented films as already discussed [10].   After the phase transformation occurred, the additional stress applied to the film, again mainly by thermal stress, was induced in the film.   However, the thermal expansion coefficient of a-axis below T_c is closed to that of CaF_2, which means the small thermal stress applied to the film from T_c to $R.T.$.   This similarity is considered to be the origin of the small residual strain in the films at $R.T.$.

Saturation polarization value obtained from polarization-electric field hysteresis loops at 1 kHz was 74 $\mu C/cm^2$ in good agreement with the estimated value for the polar-axis-oriented ones by Morioka et al. [11].   In addition, the d_{33} value measured by the scanning probe microscopy was about 50 pm/V, which is almost the same with the reported value for the polar-axis-oriented film by Nagarajan et al. taking account of the clamping effect from the substrates [12].

CONCLUSIONS

(001)-oriented $Pb(Zr_{0.35}Ti_{0.65})O_3$ films ranging from 30 – 3000 nm in thickness were grown on (100)CaF_2 substrates.   Lattice constants were almost independent of the film thickness in good agreement with reported ones for the powder with the same composition.   Obvious strain gradient was not detected in the majority of the film thickness region from the cross sectional analysis by TEM and Raman spectroscopy.   High temperature XRD data suggests that the rate of PZT in-plane, a-axis, lattice parameter change below T_c was almost the same with that of CaF_2.   This similarity is considered to be the origin of the small residual strain irrespective of the film thickness.   The similarity of the saturated polarization and the field induced strain values with reported value for the polar-axis-oriented films with the same Zr/(Zr+Ti) ratio.

ACKNOWLEDGEMENTS

A part of this study was supported by Grant-in-aid for Scientific Research on Priority Area "Nano Materials Science for Atomic Scale Modification 474" from Ministry of Education, Culture, Sports, Science and Technology (MEXT) of Japan.   Authors would thank to Canon Optron Inc. for supplying CaF_2 single crystal substrates.

REFERENCES

1.   H. Morioka, G. Asano, T. Oikawa, H. Funakubo, and K. Saito, Appl. Phys. Lett., **82**, 4761 (2003) .

2. I. Vrejoiu, G. Rhun, L. Pintilie, D. Hesse, M.Alexe, and U. Gösele, Adv.Mater., **18**(13) 1657 (2006).

3. H. Lee, S. Nakhmanson, M. Chisholm, H.Christen, K. Rabe, and D. Vanderbilt, Phys. Rev. Lett., **98**, 217602 (2007).

4. Y. Kim, H. Morioka, R. Ueno, S. Yokoyama and H. Funakubo, Appl. Phys. Lett., **86**, 212905 (2005).

5. T. Fujisawa, H. Nakaki, R. Ikariyama, T. Yamada, and H. Funakubo, Apply. Phy. Express., **1**, 085001 (2008).

6. T. Fujisawa, H. Nakaki, R. Ikariyama, T. Yamada, M. Ishikawa and H. Funakubo, J.Appl. Phys. submitted.

7. K. Kakegawa, J. Mouri, K. Takahashi, H. Yamamura, and S. Shirasaki, Chem. Soc. Jpn., **49** (1976) 717 [in Japanese].

8. J. Sanjurjo, E. Lopez-Cruz, and G. Burns, Phys. Rev. B, **28**, 7260 (1983).

9. M. Haun, E. Furman, H. Makinstry, and L. Cross, Ferroelectrics., **99**, 27 (1989).

10. K. Nishida, H. Funakubo, T. Yamamoto, and T. Katoda, Integrated Ferroelectrics., **99**, 23 (2008).

11. H. Morioka, S. Yokoyama, T. Oikawa, H. Funakubo, and K. Saito, Appl. Phys. Lett., **85**, 3516 (2004).

12. V. Nagarajan, A. Stanishevsky, L. Chen, T. Zhao, B. –T. Liu, J. Melgalis, A. L. Roytburd, R. Ramesh, J. Finder, Z. Yu, R. Droopad, and K. Eisenbeiser, Appl. Phys. Lett., **87**, 242905 (2005).

Nano-Materials and Nano-Fabrication

Mater. Res. Soc. Symp. Proc. Vol. 1129 © 2009 Materials Research Society   1129-V11-05

Investigation of the Crystalline Orientations and Substrates Dependence on Mechanical Properties of PZT Thin Films by Nanoindentation

Dan Liu, Sang H. Yoon, Bo Zhou, Barton C. Prorok and Dong-Joo Kim
Materials Research and Education Center, Auburn University, AL 36849, USA

ABSTRACT

In this paper, we investigated the effects of the substrates and crystalline orientations on the mechanical properties of Pb(Zr$_{0.52}$Ti$_{0.48}$)O$_3$ thin films. The PZT thin films were deposited by sol-gel method on platinized silicon substrates with different types of layer materials such as silicon nitride and silicon oxide. The crystalline orientations of PZT thin films were controlled by combined parameters of a chelating agent and pyrolysis temperature. A nanoindentation CSM (continuous stiffness measurement) technique was employed to characterize the mechanical properties of those PZT thin films. It was observed that (001/100)-oriented films show a higher Young's modulus compared to films with mixed orientations of (110) and (111), indicating a clear dependence on film orientation. The influence of substrates on the mechanical properties of PZT thin films was also characterized. Finally, no significant influence of the film thickness was found on the mechanical properties of films thicker than 200 nm.

INTRODUCTION

PZT thin films are widely used in microelectromechanical (MEMS) systems due to their large piezoelectric coefficients, dielectric constant, and excellent electromechanical properties [1-2]. Since the piezoelectric properties of PZT thin films are influenced by the mechanical coefficients, designing MEMS devices should understand the scaling effects of mechanical properties of PZT thin films.

In recent years, much attention has been paid to the measurement of the mechanical properties of PZT thin films. Wang et al. [3] studied the crystalline orientation dependence on the mechanical properties of PZT thin films. Delobelle et al. [4-5] measured the mechanical properties of PZT films employing different electrode materials. Fang et al. [6] investigated the influence of annealing temperature on the nanomechanical properties of PZT thin film. Xu et al. [7] and Bahr et al. [8] studied the mechanical properties of PZT thin film by nanoindentation. C.Chima-Okereke et al.[9] also investigated the mechanical properties of PZT multilayer systems by experimental ,analytical and FEA method between which good agreement was shown, and pointed out the mechanical properties of surface film can be measured in certain range of the a/t (contact radius/film thickness) value .However, a study about the effects of substrates on the mechanical properties of PZT films has not been undertaken. Since the mechanical properties of PZT thin films are often dominated by other layer structures such as substrates, this study conducted an investigation into the effects of different substrates and the crystalline orientations on the nanomechanical properties of PZT thin films. Compared with other methods of exploring the mechanical properties of thin films, including impulse acoustic method, micro-beam cantilever deflection technique, and AFM, the nanoindentation technique has a high sensitivity and does not need to remove the film from its substrate [4,7]. Hence, nanoindentation is chosen in our experiments.

EXPERIMENT AND PRINCIPLE

Thin film fabrication

The commercial PZT sol-gel solutions (110/52/48) (Inostek Inc.) were spin-coated on 10 × 10 mm^2 Pt(111)/Ta/SiO$_2$/Si and Pt(111)/Ta/SiNx/Si substrates at 4000 rpm for 20 seconds followed by drying at 150 ^{0}C for 2 minutes and pyrolysis at 300 ^{0}C for 10 minutes to remove residual organics. Such coating and pyrolysis treatments were repeated 3, 6, 9, and 12 times to achieve the desired thicknesses, which are 330, 580, 830, and 1130nm, respectively, and a pre-annealing process was carried out at 650 ^{0}C for 15 minutes every third time. After the thin films had achieved the desired thickness, a final annealing was given at 650 ^{0}C for 1 hour.

In order to deposit PZT thin films with different orientations on Pt(111)/Ti/SiO$_2$/Si(100) substrates, lead acetate trihydrate, zirconium n-propoxide , and titanium isopropoxide were used as the starting materials, and 2-methoxyethanol was used as the solvent. The precursor solution was synthesized with a Zr/Ti ratio of 52/48, and the mole concentration of the PZT precursor is 0.5M. Similarly, another solution was obtained by adding acetylacetone (AcAc) as the solvent with 2-methoxyethanol. Finally, six PZT samples (~700nm) were obtained through 7 deposition cycles by changing the pyrolysis temperature and using two PZT solutions (Table I).

Table I. Summary of the deposition parameters and measured Young's Moduli: 450C-WOAC and 450C-WAC denote 450^{0}C pyrolysis temperature without AcAc and with AcAc.

Sample number	Measured Young's Modulus (GPa)	Std.Dev	Orientation	XRD pattern
1	153.9	7.8	001,110	450C-WOAC
2	151.5	5.6	001	450C-WAC
3	154.3	11.3	001	350C-WOAC
4	135.4	11.9	110,111	350C-WAC
5	132.7	9.2	110,111	250C-WOAC
6	135.8	13.7	110,111	250-WAC

Nanoindentation measurements

Nanoindentation tests were performed using a nano-indenter (MTS Nano-indenter XP system) with a combination of a continuous stiffness method (CSM) and static method. Based upon the method developed by Oliver and Pharr [10], the hardness and Young's modulus can be determined by the indentation load-displacement data.

In order to obtain the true mechanical properties of the PZT films, the following conditions are satisfied:
1) The indentation depths were limited to be less than a tenth of the thin film thickness (Figure 1-a) (Here, we take a SiNx-based PZT sample with 12 times deposition as an example) according to the one-tenth "rule of thumb" that suggests that the measured mechanical property is very close to the true value of the films if a film has a thickness of 10 times the indentation depth [3, 11-12]. And also, microcracks and pile-ups do not occur in PZT films because the indentation

48

depths are less than 20% of the film thickness [4]. As we can see from the curves of load-displacement in Figure 1-a, no discontinuity is observed, which indicates micro-cracks do not occur in the indentation process [13-14].
2) The Young's Modulus is obtained from the flat regions in those Young's Modulus-Displacement curves [15] (Figure 1-b).

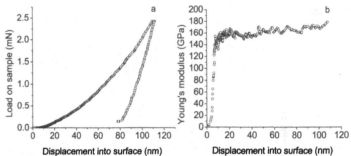

Figure 1. (a) Load-Displacement curve, (b) Young's Modulus-Displacement curve of SiNx-based PZT thin films with 12 times deposition.

RESULTS AND DISCUSSION

Crystalline orientation effects

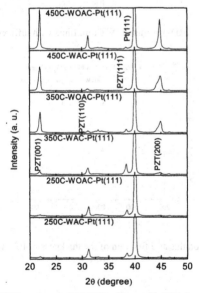

Figure 2. XRD patterns of PZT thin films with SiO₂-based substrates.

49

The PZT thin films with two different kinds of pronounced orientations were obtained (Figure 2). The average value of Young's moduli of the PZT films with strong (001) orientation are in the range of 151.5GPa-154.3GPa, which are higher than those of the samples with mixed orientations of (110) and (111) (Table I). The result is in agreement with the reports by Delobelle et al. [4] and Wang et al. [3].

Substrate effects

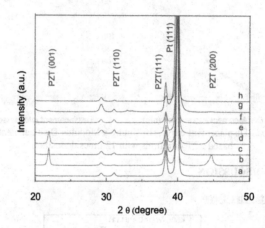

Figure 3. XRD patterns of PZT thin films with different substrates.

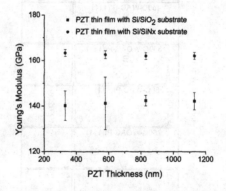

Figure 4. Young's Modulus as a function of the thickness of PZT thin films with different substrates.

The PZT thin films with four different kinds of thicknesses and two types of layer materials were prepared and characterized. Figure 3 shows the XRD patterns of PZT thin films prepared with different substrates and thin film thicknesses. The SiO_2-based PZT thin films with 3, 6, 9, and 12 times deposition are denoted as a, c, e, and g, and SiNx-based PZT thin films with 3, 6, 9, and 12 times deposition are denoted as b, d, f, and h. The PZT thin films almost show the same (111) orientation; therefore, we exclude the orientation effect on the mechanical properties of PZT thin film. Figure 4 shows the evolution of Young's modulus as a function of the thickness of PZT thin film with different substrates.

The PZT films with SiNx-based substrates show higher Young's moduli than those with SiO_2-based substrates although the penetration depths were limited to within 10% film thicknesses, which indicate the generally admitted one-tenth rule may not be completely applied for determining the mechanical properties of PZT films, otherwise we should get the same value even with different substrates.

The analysis of these results can be explained by categorizing two different structures of the PZT thin film; 1) a soft thin film on a harder substrate ($E_f<E_s$, E_f is the Young's modulus of the thin film, E_s is the Young's modulus of the SiNx or SiO_2 and 2) hard thin film on a softer substrate type structure ($E_f<E_s$). In the case of the PZT thin film on a SiNx-based substrate (soft/harder type), at small indentation depth, pronounced elastic deformation only occurs in the softer PZT thin film along vertical and longitudinal directions while the deformation in substrate is negligible, thus the measured Young's modulus of is very close to the true film modulus. And the pronounced deformation will propagate with the increasing depths until it reaches the film/substrate interface. Subsequently, the deformation is constrained vertically and only propagates laterally. However, the pronounced elastic deformation will penetrate the interface and propagate in the substrate and measured modulus will be influenced by the substrate effect if the increasing indentation depth reaches a critical value. One point needs to be stressed: the contribution of substrate will start when pronounced elastic deformation reaches the film/substrate interface; therefore, the key issue is to determine the specific indentation depth. By the finite element method, Panich etc., finding the substrate effect can be neglected as long as the indentation depth is less than 30% of the film thickness [16], and Chen etc. defined the critical value as 50% of the film thickness [17]. Based on their conclusions, the Young's modulus of the PZT thin film with SiNx-based substrate should be the intrinsic film mechanical property.

In the case of PZT thin film on SiO_2-based substrate (hard/softer type), the situation will be different. The substrate effect will be significant even for small indentation depths at which the deformation already penetrates the film/substrate interface. The critical indentation depth varies with different substrate materials, and Chen's simulation results [17] show that even an indentation depth less than 10 to 20% of the film thickness is insufficient to avoid the substrate effect, which may explain why the Young's modulus of PZT thin film with SiO_2-based substrate is smaller than that of PZT thin film with SiNx-based substrate due to the softer SiO_2 substrate. In the further research, experiments to estimate the critical indentation depth depending on hard/softer thin film structure will be conducted.

The Young's modulus of PZT films does not change evidently with the increasing film thickness, which may be attributed to the similar orientation of PZT thin films with different thicknesses.

CONCLUSIONS

(001)-oriented films show higher Young's modulus values compared to thin films with mixed orientations of (110) and (111), indicating a clear dependence on film orientation. Employing different layer materials (e.g. silicon dioxide or silicon nitride) may lead to a change in the mechanical properties of PZT films. In the case of PZT thin film on SiNx-based substrate, we may obtain the true Young's modulus of PZT thin film, which is around 160GPa. No significant influence of the film thickness above 200 nm thick PZT films was observed on the mechanical properties of those films. The generally admitted one-tenth rule may not be completely applied for determining the mechanical properties of PZT films.

REFERENCES

1. Kanno, S. Fujii, T. Kamada, and R. Takayama, "Piezoelectric properties of c-axis oriented Pb(Zr,Ti)O₃ thin films", Appl. Phys. Lett. 70, 1378 (1997).
2. B. Jaffe, W. R. Cook, and H. Jaffe, *Piezoelectric Ceramics*, London: Academic press, 1971, pp 271-280.
3. Q. M. Wang, Y. P. Ding, Q. M. Chen, M. H. Zhao, and J. R. Chen, Appl. Phys. Lett. 86, 162903 (2005).
4. P. Delobelle, E. Fribourg-Blanc, and D. Remiens, Thin Sol. Films 515, 1385 (2006).
5. P. Delobelle, G.S. Wang, E. Fribourg-Blanc, and D. Remiens, J. Eur. Ceram. Soc. 27, 223 (2007).
6. T-H Fang, S-R Jian, and D-S Chuu, J.Phys.: Condens. Matter 15, 5253 (2003).
7. X.H. Xu, P. Gu, R. jiang, G. Zhao, L. Wen and J.R. Chu, Proceedings of 1st IEEE International Conference on Nano/Micro Engineered and Molecular Systems, 18 (2006).
8. D.F. Bahr, J.S. Robach, J.S. Wright, L.F. Francis and W.W. Gerberich, Mater. Sci. Eng., A259, 126 (1999).
9. C.Chima-Okereke, A.J. Bushby, M.J. Reece, R.W. Whatmore and Q. Zhang, J. Mater. Res., 21, 409 (2006).
10. W.C. Oliver and G.M. Pharr, J. Mater. Res. 7, 1564 (1992).
11. Standard Test for Microhardness of Materials, "ASTM Standard Test Method E 384," Annual Book of Standards 3.01, American Society for Testing and Materials, p 469, 1989
12. G.N. Peggs, I.C. Leigh, Recommended procedure for microindentation Vickers hardness test, Report MOM 62, UK National Physical Laboratory, 1983.
13. X.J. Zheng, Y.C. Zhou, Y.Y. Li, Acta Mater. 51, 3985 (2003).
14. P. Delobelle, G.S. Wang, E. Fribourg-Blanc, and D. Remiens, Surf. Coat. Technol. 201, 3155 (2006).
15. Hay, J. L. and Pharr, G. M. Instrumented indentation testing. In ASM Handbook, Vol. 8: Mechanical Testing and Evaluation, 10th ed. (Kuhn, H. and Medlin, D., eds.). ASM International, 232–243, 2000.
16. N.Panich and Y.Sun, "Effect of penetration depth on indentation response of soft coatings on hard substrates: a finite element analysis", Surface and Coatings Technology 182, 342 (2004).
17. X.Chen and J.J. Vlassak, "Numerical study on the measurement of thin film mechanical properties by means of nanoindentation", J. Mater. Res.,16, 10(2001).

Mater. Res. Soc. Symp. Proc. Vol. 1129 © 2009 Materials Research Society   1129-V04-21

Carbon Nanofiber-Network Sensor Films for Strain Measurement in Composites

Nguyen Quang Nguyen, Sangyoon Lee, and Nikhil Gupta
Composite Materials and Mechanics Laboratory
Mechanical and Aerospace Engineering Department
Polytechnic Institute of New York University
Brooklyn, NY 11201
Phone: 718-260 3080, Fax: 718-260 3532, Email: ngupta@poly.edu

ABSTRACT

A carbon nanofiber-based sensor film is designed and calibrated for force measurement. The sensor is designed for use in structural health monitoring of composite materials. The sensing scheme is based on creating a network of carbon nanofibers on the surface of the composite material. In the experimental scheme a patch of nanofiber reinforced epoxy resin film is developed and adhesively bonded to the laminate. The extension of the sensor film due to the applied force leads to a change in the connectivity of carbon nanofibers in the film, resulting in the change in the resistance of the network. Results show that such sensing schemes have high sensitivity and repeatability. Use of nanofibers can provide a low cost and more efficient alternative to other sensor films that rely on carbon nanotubes.

INTRODUCTION

Carbon nanotube based sensors are being explored in the recent years [1, 2]. In nanostructures electron transport through one nanotube and through the network are significant issues [3]. In addition contact resistance plays a dominating role because of losses at each junction. Some of these effects can be reduced if the conducting network is made of high aspect ratio structures, having fewer connections but still very high surface area. Carbon nanofibers (CNFs), which possess mechanical and electrical properties similar to those of carbon nanotubes (CNTs), are cylindrical nanostructures with stacks of graphene layer perpendicular to the nanofiber axis [4]. Compared to carbon nanotubes, CNFs are much cheaper than the multi-walled or single-walled carbon nanotube [5]. Therefore, carbon nanofibers are used in the present study to explore the possibility of developing nanofiber network based sensing schemes.

Structural health monitoring techniques based on CNF sensors have potential application in civil, mechanical, and aerospace systems [6]. CNFs have been extensively used as biosensor and gas sensors in published studies [4, 7-9]. However, their application in force, displacement or temperature sensors are not yet fully developed despite having great potential in this area.

Normally the nanotube/nanofiber networks are connected to the electrical measurement systems using conductive silver paint. However, the properties of the adhesive paint in bonding the electrodes with the network dominate the measurement accuracy. In our sensors we expect displacements of the order of several mm. At this displacement scale the measurements are observed to have significant fluctuations, resulting in loss of sensitivity. In present work embedded electrodes are used to improve the connection quality to obtain consistent measurement of resistivity.

Published studies have shown high resistivity below 1.5 wt.% of CNFs in epoxy resin matrix composites [10]. Therefore, the CNF weight fraction is maintained between 1.5 and 2% in the present study. This weight fraction of CNFs is also known to enhance mechanical properties of composites (or composite films) [11].

The CNF films can be also used as a temperature sensor due to the difference of thermal expansion coefficients between CNFs and the epoxy matrix [12]. As the temperature increases the thermal expansion mismatch between fibers and resin introduces stresses in the film and leads to a change in the resistance measured between two ends of the film. The film is used as the temperature sensor based in this principle.

FABRICATION OF THE SENSOR FILM

Carbon nanofiber films are fabricated by mechanically mixing nanofibers in an epoxy resin. The nanofibers of type PR19 XT PS are obtained from Pyrograph Inc. Epoxy resin DER 332, manufactured by DOW Chemical Co. is used as the matrix resin. An amine based hardener DEH 24, also manufactured by DOW Chemical Co., is used as the curing agent for the resin. The CNF/epoxy mixture is spread uniformly on an aluminum plate and compressed between two aluminum platens coated by a release agent. The CNF film is then allowed to cure at room temperature for 10 hours. After curing, the film has about 750 μm thickness. Table 1 shows compositions of two types of films fabricated in the experiments.

In order to measure the resistance, conductive wires of diameter 160 μm are placed at each 24.5 mm distance along the length of the film during the fabrication process. The wires have two ends extended out of the film to facilitate the resistance measurement. The wires keep the contact fixed with the carbon nanofiber structure inside the film during the test; and small cross sectional area allows precise measurement of the resistance. This measurement scheme helps in substantially reducing the variation in resistance measurement due to the quality of the electrical contact.

Glass fiber/epoxy resin laminates are used as the base material for testing the CNF sensor films. Two different schemes are used for sensor placement on the laminate. In the first scheme a sensor film is fabricated and adhesively bonded to the surface of the laminate. In the second approach the sensor film is directly fabricated on the surface of the laminate. The laminates are constituted of four layers of glass fabrics. Epoxy resin DER 332 is used as the matrix resin. The glass fabrics are oriented in ±60° orientation in order to obtain large deformation in the test.

Table 1. Composition of carbon nanofiber sensor films.

Film type	Components	Quantity	
		Weight (%)	Volume* (%)
1	Carbon Nanofibers	1.7%	1%
	Matrix resin	98.3%	99%
2	Carbon Nanofibers	1.4%	0.8%
	Matrix resin	98.6%	99.2%

* Constituent properties obtained from [13, 14]

The laminates are fabricated using hand lay-up technique. Subsequently, a hydraulic compression molding press is used to apply a load of 1000 kg and remove the excess resin from the laminate. The resulting laminates are about 1.3 mm thick. After compression, the laminate is allowed to cure at room temperature for at least 1 hour and then cut into the flexural test specimens. Two specimens of 24.5×122.5 mm^2 size are selected for conducting the bending tests. A thin layer of epoxy resin is used to adhesively bon the CNF films on the laminate surface as shown in Figure 1(a). To improve the adhesion, the surface of the laminate is scratched and sanded before bonding the film.

EXPERIMENTAL SETUP

The laminate with adhesively bonded sensor film is tested under flexural loading conditions using a three-point bend test configuration on a computer controlled INSTRON 4467 test machine. Two fixed supports are set 101 mm apart and a cross head applied load at the center of the laminate. The CNF film is attached on the lower surface of the specimens, which is under tension in the present test configuration as shown in Figure 1(b). The resistance across a desired length is measured using a multimeter, while the load and deflection are detected by the INSTRON machine using Bluehill software. The experiment is conducted at room temperature.

Temperature sensor consists of a 24.5 mm long carbon nanofiber film as shown in Figure 2. The thin film configuration allows bending due to the temperature changes. The film is kept into a furnace and the temperature is increased from the room temperature to 160°C in 20°C steps. A thermocouple is also used to independently measure the temperature in the same region inside the furnace.

RESULTS AND DISCUSSION

Force and displacement sensor

Load-deflection curves obtained in three different flexural tests performed on a laminated specimen containing the sensor film attached to it are presented in Figure 3(a). In the same tests the resistance data are also obtained, which are presented in (Figure 3b). The resistance change is presented in terms of a difference between the resistance of the film at a given load and the resistance in the unloaded condition. For specimen 1 the change in resistance is measured across two different lengths of the sensor film, 122.5 and 25.4 mm. For the specimen 2 the resistance change is measured across the film length of 122.5 mm. This scheme allows comparing the results for two different sensor film lengths and also for two specimens of the same sensor film length. Results in Figure 3(a) show that the load-displacement curves obtained in all three tests are comparable to each other. Figure 3(b) shows that the change in resistance with respect to the deflection is almost linear. The linear trend allows directly relating the displacement or load with the resistance values. It is also observed that for two different specimens of the same length the resistance values are different. The CNFs are randomly dispersed in these sensor films. Therefore the connectivity of fibers in the network is different for two film of similar composition. However, it is expected that an increase in the volume fraction of carbon fibers in the network will make their resistance values more uniform and within a narrow range.

55

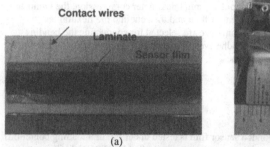

(a)   (b)

Figure 1. (a) Glass fiber/epoxy laminates with attached sensor films and (b) experimental setup for testing a sensor film attached to a laminate.

Figure 2. Carbon nanofiber film used as temperature sensor.

The change in resistance with respect to the displacement and force depends on the components of composites, mixing quality, as well as the size of the specimen. These parameters can be used to tailor the properties and sensitivity of the sensor film. Types of materials used in the composites also can be used to adjust the change and extend the application of the sensor in other fields. It shoud be noted that the presence of strong interfacial bonding between CNFs and epoxy resin may reduce the sensitivity of these films because of complete wetting and increase in the contact resistance between various CNFs.

Temperature sensor

Two sensor films, of 25.4 and 76.2 mm length, respectively, are tested for temperature sensitivity. The change in resistance with respect to the temperature is plotted in Figure 4 for both these films. The experimental results show that the temperature increase leads to a decrease in the resistance of the film.

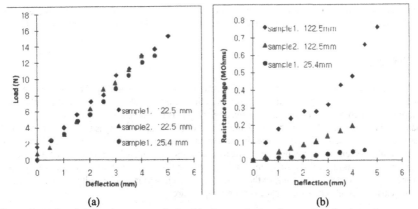

(a)   (b)

Figure 3. (a) Load-deflection curves obtained in three-point bend tests of the laminated composites attached with the sensor films and (b) change in the resistance of the sensor film with respect to the deflection.

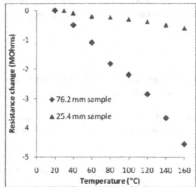

Figure 4. Change in the relative resistance of the sensor film with temperature.

The difference in the thermal expansion coefficients of the carbon nanofibers and epoxy resin causes significant thermal stresses in the film.Since, the microstructure of the film contains random distribution of nanofibers, the thermal stresses induce bending in the film. These two effects, change in the connectivity of the CNFs in the network due to thermal expansion/contraction and bending, both increase the resistance of the film as the temperature increases. It is observed that the decrease is steeper in the longer film at the same temperature range because of the higher deflection in the longer film. Compared to the sensitivity of the sensor films for strain observed in Figure 3(b) the variation of resistance with respect to the temperature is about 8 times higher, showing a promising application of these films in temperature measurement. The temperature is detected in real time domain showing a potential

57

application under dynamic test conditions. Moreover, the sensitivity can be tailored by means of properties, volume fraction, and contact resistance of CNFs used in preparing these sensors.

CONCLUSIONS

Carbon nanofiber based films are fabricated and tested for potential applications in stress, strain, and temperature measurement. The applications are focused on the structural health monitoring of composite materials. The sensing scheme is based on creating network of nanofibers in the film. Applied force, displacement or temperature changes the connectivity (and resistance) of the network. The change in the resistance can be correlated to the applied stimulus to calibrate the sensor. The experimental results have shown that such sensors can be created and used in the structural health monitoring applications. The nanofiber volume fraction and wetting properties can be changed to change the sensitivity of such sensors. The results also show that these sensors can also be used for measuring temperature.

ACKNOWLEDGMENTS

This research is supported by the National Science Foundation Grant #CBET 0619193, through the Major Research Instrumentation – Development program.

REFERENCES

1. Thostenson, E.T. and T.-W. Chou, *Carbon nanotube-based health monitoring of mechanically fastened composite joints.* Composites Science and Technology, 2008. **68**(12): p. 2557-2561.
2. Thostenson, E.T., Z. Ren, and T.-W. Chou, *Advances in the science and technology of carbon nanotubes and their composites: a review.* Composites Science and Technology, 2001. **61**(13): p. 1899-1912.
3. Li, C.Y., E.T. Thostenson, and T.W. Chou, *Dominant role of tunneling resistance in the electrical conductivity of carbon nanotube-based composites.* Applied Physics Letters, 2007. **91**(22): p. 223114.
4. Lu, X., et al., *Carbon nanofiber-based composites for the construction of mediator-free biosensors.* Biosensors and Bioelectronics, 2008. **23**: p. 1236-1243.
5. Allaoui, A., S.V. Hoa, and M.D. Pugh, *The electronic transport properties and microstructure of carbon nanofiber/epoxy composites.* Composites Science and Technology, 2008. **68**(2): p. 410-416.
6. Kang, I., et al., *Introduction to carbon nanotube and nanofiber smart materials.* Composites: Part B, 2006. **37**: p. 382-294.
7. Vaseashta, A. and D. Dimova-Malinovska, *Nanostructured and nanoscale devices, sensors and detectors.* Science and Technology of Advanced Materials, 2005. **6**: p. 312-318.
8. Vamvakaki, V., K. Tsagaraki, and N. Chaniotakis, *Carbon Nanofiber-Based Glucose Biosensor.* Analytical Chemistry, 2006. **78**(15): p. 5538-5542.

9. Baker, S.E., et al., *Fabrication and characterization of vertically aligned carbon nanofiber electrodes for biosensing applications*. Diamond and Related Materials, 2006. **15**(2-3): p. 433-439.

10. Higgins, B.A. and W.J. Brittain, *Polycarbonate carbon nanofiber composites*. European Polymer Journal, 2005. **41**(5): p. 889-893.

11. Zhou, Y., et al., *Microstructure and resistivity of carbon nanotube and nanofiber/epoxy matrix nanocomposite*. Journal of Materials Processing Technology, 2008. **198**: p. 445–453.

12. Yu, C., et al., *Thermal Contact Resistance and Thermal Conductivity of a Carbon Nanofiber*. Journal of Heat Transfer, 2006. **128**: p. 234- 239.

13. Pyrograf-III, *http://www.apsci.com/ppi-pyro3.html*. 2008.

14. Dow Plastics Co., *http://test.newpolystar.com/tds/211/332.pdf*. 2008.

Mater. Res. Soc. Symp. Proc. Vol. 1129 © 2009 Materials Research Society　　　1129-V05-05

Absorption-induced deformations of nanofiber yarns and webs

Daria Monaenkova, Taras Andrukh, and Konstantin G.Kornev
School of Materials Science and Engineering, 161 Sirrine Hall, Clemson University, Clemson, South Carolina 29634, kkornev@clemson.edu

ABSTRACT

Current advances in manufacturing of nanotubular and nanofibrous materials with high surface- to - volume ratios call for the development of adequate characterization methods and predictive estimates of the materials absorption capacity. Extremely high flexibility of these materials poses a challenge: their pore structure easily changes upon contact with the fluid in question. This paper sets a physical basis for analyses of absorption processes in nanotubular and nanofibrous materials. As an example, we study absorption of droplets by yarns and webs made of nanofibers and microfibers. We show that absorption can induce different types of deformations: visible deformations of the sample profile and deformations of the yarn diameter/length caused by the capillary forces. Using experimental data and theory, we estimate elastic and transport characteristics of the nanofibrous materials. The reported experiments and proposed theory open a new area of research on absorption-induced deformations of nanotubular and nanofibrous materials and show their potential applications as sensors to probe minute amount of absorbable liquids.

INTRODUCTION

Due to recent progress in electrospinning, the number of applications of electrospun nanofibers and webs grow exponentially (1-3). Varying the density of nanofibers in the web, one can significantly decrease the weight of electrospun mats. In addition to that, the nanofibers can be made porous (4). These materials are very important for some micro and nanofluidic applications which require a hierarchical pore structure with the range of pore sizes varying from nanometers to micrometers (5, 6). In Figure1, we show some typical electrospun nanofibrous structures from polyvinylidene difluoride/polyethylene oxide (PVDF/PEO) blends which we use to control fluid release/absorption. These 0.5 μm - 2 μm diameter fibers have about 83% porosity (7)!

The characterization of nanoweb properties is a challenging task, especially when the web thickness is very small; in our case it is comparable with the nanofiber diameter.

In this paper we report on development of a new method for characterization of fiber-based materials. The proposed method allows us to measure the permeability of the fiber-based materials and the internal stresses induced by the capillary forces and the weight of absorbed liquids.

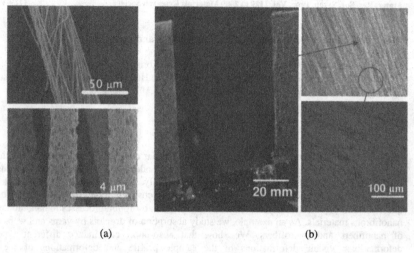

(a)   (b)

Figure 1. a) Yarn made of nanoporous PVDF nanofibers produced by electrospinning, b) electrospun PVDF webs.

EXPERIMENTAL

In the proposed method, a sample of fiber-based material (yarn or film) hangs freely under its own weight between two posts. One end of the specimen is immersed into a vessel with a wetting liquid. If the sample is highly-porous, the liquid invades the pores spontaneously thanks to the wetting forces. When the liquid propagates through the sample, the flow induces visible deformations: one can see how the sag moves. The ends of the sample are firmly fixed, therefore the sample doesn't slide during absorption experiments. Since the sag profile depends on the position of the liquid front, for full description of the sample deformations one needs to follow the front movement.

Therefore, we placed a mirror to image the front position from the top. A monochromatic camera (1.0 Mp, Diagnostic Instruments Inc., MI) captures simultaneously the sag profile and the image of the wet area reflected from the mirror. Experiments showed that the length of the sag did not change appreciably. Hence we assumed that the elastic deformations imposed by the flow were negligible and the sample initial length L, diameter R (for yarns), and the width W (for planar substrates) stayed the same. We checked this assumption with the nanofiber yarn, which, morphology is shown in Figure 1 a). A sequence of frames shown in Figure 2 illustrate the process of droplet absorption by the nanofiber yarn. When the drop of ethylene glycol invades the pores in the yarn, the yarn diameter does not change appreciably, i.e. we cannot distinguish these deformations analyzing the images.

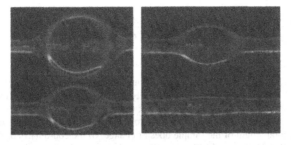

Figure 2. The drop of ethylene glycol on the yarn composed of PVDF/PEO electrospun fibers.

THEORY

As follows from these experiments, elastic deformations are not significant, and the sag profile is controlled only by the weight of the absorbed liquid. The sample weight is balanced by the total tension σ which can be represented as the difference between the tension on fibers T and the hydrostatic pressure in the liquid P, i.e. $\sigma = T-P$ (Figure 3a).

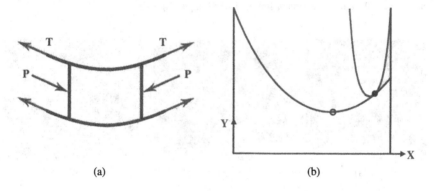

(a)   (b)

Figure 3. a) The model of a porous material as a deformable tube filled by a liquid. For any point of the sample, the tension T is supported by the fibers, and the pressure P is supported by the liquid. The total tension is therefore $\sigma = T-P$; b) theory predicts that the suspended partially saturated sample is described by two catenaries, the red catenary describes the wet part, while the blue one describes the dry part. The empty circle with coordinates $(X_{dry}, Y_{dry}+a_{dry})$ marks the minimum for dry catenary, the solid circle with coordinates $(X_{wet}, Y_{wet}+a_{wet})$ marks the minimum for wet catenary.

Writing the force balance for each point of the sample, one can show (8) that the sample profile can be described by a combination of two catenaries corresponding to wet and dry parts of the sample. At each instant of time, these profiles are described by the following equations

63

$$y = a_{wet} \cosh\left(\frac{x - X_{wet}}{a_{wet}}\right) + Y_{wet} \quad , \quad y = a_{dry} \cosh\left(\frac{x - X_{dry}}{a_{dry}}\right) + Y_{dry} \quad ,$$

$$(1)$$

where $a_{wet} = \rho_{wet}g/\Sigma$, $a_{dry} = \rho_{dry}g/\Sigma$; ρ_{dry} and ρ_{wet} are the densities of dry and wet parts of the sample, respectively; g –acceleration due to gravity; Σ – horizontal component of the total stress in the sample. This component is the same for wet and dry parts.

All unknown parameters can be found from the boundary and initial conditions which state that a) the total length of the sample doesn't change; b) the coordinates of the sample ends are known; c) the forces are balanced at the propagating front and the sample profile is continuous. After solution of this system of equations for unknown parameters a_{wet}, a_{dry}, X_{dry}, Y_{dry}, X_{wet}, Y_{wet}, the model allows us to predict the sag profile with respect to the sample saturation level.

RESULTS

Typical configurations of the fibrous samples are shown in Figure 4. As seen from these pictures, the deformations of these nano and microfibrous materials are visible. Using our model, we were able to describe the distribution of stresses in the sample, Figure 5.

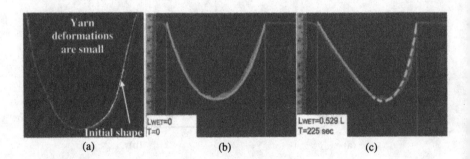

Figure 4. a) Deformation of nanofiber yarn after partial saturation with water. White line is the profile of dry yarn, and red line is the profile of wet yarn. b)-c) The results of the model profile over-imposed on the experimental configurations of the strips of paper towel saturated with water: b) dry paper towel, c) partially saturated paper towel (wet part is shown as dashed line, dry part is shown as solid line)

According to the proposed model, when the wet part of the sample is almost vertical, it supports a stronger tension compared to that produced in the dry part, Figure 5. The model also reveals that the tension on fibers in the sample stays almost constant in the wet part. Augmented with the Darcy's law and mass balance (9,10), the model allows us to characterize the sample permeability.

To appreciate the significance of the capillary forces, we show the results of experiments on drop spreading over a channel comprised of two fibers. The capillary pressure P forces the fibers to snap off the gap between them, Figure 6. The tension T acting on the fibers opposes this meniscus-induced contraction. For tightly twisted nanofiber yarns this deformation effect is not visible: the internal elastic stresses in the fibers are able to withstand the capillary pressure without appreciable deformations of the nanofibers. However, the absolute value of the internal tension T, is expected to be appreciable as shown in Figure 5.

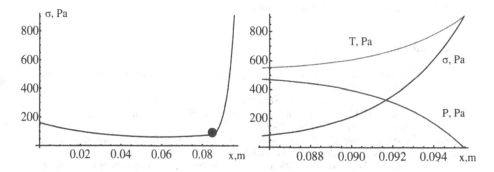

Figure 5. (a) The tension distribution along the sample. The jump in the slope marked by the red circle corresponds to the front position. In this example, the wet part occupies (3/10) the total sample length, (b) The distribution of the tension components in the wet part of the sample. The density of the wet sample is assumed to be 10 times greater than the density of the dry sample. The distance between posts is 9.546 cm.

Figure 6. Droplet of ethylene glycol absorbed by the fiber rail made of two metal wires. Observe the gap contraction when the drop got sacked into the interfiber channel.

CONCLUSION

We propose a new method for characterization of nanofibrous materials with respect to absorption-induced deformations and permeability. This method allows one to estimate the internal stresses exerted by the capillary forces on nanofiber yarns and webs. From a single experiment, one can extract the internal stresses, materials permeability and absorption capacity.

Experimental and theoretical analyses of model fibrous systems reveal highly inhomogeneous stress distribution in the materials.

ACKNOWLEDGMENTS

This work is supported by National Textile Center, project M08-CL10 and National Science Foundation, project CMMI-0826067.

REFERENCES

1. Reneker, D. H., and Yarin, A. L., Electrospinning jets and polymer nanofibers, *Polymer, 49*, 2387 (2008).
2. Dzenis, Y., Materials science - Structural nanocomposites, *Science, 319*, 419 (2008).
3. Rutledge, G. C., and Fridrikh, S. V., Formation of fibers by electrospinning, *Advanced Drug Delivery Reviews, 59*, 1384 (2007).
4. Zhang, L. F., and Hsieh, Y. L., Nanoporous ultrahigh specific surface polyacrylonitrile fibres, *Nanotechnology, 17*, 4416 (2006).
5. Gibson, P. W., Schreuder-Gibson, H. L., and Rivin, D., Electrospun fiber mats: Transport properties, *Aiche Journal, 45*, 190 (1999).
6. Neimark, A. V., Ruetsch, S., Kornev, K. G., Ravikovitch, P. I., Poulin, P., Badaire, S., and Maugey, M., Hierarchical Pore Structure and Wetting Properties of Single Wall Carbon Nanotube Fibers, *Nano Letters, 3*, 419 (2003).
7. Andrukh, T., Few, C., Burtovyy, O., Burtovyy, R., Luzinov, I., and Kornev, K. G., Preparation and characterization of highly porous PVDF nanofibers and yarns, in *Nanotechnology 2008: Life Sciences, Medicine & Bio Materials - Technical Proceedings of the 2008 NSTI Nanotechnology Conference and Trade Show*, Vol. 2, NSTI, Nanotech Conference Publications, Boston, MA, pp. 763 (2008).
8. Freeman, I., A general form of the suspension bridge catenary *Bulletin of the American Mathematical Society, 31*, 425 (1925).
9. Lucas, R., Ueber das Zeitgesetz des kapillaren Aufstiegs von Flussigkeiten, *Kolloid Zeitschrift, 23*, 15 (1918).
10. Washburn, E. W., The dynamics of capillary flow, *Physical Review, 17*, 273 (1921).

Mater. Res. Soc. Symp. Proc. Vol. 1129 © 2009 Materials Research Society   1129-V04-20

Study of the conduction mechanism and the electrical response of strained nano-thin 3C-SiC films on Si used as surface sensors

Ronak Rahimi[1], Christopher M. Miller[1], Alan Munger[2], Srikanth Raghavan[1], C. D. Stinespring[3] and D. Korakakis [1, 4]

[1]Lane Department of Computer Science and Electrical Engineering, West Virginia University, Morgantown, WV 26506-6109
[2]Department of Mechanical and Aerospace Engineering, West Virginia University, Morgantown, WV 26506-6106
[3]Department of Chemical Engineering, West Virginia University, Morgantown, WV 26506-6102
[4]National Energy Technology Laboratory, 3610 Collins Ferry Road, Morgantown, WV 26507-0880

ABSTRACT

Various superior properties of SiC such as high thermal conductivity, chemical and thermal stability and mechanical robustness provide the basis for electronic and MEMS devices of novel design [1]. This work evaluates heterostructures that consist of a few nanometers-thick 3C-SiC films on silicon substrates. Nano-thin SiC films differ significantly in their electrical behavior compared to the bulk material [2], a finding that gives rise to a potential use of these films as surface sensors. To gain a better understanding of the effect of surface states on the electrical response of these thin, strained films, several metal-semiconductor-metal heterostructures have been examined under variable conditions. The nano-thin, strained films were grown using gas source molecular beam epitaxy. Reflection high-energy electron diffraction patterns obtained from several 3C-SiC films indicate that these films are strained nearly 3% relative to the SiC lattice constant. Al, Cr and Pt contacts to a nano-thin film 3C-SiC were deposited and characterized. I-V measurements of the strained nano-thin films demonstrate metal-semiconductor-metal characteristics. Band offsets due to biaxial tensile strain introduced within the 3C-SiC films were calculated and band diagrams incorporating strain effects were simulated. Electron affinity of 3C-SiC has been extracted from experimental I-V curves and is in good agreement with the value that has been calculated for a strained 3C-SiC film [3]. On the basis of experimental and simulation results, an empirical model for the current transport has been proposed. Fabricated devices have been characterized in a controlled environment under hydrogen flow and also in a reactive ambient, while heating the sample and oxidizing the surface, to investigate the effects of the environment on the surface states. Observed changes in I-V characteristics suggest that these nano-thin films can be used as surface sensors.

INTRODUCTION

In multilayer semiconductor structures, stress and strain may be introduced due to differences in the lattice constants and thermal expansion coefficients of materials. As long as the thickness of the grown layers remains below the critical thickness [4], the grown films will be strained. In contrast, if the thickness of the layers is beyond the critical thickness, strain will be relaxed. Misfit dislocations which are the result of strain relaxation in the film appear increasingly beyond the critical thickness [4]. Studies of strained heterostructures have shown that strain in the films creates changes in the electronic band structure of the layers [5]. Some of these changes include band gap shift and the removal of degeneracy of light and heavy holes at the valence-band maximum. In order to obtain the desired performance from devices based on strained heterostructures, the aforementioned effects of the strain should be considered and incorporated in the design parameters. In this work, a thin film of 3C-SiC has been grown on Si substrate and the effects of strain on the electrical properties of this heterostrcuture have been studied. In the 3C-SiC/Si heterojunction, since the thermal expansion coefficient of 3C-SiC is greater than that of Si, the introduced strain to the epitaxial films will be biaxial tensile strain, as attested by various measurement techniques and calculations [6, 7].

EXPERIMENT

Nano-thin 3C-SiC films evaluated were grown on n-type Si (001) using gas source molecular beam epitaxy (GSMBE). Details about the GSMBE system and the growth conditions have been reported elsewhere [8, 9]. Reflection high-energy electron diffraction (RHEED) patterns obtained from several SiC films show that the SiC films are about 3% strained relative to the SiC lattice constant. One of the key issues that has been investigated involves an understanding of the overall effect of lattice strain on the semiconductor-metal interface states associated with the electrical contacts. Due to the lattice mismatch between 3C-SiC and Si, a high density of planar lattice defects will be introduced in the grown films. These defects, manifesting as vacancies, can become filled with electrically active nitrogen atoms, ultimately resulting in n-doped SiC films [10]. In this research doping concentration of the unintentionally doped films has been evaluated from the experimental results.

Metal semiconductor metal (MSM) structures were used to study the conduction mechanism and the electrical properties of the strained nano-thin 3C-SiC films on Si. MSM structure is a two terminal device consisting of 3C-SiC film and two Schottky contacts. MSM devices exhibit various transport properties considerably different from a single Schottky diode. Schottky barriers formed at the interface of the metal and the semiconductor have an essential role in the electrical transport in the MSM structure [2]. To fabricate the MSM device, 3C-SiC nano-thin films were patterned using a mask consisting of a series of different-sized contact pads at different spacings. Several metal contacts (Al, Cr and Pt) were deposited on the patterned samples using a CVC 610 dc magnetron sputter deposition system.

RESULTS AND DISCUSSION

In this research, electrical properties of the grown nano-thin 3C-SiC films have been characterized under different conditions. Several samples were made using the procedure described in the previous section. Samples have been characterized after using hydrofluoric (HF) acid to modify the surface treatment, during exposure to hydrogen in a gas test cell and at elevated temperatures using a hot chuck probe. Due to the HF treatment, the original Schottky characteristics change to Ohmic and the results of this test have been presented and explained in detail elsewhere [9]. Fabricated samples were also studied by placing them in a reactive ambient (high temperature in atmosphere) to induce slow surface changes.

Figure 1 shows the I-V characteristics at different temperatures, from room temperature to 100°C and then measuring again in room temperature after heating the samples up to 200°C. As shown in this figure, changing temperature has essentially rendered the electrical behavior of the contacts to be near Ohmic versus their original Schottky state. To analyze the effect of the environment on the surface states and the conduction mechanism, electrical characteristics of the

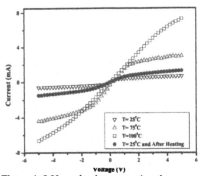

Figure 1. I-V results demonstrating the effect of temperature on nano-thin 3C-SiC/n-type Si

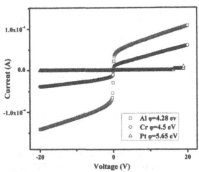

Figure 2. Schottky characteristics demonstrated by Al, Cr and Pt on nano-thin film 3C-SiC/Si.

structures have been evaluated in a hydrogen test cell. For this experiment, three different metals (Al/Ti/Au (100/30/600nm)) were deposited on the grown films. Samples were bonded to a multi-pin mounting and connected to a wire feed-through which provided the necessary connections for the electrical characterization of the devices. Results of the hydrogen tests show that this structure can be used as a gas sensor. Preliminary results of this structure as a gas sensor and its response to hydrogen have been discussed elsewhere [9].

Nano-thin films differ significantly in their electrical behavior compared to the bulk material [2]. Therefore, to completely understand and explain the results obtained under different conditions, electrical properties of nano-thin 3C-SiC/Si need to be further evaluated. In this work, back-to-back Schottky diode characteristics were reproduced consistently for as-deposited contacts (Al, Cr, and Pt) on the n-type 3C-SiC. I-V characteristics of these different contacts are shown in Figure 2. In this figure, for Pt contacts with the higher work function, current is much smaller than Al with the lower work function.

As mentioned before, band alignment and Schottky barriers at the interface of the metal and the semiconductor affect the electrical transport considerably. Studies of strained heterostructures have also shown that strain in the films create changes in the electronic band structure of the layers [3, 5]. Band offsets due to biaxial tensile strain introduced within the 3C-SiC films were calculated and band diagrams incorporating strain effects were simulated. Figure 3 shows the simulation results of Al contacts on 3C-SiC/Si for strained and unstrained SiC films. Parameters such as band gap shift and electron affinity change can also be extracted from the I-V curves. For the MSM structures the current of the reverse-biased Schottky barrier dominates the total current [11]. This current is due to barrier lowering and can be calculated from the following equation,

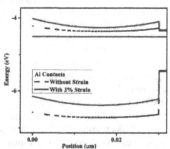

Figure 3. Band structure simulation using ADEPT [15] for Al contacts on 3C-SiC/Si for strained and unstrained SiC films.

69

$$I = AA^*T^2 \exp(\frac{-q\phi_b}{kT})\exp(\frac{-q\Delta\phi_b}{kT})\{\exp(\frac{qV}{kT})-1\} \tag{1}$$

where A is the contact area, A* is the Richardson's constant, ϕ_b is the barrier height of the reverse-Schottky barrier, k is the Boltzmann constant, V is the applied bias, $\Delta\phi_b$ is the barrier lowering, E is the electric field and V_{bi} is the built-in potential [12].

$$A^* = \frac{4\pi qkm^*}{h^3} = 1.201\times10^2(\frac{m^*}{m_0})[A/cm^2K^2], \quad \Delta\phi_b = \sqrt{\frac{qE}{4\pi\varepsilon_s}}, \quad E = \sqrt{\frac{2qN_D(V+V_{bi}-\frac{kT}{q})}{\varepsilon_s}}, \quad \varphi_b = \varphi_M - \chi_S$$

After substituting and simplifying all the above values in equation (1), an exponential relationship between current and $Z(V,N_D,\chi_s)$ can be observed as follows.

$$I = \exp((Ln(Area\times A^*\times T^2)\times(-\frac{q}{kT}\varphi_M))+\frac{q}{kT}\chi_s+\frac{q}{kT}\sqrt{\frac{q}{4\pi\varepsilon_s}}\sqrt[4]{\frac{2q}{\varepsilon_s}}\sqrt[4]{N_D}Z(V,N_D,\chi_s)) \tag{2}$$

$$Z(V,N_D,\chi_s) = \sqrt[4]{qV+\varphi_M-\chi_s-\frac{kT}{q}-\frac{kT}{q}Ln(\frac{N_c}{N_D})}$$

Figure 4 shows a semi-logarithmic plot of current as a function of Z (V, N_D, χ_s) for Al contacts for four different samples. For each sample, a curve fit has been done by minimizing the mean square error with respect to N_D and χ_s. Using a numerical analysis, the donor carrier concentration (N_D) and the electron affinity (χ_s) of the strained 3C-SiC have been extracted from these curve fits and summarized in Table I. The experimental values for the electron affinity of strained 3C-SiC (3.76 - 3.84 eV) are very close to the calculated value of the electron affinity for strained 3C-SiC films (3.8 eV) used in the simulations [6, 13]. It is also known that bulk 3C-SiC has no strong Fermi level pinning, and hence, it can be assumed that the metal-semiconductor barrier depends on the 3C-SiC electron affinity and Fermi level [14]. Experimental and fitted semi-logarithmic plot of the current as a function of Z (V, N_D, χ_s) for Sample 1 is shown in Figure 5. As presented in this figure, three different regions can be identified. The values for N_D and χ_s of 3C-SiC have been calculated from the curve fit in the Schottky barrier lowering (SBL) region. In this region, with increasing applied voltage, current does not change much and the current depends more on carrier concentration than on voltage. Therefore, the

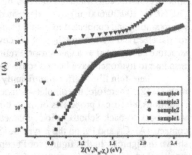

Figure 4. Semi-logarithmic plot of reverse current vs. Z (V, N_D, χ_s) for Al contacts on four different samples.

	N_D (cm^{-3})	χ_s (eV)
Sample 1	1.0×10^{16}	3.762
Sample2	1.0×10^{15}	3.789
Sample3	3.0×10^{15}	3.821
Sample4	2.2×10^{15}	3.840

Table I. N_D and χ_s values for four different samples consisting of Al contacts on 3C-SiC/Si heterostructure.

shift in current at a constant voltage in this region provides the basis for a sensor signal. Fabricated device use this relationship to sense different concentrations of hydrogen. When hydrogen is absorbed at the surface, it changes the donor carrier concentration, which will shift up the Fermi level, and as a result, the SiC surface will be metalized. As the surface becomes increasingly metalized, more current will flow as hydrogen increases, which provides the basis for sensing the change in hydrogen concentration. The application of this structure as a hydrogen sensor and its results has been presented elsewhere [9].

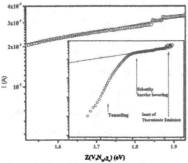

Figure 5. Experimental and fitted semi-logarithmic plot of I vs. Z (V, N_D, χ_s) for Sample 1.

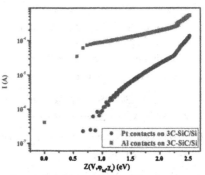

Figure 6. Semi-logarithmic plot of reverse current vs. Z (V, N_D, χ_s) for Al and Pt contacts.

Semi-logarithmic plots of current as a function of Z (V, N_D, χ_s) for Al and Pt contacts are shown in Figure 6. For Pt with a higher work function compared to Al, a tunneling mechanism dominates up to much higher voltages and the conduction mechanism switches directly from tunneling to thermionic emission. This switch can be explained intuitively since, with a higher energy barrier, any lowering will not be as evident as with a lower Schottky barrier.

CONCLUSIONS

In this work, the effects of strain and stress generated because of the lattice mismatch as well as the difference in thermal expansion coefficients between the 3C-SiC film and the thick Si substrate have been studied. Experimental I-V measurements show that a back-to-back Schottky diode characteristic is reproduced consistently. In this structure, based on the simulation and experimental results, Schottky barrier lowering current is believed to be the dominant current mechanism and it depends more on the donor carrier concentration than on voltage. In the gas sensors, due to the surface metallization, donor carrier concentration increases which will shift the current at a constant voltage in the Schottky barrier lowering region. This shift of the current provides the basis for a sensor signal.

ACKNOWLEDGMENTS

This technical effort was performed in support of the National Energy Technology Laboratory's under DOE / RDS, LLC Contract No. DE-AC-04NT41817.606.02.09. R.R. has been partially supported through the WVNano Bridge Award at West Virginia University.

REFERENCES

1. R.G. Azevedo, D.G. Jones, A.V. Jog, B. Jamshidi, D.R. Myers, L. Chen, X. Fu, M. Mehregany, M.B. Wijesundara, A.P. Pisano, IEEE Sens. J. **7**, 801-803 (2007).

2. W.T. Hsieh, Y.K. Fang, K.H. Wu, W.J. Lee, J.J. Ho and C.W. Ho, IEEE Trans. Electron Devices **48**, 801-803 (2001)

3. W.J. Choyke, Z.C. Feng and J.A. Powell, Jour. of Appl. Phys **64**, 3163 (1988).

4. J.W. Matthews and A.E. Blakeslee, *Jour. of Crys.Growth* **27**, 118 (1974).

5. C. Ohler, C. Daniels, A. Forster and H. Luth, *Phy. Rev.*B **58**, 7864 (1998).

6. Z.C. Feng, A. Mascarenhas, W.J. Choyke and J.A. Powell, *J. Appli. Phys.* **64**, 3176 (1988).

7. H. Mukaida, H. Okumura, J.H. Lee, H. Daimon, E. Sakuma, K. Endo and S. Yoshida, *J. Appl. Phys.* **62**, 254 (1987).

8. K. S. Ziemer, A.A. Woodworth, C.Y. Peng and C.D. Stinespring, *Diamond and Related Mat.***16**, 486 (2007).

9. R. Rahimi, S. Raghavan, N.P. Shelton, D. Penigalapati1, A. Balling, A.A. Woodworth, T. Denig, C.D. Stinespring and D. Korakakis, *Mater. Res. Soc. Symp. Proc.***1056**, (2008).

10. R.F. Davis, G. Kelner, M. Shur, J.W. Palmour, J.A. Edmond, Proc. of the IEEE **79**, 677-701 (1991).

11. Z.Y. Zhang, C.H. Jin, X.L. Liang, Q. Chen and L.M. Peng, *Appl. Phys. Lett.* **88**, 73102 (2006).

12. S.M Sze, "Metal-Semiconductor Contacts" Physics of Semiconductor Devices New York: John Wiley & Sons, (1981).

13. C.P. Kuo, S.K. Vong, R.M. Cohen and G.B. Stringfellow, *J. Appli. Phys.* **57**, 5428 (1985).

14. J.S. Foresi and T.D. Moustakas, *Applied Physics Letters* **62**, 2859-61 (1993).

15. www.nanohub.org.

Mater. Res. Soc. Symp. Proc. Vol. 1129 © 2009 Materials Research Society   1129-V11-14

Electrical and Optical Properties of Gold-Strontium Titanate Nano-Composite Thin Films

S. Ganti, Y. Dhopade, R.K. Gupta, K. Ghosh, P.K. Kahol
Department of Physics, Astronomy, and Materials Science, Missouri State University,
Springfield, MO 65897, U.S.A.

ABSTRACT

Thin films based on nano-composites have attracted considerable attention for their possible applications in devices and sensors. These nano-composite thin films are formed by embedding metal or semiconductor nano-particles in a host material and they exhibit interesting electrical transport properties. Using pulsed laser deposition technique, we have prepared nano-composite thin films of gold-strontium titanate on quartz substrate. Gold and strontium titanate were used as targets for pulsed laser deposition. Thin films having different compositions were grown. The effect of different composition on their electrical and optical properties has been studied in details. The structural characterizations of the films were done by x-ray diffraction, transmission electron microscopy, scanning electron microscopy, and atomic force microscopy. Transmission electron microscopy as well as atomic force microscopy shows the presence of gold nano-particles in these films. X-ray diffraction and energy dispersive x-ray spectroscopy shows the existence of strontium titanate and gold. Current-voltage characteristics and temperature dependent resistivity measurements were made to characterize electrical properties of these films. Electrical properties can be manipulated from metal to insulator through semiconductor by varying the composition. In addition, it is observed that the absorption of visible light increases with increase in gold percentage. This indicates that these nano-composites could also use as active materials for many electronic as well as optical sensors.

INTRODUCTION

Pulsed laser deposition (PLD) technique is widely used for deposition of thin films of advanced materials having different properties [1]. The PLD has many advantages such as epitaxial growth of thin films [2], retention of target composition in the deposited films [3], and formation of multilayers [4]. To grow composite films by PLD, two approaches can be envisaged. The first is the combinatorial approaches, which is based on the use of multitarget systems. The second approach is to use the single target having the desired composition. In this paper, we report the properties of gold (Au)-strontium titanate (STO) nano-composites using combinatorial approach.

STO is an insulator with a band gap of ~3.2 eV having perovskite crystal structure [5]. STO is widely used for ferroelectric applications because of its high dielectric constant [6]. The dielectric constant of STO is nearly 300 at room temperature and 10000 at low temperature. Rao et al have studied the effect of growth parameters such as ablation fluence, pressure, and substrate temperature on the properties of PLD grown STO films [7]. Several researchers have used STO as a dielectric in metal-oxide-semiconductor field effect transistor [8]. It is reported that the electrical properties of STO could be modify from insulator to semiconductor by introducing oxygen vacancies [9]. Bellingeri et al have used semiconducting and insulating STO to fabricate field effect devices using PLD technique [10]. In this communication, we report synthesis and characterization of gold-STO nano-composites using PLD. The effect of composition on structural, optical, and electrical properties were studied.

EXPERIMENTAL DETAILS

Gold target for the pulsed laser deposition is purchased from Kart J. Lesker, USA. The STO powder is purchased from Alfa-Aesar, USA. The STO target is made by cold pressing the STO powder at 6×10^6 N/m^2 load, followed by sintering at 950 °C for 12 hours. The nano-composite of gold and STO is prepared using KrF excimer laser (Lambda Physik COMPex, λ = 248 nm and pulsed duration of 20 ns). The laser was operated at a pulse rate of 10 Hz, with an energy of 200 mJ/pulse. The laser beam was focused onto a rotating target at a 45° angle of incidence. All the composites were deposited at room temperature under vacuum of base pressure 1.0×10^{-6} mbar. The structural and optical characterizations of the films were done by x-ray diffraction, transmission electron microscopy, scanning electron microscopy, atomic force microscopy, and uv-visible spectroscopy. The current voltage (I-V) characteristics and temperature dependence resistivity were measured by standard four-probe method using programmable electrometer (model 617, Keithley) and programmable voltage source (model 230, Keithley).

RESULTS AND DISCUSSION

Figure 1 shows the x-ray diffraction patterns of pure gold, STO, and their composite. All the films were grown at room temperature on quartz substrate. It is observed that deposited pure gold film is highly oriented along (111) direction. The (111) preferred orientation of gold nano particles is also observed in the composite films. The pure STO film is amorphous in nature. It is seen that intensity of gold peak decreases with increase in STO content in the films. The film composition is determined with the help of energy dispersive x-ray spectroscopy. The observed percentage composition of the film is almost same to that of target composition. The small variation in composition is observed that may be due to difference in melting points and vapor pressures of gold and STO.

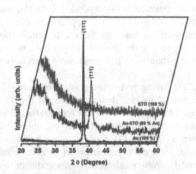

Fig. 1. XRD patterns of Au-STO nano-composite.

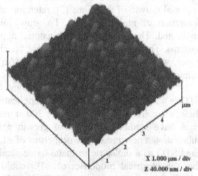

Fig. 2. AFM image of Au-STO (40 % Au) nano-composite.

74

The atomic force microscopic image of Au-STO (40 % Au) nano-composite grown at room temperature under vacuum is shown in Figure 2. The scan was carried out in tapping mode. The spring constant of the cantilever was ~42 N/m. The cantilevered tip was oscillated close to the mechanical resonance frequency of the cantilever (typically, 200–300 khz) with amplitudes ranging from 10 to 30 nm. The surface morphology of the film suggests granular growth of the composite. The root mean square (rms) roughness and average roughness of the film is 2.9 nm and 2.2 nm respectively. The peak to valley roughness of the film is observed to be 20.2 nm.

The optical absorbance spectra of Au-STO nano-composites are discussed below. Figure 3 shows absorbance spectra of the composites films recorded in the 250-950 nm

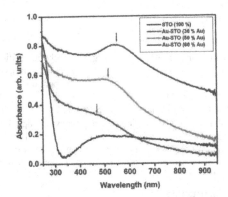

Fig. 3. UV-Visible spectra of Au-STO nano-composite.

range. It is evident from the optical absorbance spectra that the absorbance of visible light increases with increase in gold percentage in the composites. These composites could be use as active absorbance materials in solar cells. The appearance of peak around 510 nm indicates the presence of gold nano-particles. The peak shifts toward higher wavelength with increase in gold composition in the films suggesting formation of the gold clusters [11]. Buso et al have also reported increase in optical absorbance of titania (TiO_2) films by incorporation of gold nano-particles [12].

The dc electrical characterizations of the films were discussed next. Current-voltage characteristics of all the films are linear suggesting Ohmic nature of these films. Figure 4 shows the variation of electrical resistivity of Au-STO nano-composites with composition. It is observed that the resistivity of the films decreases with increase in concentration of gold in the composite. We do not observe any sudden change in resistivity with gold concentration suggesting that no percolation model could apply here. Figure 5 shows the effect of temperature on current-voltage characteristics of Au-STO (30 % Au) nano-composite. It is seen that the measured current of the film increases (resistivity decreases) with increase in temperature, suggesting semiconducting nature of the film. The composite films having lower percentage of gold show semiconducting nature while films with higher gold percentage show metallic behavior.

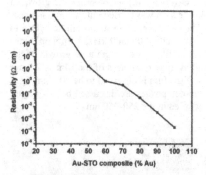

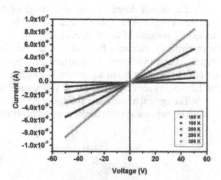

Fig. 4. Resistivity vs. composition plot of Au-STO nano-composites.

Fig. 5. I-V characteristics Au-STO (30% Au) nano-composite at different temperature.

Figure 6 shows the temperature dependent resistivity measurement for Au-STO nano-composites. These films show similar to semiconducting behavior: $d\rho/dT$ is less than zero. To better understand the electrical transport mechanism, various electrical transport models such as granular metal, ionized impurity scattering, and hopping conduction were considered. Different kinds of conduction mechanism can be represented by an exponential temperature dependent resistivity $\rho(T)$ of the form [13]

$$\rho(T) = B_1 \exp [(B_2/T)^{1/q}]$$

where B_1, B_2, and q are constants. These gold-STO nano-composites may be considered as the granular metal system, where small metallic grains are separated by thin insulating layers. For such type of system, a linear fit between log ρ vs. $T^{-1/2}$ (q=2) would indicate that this theoretical model applies. Unfortunately our data also do not fit with this model (Figure 7).

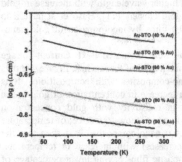

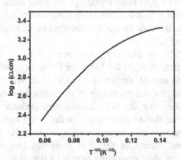

Fig.6. Resistivity vs. temperature plot for Au-STO nano-composites.

Fig. 7. log ρ vs. $T^{-1/4}$ plot for Au-STO (40 % Au) nano-composite.

76

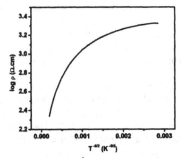

Fig. 8. log ρ vs. T^{-1} plot for Au-STO (40 % Au) nano-composite.

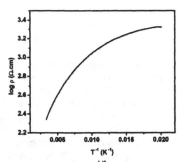

Fig. 9. log ρ vs. $T^{-1/2}$ plot for Au-STO (40 % Au) nano-composite.

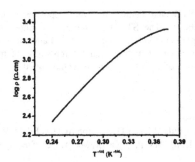

Fig. 10. log ρ vs. $T^{-3/2}$ plot for Au-STO (40 % Au) nano-composite.

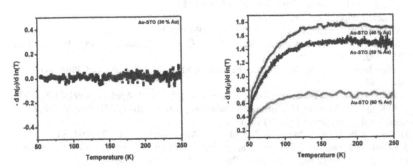

Fig. 11. W plots for Au-STO nano-composites.

The non linear plot between log ρ vs. $T^{-3/2}$ (q=2/3) (Figure 8) suggests that electrical transport is also not due to ionized impurity scattering [13]. Next we consider hopping conduction mechanism for these composites. The straight line between log ρ vs. T^{-1} (q=1) would represent nearest neighbor hopping conduction (Figure 9), while linear relation between log ρ vs. $T^{-1/4}$ (q=4) would suggest variable range hopping (NNH) conduction mechanism. Our data do not fit perfectly with nearest neighbor hopping mechanism, but variable range conduction mechanism seems to apply at higher temperature range.

From $\ln(\rho)$ vs. T^{-1} we see that these composites are not typical semiconductor. To further confirms the nature of composites, we used W-plot (W=-d(lnρ)/d(lnT)) [14]. W-plots are particularly useful when a system is neither in the metallic region and nor in the insulating/semiconducting region, i.e, the system happens to be in critical region. The W-plot reveals that these nano-composites are more towards the metallic regime at higher concentration of gold while composite having 30 % Au is in the critical region.

CONCLUSIONS

Thin films of Au-STO nano-composites were deposited using pulsed laser deposition technique at room temperature. The STO film shows amorphous nature, while gold shows preferred orientation along (111) direction. The current-voltage characteristics of the composites were studied. The W-plot reveals that these nano-composites are more towards the metallic regime at higher concentration of gold while composite having 30 % Au is in critical region.

REFERENCES

[1] D.B. Chrisey, G.K. Hubler, Pulsed Laser Deposition of Thin Films, Wiley, New York, 1994.
[2] R.K. Gupta, K. Ghosh, S.R. Mishra, P.K. Kahol, Mater. Res. Soc. Symp. Proc. 1035 (2007) 1035L-11-17.
[3] R.P. Casero, J. Perriere, A.G. Llorente, D. Defourneau, Phys. Rev. B 75 (2007) 165317.
[4] F.F. Ngaffo, A.P. Caricato, M. Fernandez, M. Martino, F. Romano, Appl. Surf. Sci. 253 (2007) 6508.
[5] N. Shanthi, D.D. Sharma, Phys. Rev. B 57 (1998) 2153.
[6] J.H. Barrett, Phys. Rev. 86 (1952) 118.
[7] G.M. Rao, S.B. Krupanidhi, J. Appl. Phys. 75 (1994) 2604.
[8] X.X. Xi, C. Doughty, A. Walkenhorst, S.N. Mao, Q. Li, T. Venkatesan, Appl. Phys. Lett. 61 (1992) 2353.
[9] I. Pallecchi, G. Grassano, D. Marre, L. Pellegrino, M. Putti, A.S. Siri, Appl. Phys. Lett. 78 (2001) 2244.
[10] E. Bellingeri, L. Pellegrino, D. Marre, I. Pallecchi, A.S. Siri, J. Appl. Phys. 94 (2003) 5976.
[11] R.B. Konda, R. Mundle, H. Mustafa, O. Bamiduro, A.K. Pradhan, U.N. Roy, Y. Cui, A. Burger, Appl. Phys. Lett. 91 (2007) 191111.
[12] D. Buso, J. Pacifico, A. Martucci, P. Mulvaney, Adv. Funct. Mater. 17 (2007) 347.
[13] J. Ederth, P. Johnsson, G.A. Niklasson, A. Hoel, A. Hultaker, P. Heszler, C.G. Granqvist, A.R. Van Doorn, M.J. Jongerius, D. Burgard, Phys. Rev. B 68 (2003) 155410.
[14] A.G. Zabrodskii, Sov. Phys. Semicond. 11 (1977) 345.

Mater. Res. Soc. Symp. Proc. Vol. 1129 © 2009 Materials Research Society   1129-V07-08

Pb(Zr,Ti)O₃ nanofibers produced by electrospinning process

Ebru Mensur Alkoy, Canan Dagdeviren and Melih Papila
Faculty of Engineering and Natural Sciences, Sabanci University, 34956,Tuzla, Istanbul, Turkey

ABSTRACT

Lead zirconate titanate (PZT) nanofibers are obtained by electrospinning a sol-gel based solution and polyvinyl pyrrolidone (PVP) polymer, and subsequent sintering of the electrospun precursor fibers. The PVP content of the precursor solution is critical in the formation of the fully fibrous mats. Scanning electron microscope (SEM) is used to examine the morphology of the precursor fibers and annealed PZT nanofibers. The diameters of the precursor PZT/PVP green fibers have increased with the aging of the precursor solution along with an increase in the viscosity. The viscosity of 500 mPa results in successful fibrous mats, yielding green PZT/PVP fibers with a diameter of 400 nm. The fiber mats are then sintered at 700°C. X-ray diffraction (XRD) pattern of the annealed PZT fibers exhibits no preferred orientation and a pure tetragonal perovskite phase. Preparation of piezocomposites by infusion of epoxy into the nanofiber mat facilitates successful handling of the fragile mats and enables measurements of dielectric properties.

INTRODUCTION

PZT is the most widely used ferroelectric material in ultrasonic transducers, non-volatile random access memory devices, microelectromechanical devices, sensor and actuator applications due to its high dielectric constant, high electromechanical coupling coefficient, and large remnant polarization [1,2]. This material can be processed into various forms such as bulk ceramics, thin films and fibers depending on application area. PZT in fiber form is appealing because of its increased anisotropy, improved flexibility and strength over monolithic PZT ceramics. Micron-scale PZT fibers are usually incorporated into a polymer matrix to obtain smart piezocomposite structures [3]. They can also be used as individual fibers in novel actuator and sensor devices, such as energy harvesting and self-powered in-vivo medical devices, high-frequency transducers, non-volatile ferroelectric memory devices [4]. Nano-scale PZT fibers are also expected to find wide applications particularly in nano-electronics, photonics, sensors and actuators [5].

There are a few methods to obtain PZT in fiber form. Sol-gel, dicing of the bulk ceramic, and extrusion, for instance, are applied to produce PZT fibers typically in the micron scale [6,7,8]. Electrospinning technique has recently gotten attention because the fibers at micro- and even nanoscale can be produced by this method. Nano-scale PZT fibers have been also produced by electrospinning method [9,10,11]. Zhou et al. [12] have produced PZT nanofibers and found that these fibers exhibit significant reversible piezoelectric strains under applied electric field. The level of this strain was measured to be about 4.2% which is reportedly six times larger than that observed in thin films.

The objectives of this study are to investigate the processing conditions in electrospinning of polymeric pre-cursor fiber mats followed by annealing for PZT nanofibers, to examine the phase and morphology of these nanofibers prior and posterior to annealing process and to characterize the dielectric properties of the resultant PZT/epoxy composite.

EXPERIMENTAL

The PZT sol-gel precursor solution was prepared from lead acetate trihydrate, titanium isopropoxide and zirconium n-butoxide in n-butanol. 2-methoxyethanol was used as the main solvent [13]. Final concentration of the solution was 0.4 M and Zr/Ti ratio was 0.5:0.5. The process flow chart for the preparation of the sol-gel solution is given in Figure 1(a). Addition of two alternative polymers, namely, polyvinyl pyrrolidone (PVP) and polyvinyl alcohol (PVA) was investigated. Each polymer choice at various concentrations was added to the sol-gel solution. A home-made, electrospinning set-up allowing computer controlled polymer solution flow rate, as depicted in Figure 1(b), was used to prepare the pre-cursor nano-fiber mats. The PZT/ polymer precursor solution was electrospun on an aluminum foil collector and randomly oriented green fibers were collected. The applied voltage (12 kV) and the distance between the tip of the needle and the collector (10 cm) resulting in electrostatic field of 1.2 kV/cm were kept fixed after preliminary screening experiments. The effect of electrospinning parameters, such as the type of the polymer (PVP versus PVA) and associated polymer solution concentration (6 – 34 wt%), on the morphology of the green precursor fibers and mats were investigated in detail. Annealing regime (T=700°C, heating rate=0.5-5°C/min) was another crucial parameter studied in order to control the crystallinity and the final sintered morphology of the mats.

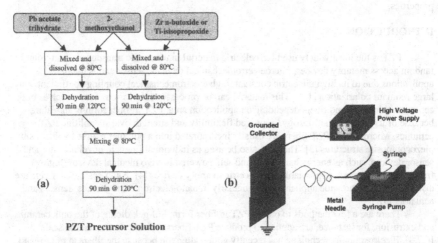

Figure 1. a) The process flow chart for the preparation of the precursor sol-gel solution and (b) schematic representation of the electro-spinning setup.

Thermo Gravimetric Analysis (TGA) was used to identify the crucial steps in the pyrolysis and sintering processes. Sintered PZT mats were characterized by X-ray diffraction

(XRD) and scanning electron microscopy (SEM). Dielectric properties of the PZT/resin composites were also measured by Impedance Analyzer.

DISCUSSION

Microstructural features of green fiber mats

The as-prepared green and sintered nanofiber mats were examined by scanning electron microscope (SEM). Images of the green PZT/PVA mats corresponding to various PVA ratios indicated that no uniform fibers could be collected; instead the mats were predominantly composed of nanoscale beads. In fact, using PVA as a transfer polymer has only led to a spray coating. As a result, PVA was not used as a carrier polymer in the rest of the study.

Figure 2 shows the PZT/PVP green mats by electrospinning at various PVP content in the solution. From Figure 2 (a) and Figure 2(b), small amount of PVP, such as 6wt% or 12wt%, was clearly seen to be insufficient to obtain fiber formation. On the other hand, the concentration of the PVP polymer in the solution exceeding 22wt% appears to result in fibrous mat with reasonable amount of beads. In particular, 28wt% PVP ratio resulting viscosity of 290 mPa was considered to be optimal as the bead-free uniform green PZT/PVP fibers were collected (Figure 2 (d)).

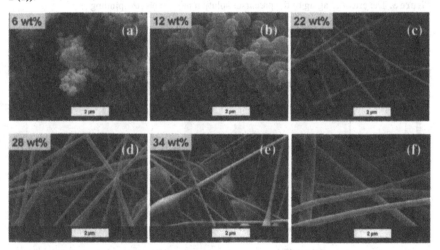

Figure 2. PZT/PVP green fibers from as-prepared solution with (a) 6wt%, (b) 12wt%, (c) 22wt%, (d) 28wt%, (e) 34 wt% PVP content and (f) PZT/PVP green fibers from 72 hour aged solution with 22wt% PVP content.

Additional viscosity measurements and investigation were carried out on the PZT precursor solution of 22wt% PVP in order to explore the effect of aging. The average diameters of the fibers prepared from aged solutions were also measured by SEM. Figure 3 shows the diameter of the PZT/PVP green fibers and viscosity of the precursor solution as a function of aging time. It appears that aging the electrospinning solution for several days led to an increase in viscosity along with an increase in fiber diameters. From Figure 2(c), (f) due to correlation between the viscosity and the PVP content along with Figure 3, it is clearly seen that the viscosity of the solution has significant effect on formation of fibers and their size.

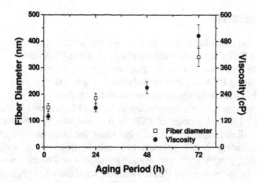

Figure 3. The effect of aging of the precursor solution prior to electrospinning

XRD study and microstructure of the sintered PZT fibers

The phase identification of the PZT nanofibers was performed using XRD. The XRD pattern of the PZT fibers annealed at 700°C for 1 hr indicates that these nanofibers are crystallized in the tetragonal pure perovskite phase with no preferred orientation (see figure 4).

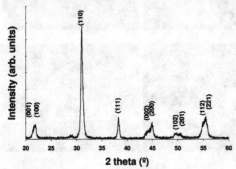

Figure 4. XRD pattern of the PZT nano-fiber mat sintered at 700°C for 1 hour.

Various annealing regimes (temperature = 600-700°C and heating rate = 0.5 – 5°C/min) were investigated to obtain dense PZT nanofiber mats. As depicted from Figure 2, precursor fibers have uniform cylindrical geometry with average diameter varying in the range of 200-350 nm, depending on electrospinning process parameters. After sintering, however, branching occurred between the fibers and the typical diameter shrunk to 100-250 nm range. **Figure 5**(a) and (b) present the morphology of the annealed fibers prepared from the PZT precursor solution with 28 wt% PVP content sintered at 700°C using 5°C/min and 0.5°C/min heating rates, respectively. It is clear that heating regime has significant effect on the microstructure of the fibers.. At the slow heating regime (0.5°C/min) all organics are removed, leaving behind a uniform microstructure with fiber diameters varying between 120-160 nm, whereas fast heating regime (5°C/min) yields a more complex microstructure with a mixture of fibers and platelet-like grains.

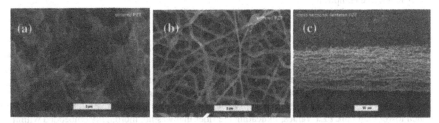

Figure 5. The micrographs of PZT fibers sintered at 700°C for 1 hour with (a) 5°C/min, (b) 0.5°C/min heating regime and, (c) Cross sectional view of the sintered PZT fiber mat.

Dielectric properties of the PZT nano-fiber/resine composites

After obtaining the PZT fibers, PZT/polymer matrix composite samples were prepared. T676NA vinyl ester, Accelerator D (styrene 10wt%, N-N dimethylaniline with >89 wt %), Accelerator G (styrene >80wt %) and Butanox LPT (methyl-ethyl ketone peroxide in diisobutyl phthalate) were mixed to obtain the polymer matrix of the composite. This solution was poured on to the sintered PZT fiber mat and the sample was dried 10 hours at room temperature. The dielectric constant and loss of the PZT/vinylester composites were then measured from 10 kHz to 1 MHz using an HP 41494A Impedance Analyzer (Figure 6).

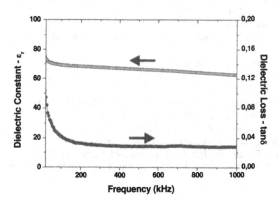

Figure 6. Dielectric constant and of the PZT/resin sample as a function of frequency at the room temperature.

Figure 6 shows dielectric behavior of the PZT/vinylester composite sample. The dielectric constant was found to be fairly stable and vary from 72 to 62 within the measurement range. This value is more that an order of magnitude higher than the dielectric constant of the resin itself. Comparing with bulk PZT [14], however, the composite has much lower dielectric constant. This is expected due to the low PZT fiber ratio in the composites. It may also be related to insufficient embedding or wetting of the fibers by the polymer matrix that may lead to voids within the composite.

CONCLUSIONS

In summary, PZT nanofibers were successfully obtained by electrospinning based process. Processing conditions and the annealing regime was found to have significant influence on the morphology of the PZT nanofiber mats. The dielectric constant and dielectric loss measured at room tempretaure is order of magnitude higher than polymer matrix. It is however much lower than bulk PZT values, indicating the fiber mat was not thoroughly embedded within the matrix. Improving the composite making procedure will further be studied.

ACKNOWLEDGMENTS

We kindly acknowledge the support for this work from The Scientific and Technological Research Council of Turkey - TÜBİTAK Grant 106M364 and post doctoral fellowship of Turkish Academy of Sciences.

REFERENCES

1. R. Maeda, J.J. Tsaur, S.H.Lee and M. Ichiki, *J. Electroceram.* **12**, 89 (2004).
2. N. Setter and R. Waser, *Acta Mater.* **48**,151 (2000).
3. J. Ouellette, *The Industrial Physicist* **2**, 4, 10 (1996).
4. S. Xu, Y. Shi and S. Kim, *Nanotechnology* **17**, 4497 (2006).
5. X.Y. Zhang, X. Zhao, C.W. Lai, X.G. Tang, and J.Y. Dai., *Appl. Phys. Lett.* **85**, 18, 4190 (2004).
6. R. Meyer Jr., T. Shrout and S. Yoshikawa, *J. Am. Ceram. Soc.* **81**, 861-868 (1998)
7. M. Zhang, I.M.M. Salvado and P.M. Vilarinho, *J. Am. Ceram. Soc.* **90**, 358-363 (2007)
8. K. Kitaoka, *J. Am. Ceram. Soc.* **81**, 1189-1196 (1998)
9. Y.Wang, R. Furlan, I.Ramos, J.J. Santiago-Aviles, *Appl. Phys. A* **78**, 1043-1047 (2004)
10. N. Dharmaraj, C.H. Kim, H.Y. Kim, *Mat. Lett.* **59**, 3085-3089 (2005)
11. S. Xu, Y. Shi, S. Kim, *Nanotechnology* **17**, 4497-4501 (2006)
12. Z.H. Zhou, X.S. Gao and J. Wang, *Appl. Phys. Lett.* **90**, 052902 (2007)
13. E. Mensur Alkoy, S. Alkoy and T. Shiosaki, Jpn. J. Appl. Phys., **44**, 12, 8606 (2005).
14. P. K. Sharma, Z. Ounaies, V. V. Varadan and V. K. Varadan, Smart Mat. And Str., 10, 878-883, (2001).

84

Mater. Res. Soc. Symp. Proc. Vol. 1129 © 2009 Materials Research Society   1129-V11-25

Thermal, Mechanical, and Electric Properties of Exfoliated Graphite Nanoplate Reinforced Poly(vinylidene fluoride) Nanocomposites

Fuan He, Jintu Fan*
Institute of Textiles and Clothing, The Hong Kong Polytechnic University,
Hung Hom, Kowloon, Hong Kong, China

ABSTRACT

Poly(vinylidene fluoride) (PVDF)/exfoliated graphite nanoplate (xGnP) nanocomposites were prepared by a solution mixing method for the first time. The thermal, mechanical and electric properties of these nanocomposites were studied by differential scanning calorimetry (DSC), dynamic mechanical analysis (DMA), and an impedance analyzer, respectively. The DSC results indicated that xGnP might act as the nucleating agents and accelerated the overall non-isothermal crystallization process of PVDF. Meanwhile, the incorporation of xGnP also significantly improved the storage modulus and conductivity of the PVDF/xGnP nanocomposites with an increment in the graphite nanoplate content, respectively.

INTRODUCTION

Exfoliated graphite nanoplate (xGnP) has good electrical and thermal conductivity, high mechanical strength, and large aspect ratio [1]. Because of these outstanding features, xGnP has been considered as a promising reinforced agent for the preparation of polymeric nanocomposites [2-8]. Poly (vinylidene fluoride) (PVDF) has received considerable attention because of its good mechanical properties, resistance to chemicals, high dielectric permittivity, and unique pyroelectric and piezoelectric properties [9,10]. In the present work, we attempt to develop new nanocomposites consisting of poly(vinylidene fluoride) (PVDF) and exfoliated graphite nanoplates (xGnPs) by solution mixing method. The effect of different concentrations of graphite nanoplates on the non-isothermal crystallization behavior, dynamic mechanical properties, and conductivity of PVDF/xGNP nanocomposites was investigated.

EXPERIMENT

xGnPs were obtained from subjecting natural graphite flake to acidic intercalation, rapid thermal treatment, and ultrasonic powdering in sequence (see Figure 1). The PVDF/xGnP nanocomposites were prepared by mixing desired amount of xGnP and PVDF in 100 ml of DMF solution at 80 °C under stirring for 2h and then treated with ultrasonic powdering for another 2h. The resultant products were dried at 70 °C for 3 days.

DSC thermal analysis was carried out with a Perkin Elmer DSC-7 differential scanning calorimeter calibrated under a nitrogen atmosphere. Dynamic mechanical property was measured by a Perkin Elmer diamond DMA lab system at a frequency of 1 Hz in a nitrogen atmosphere, with a temperature range from −80 to 80 °C at a scan rate of 5 °C /min. Conductivity of the samples were measured using an Agilent 4294 A impedance analyzer in the frequency range of 50-10^7 Hz at room temperature.

DISCUSSION

Figure 1(b) presents the SEM image of xGnPs, the obtained graphite nanoplates have a size of 20–60 nm in thickness and 0.5–25 μm in diameter. According to the FTIR results reported by other groups [1, 4], there are some functional groups, such as C–O–C, C–OH and COOH, on the surface of xGnP. Figure 1(c) depicts the TEM image of the PVDF/xGnP nanocomposite obtained from solution casting. It can be seen that graphite nanoplates were well dispersed in the PVDF matrix without any aggregation. The good dispersion stability of xGnP in PVDF matrix may be attributed to the interaction between the polar group on the surfaces of graphite nanoplates and the $-CF_2-$ group of PVDF.

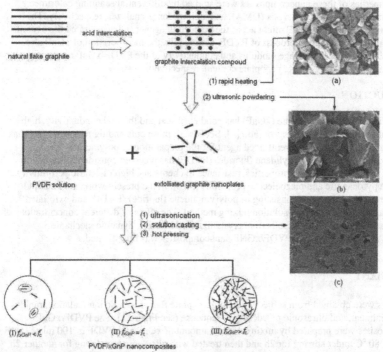

Figure 1. Schematic illustration of the PVDF/xGnP nanocomposites formation

Crystallization behavior

From the DSC curves of melting crystallization for PVDF and its nanocomposites (not shown here), the values of X_t at different time t can be calculated according to the following equations:

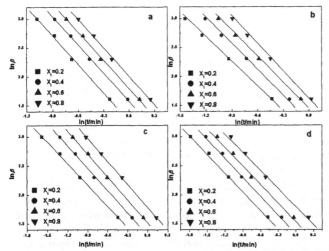

Figure 2 Plots of ln β against ln t for crystallization of(a) PVDF, (b) PVDF nanocomposite with the xGnP loading of 1 wt%, (c) PVDF nanocomposite with the xGnP loading of 5 wt% and (d) PVDF nanocomposite with the xGnP loading of 10 wt%.

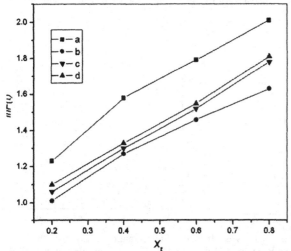

Figure 3 The lnF(t) values at different crystallization degrees for (a) PVDF, (b) PVDF nanocomposite with the xGnP loading of 1 wt%, (c) PVDF nanocomposite with the xGnP loading of 5 wt% and (d) PVDF nanocomposite with the xGnP loading of 10 wt%.

87

$$X_T = \frac{\int_{T_0}^{T} (dH_c / dT) dT}{\int_{T_0}^{T_\infty} (dH_c / dT) dT} \qquad (1)$$

$$t = (T_0 - T) / \beta \qquad (2)$$

where X_T is the relative degree of crystallinity at temperature T, T_0 and T_∞ represent the onset and end of crystallization temperatures, respectively, dH_c/dT is the heat flow rate. By combing Equation (1) with Equation (2), the development of relative crystallinity X_t of PVDF and its nanocomposites with time t at different cooling rates can be obtained. Mo et al. proposed that the relation between β (cooling rate) and t could be defined under a certain crystallinity degree because X_t or X_T is related to β and the crystallization time t or temperature T) [11]. Therefore, the following equation can be obtained for a given degree of crystallinity:

$$\log \beta = \ln F(T) - \alpha \ln t \qquad (3)$$

where $F(T) = [K(T)/k]^{1/m}$, which refers to the cooling rate that must be selected within an unit of crystallization time when the measured system reaches a certain degree of crystallinity, and α is the ratio of the Avrami exponent (n) to the Ozawa exponent (m), that is, n/m. As presented in Figure 2, the plots of $\ln \beta$ vs. $\ln t$ at different degree of crystallinity for PVDF and its nanocomposite show a good linearity. The values of $\ln F(T)$ and α are shown in Figure 3. It can be seen that the values of $\ln F(T)$ for PVDF/xGnP nanocomposites are smaller than that for pure PVDF at the same degree of crystallinity, indicating that the addition of graphite nanoplates can accelerate the overall crystallization process of PVDF. Moreover, the $\ln F(T)$ values of PVDF/xGnP nanocomposites appear to decrease with an increment in the graphite nanoplate content.

Mechanical Property

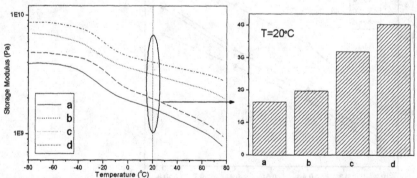

Figure 4. The variation of storage modulus as a function of temperature for (a) PVDF, (b) PVDF nanocomposite with the xGnP loading of 1 wt%, (c) PVDF nanocomposite with the xGnP loading of 5 wt% and (d) PVDF nanocomposite with the xGnP loading of 10 wt%. The right-side image shows the data of storage modulus for PVDF and its nanocomposites at 20 °C.

From Figure 4 it can be seen that the addition of xGnP could significantly increase the storage modulus of PVDF over the whole temperature range. Moreover, the storage modulus of the composites increased with increasing xGnP content. For example, when 20 °C was selected as a point for comparison, the storage modulus for pure PVDF and its nanocomposites with the xGnP loading of 1, 5 and 10 wt% are determined as 1.63, 1.97, 3.19 and 4.03 GPa, respectively. In particular, when the xGnP loading was 10 wt%, the storage modulus at 20 °C was approximately 150% higher than that of pure PVDF.

AC conductivity

The effective AC conductivity as a function of xGnP volume fraction at 1000 Hz is shown in Figure 5. It shows that the conductivity increases only slightly with the increase of xGnP volume fraction (f_{xGnP}), when the xGnP volume fraction is low. This is followed by an obvious insulator-to-conductor transition (known as percolation transition) at $f_{xGnP} \approx 0.76$–1.55 vol.%.

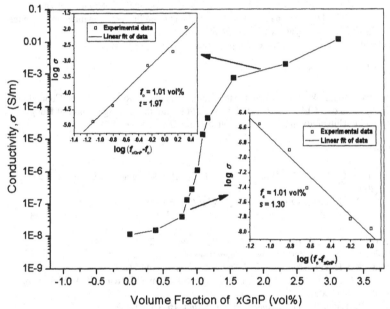

Figure 5. Effective AC conductivity of the PVDF/xGnP nanocomposites as a function of the xGnP volume fraction, measured at 1000 Hz and room temperature. The insets show the best fits of the conductivity to Eq. (4).

The conductivity (σ) of conductor-insulator composites near the percolation threshold can be predicated by the power laws as follows:

$$\sigma(f_{xGnP}) \propto (f_c - f_{xGnP})^{-s'} \text{ for } f_{xGnP} < f_c \quad (4a)$$

$$\sigma(f_{xGnP}) \propto (f_{xGnP} - f_c)^t \quad \text{for } f_{xGnP} > f_c \quad (4b)$$

where f_c is the percolation threshold, f_{xGnP} is the volume fraction of xGnP, s and t are the critical exponents in the insulating ($f_{xGnP} < f_c$) and conducting ($f_{xGnP} > f_c$) region, respectively. The best fits of experimental conductivity values to the log–log plots of the power laws give $f_c = 1.01$ vol%, $t = 1.97$, and $s' = 1.30$ (see the insets in Figure 5).

CONCLUSIONS

PVDF/xGnP nanocomposites were prepared by a solution mixing method for the first time. DSC results showed that the xGnP might act as a nucleating agent and accelerated the overall non-isothermal crystallization process of PVDF. The PVDF/xGnP nanocomposites exhibited much higher storage modulus compared to that of pure PVDF, especially when the xGnP loading amount was 10 wt%. Conductivity of the PVDF/xGnP nanocomposites film increased with the increment in xGnP loading amount and this enhancement can be explained by the percolation theory. These novel PVDF/xGnP nanocomposites can be applied in electromagnetic interference shielding, thin-film transistors and static-charge dissipation.

Part of this work has been accepted for publication in *Advanced Materials*.

ACKNOWLEDGMENTS

The authors would like to acknowledge the funding support of Research Grant Council of HKSAR in the form of a CERG project (PolyU 5164/06E) and the funding support of Hong Kong Polytechnic University in the form of a Niche Area Project (1-BB82).

REFERENCES

1. G.H. Chen, W.G. Weng, D.J. Wu, C.L. Wu, J.R. Lu, P.P. Wang and X.F. Chen, *Carbon* **42**,753 (2004).
2. G.H. Chen, W.G. Weng, D.J. Wu and C.L. Wu, *Eur. Polym. J.* **39**,2329 (2003).
3. D.G. Miloaga, H.A.A. Hosein, M. Misra and L.T. Drzal, *J. Appl. Polym. Sci.* **106**, 2548 (2007).
4. G.H. Zheng, J.S. Wu, W.P. Wang and C.Y. Pan, *Carbon* **42**, 2839 (2004).
5. W.G. Weng, G.H. Chen and D.J. Wu, *Polymer* **44**, 8119 (2003).
6. K. Kalaitzidou, H. Fukushima and L.T. Drzal, *Carbon* **45**, 1446 (2007).
7. H. Yang, M. Tian, Q.X. Jia, J.H.Shi, L.Q. Zhang, S.H. Lim Z.Z.Yu and Y.W. Mai, *Acta. Mater.* **55**, 6372 (2007).
8. L.W. Li, Y.L. Luo and Z.Q. Li, *Smart. Mater. Struct.* **16**, 1570 (2007).
9. W.A. Yee, M. Kotaki, Y. Liu and X.H. Lu, *Polymer* **48**, 512 (2007).
10. M. Wang, J.H. Shi, K.P. Pramoda and S.H. Goh, *Nanotechnology* **18**, 235701 (2007).
11. T.X. Liu, Z.S. Mo, S.G. Wang and H.F. Zhang, *Polym. Eng. Sci.* **37**, 568 (1997).

Mater. Res. Soc. Symp. Proc. Vol. 1129 © 2009 Materials Research Society   1129-V03-02

Controlled assemble and microfabrication of zeolite particles on SiO₂ substrates for potential biosensor applications

S. Ozturk[1,4], K. Kamisoglu[2], R. Turan[1,3], B. Akata[1,4]
[1] Micro and Nanotechnology Department, [2] Chemical Engineering Department,
[3] Department of Physics, [4] Central Laboratory
Middle East Technical University, Ankara, 06531 Turkey

ABSTRACT

Zeolite nanoparticles were organized into functional entities on SiO₂ substrates and microfabrication technique was tested to form patterns of zeolite nanoparticles on SiO₂ using the electron beam lithography (EBL). The effect of different techniques for efficient zeolite assembly on the SiO₂ substrates was investigated. For this purpose, three different assembly techniques were tested. The first two methods are spin-coating (SC) and ultrasound aided strong agitation (US) methods, which were tested using bare and silanized zeolite nanoparticles. The third technique is the manual assembly method, which was also investigated using bare zeolites. All methods were facile in terms of experimental approach. Full coverage of the substrate was obtained after all three methods, however strong agitation (US) leads to better organization of zeolite nanoparticles. Among all techniques, manual assembly method lead to the most organized zeolite nanoparticles with full coverage. Although strong agitation (US) also results in organized zeolite entities, it was not found to be a suitable technique for EBL studies. Using the manual assembly method, it was possible to form monolayers of zeolite nanoparticles on SiO₂ and to make patterns of zeolite nanoparticles by EBL, which offers a simple technique to engineer the surfaces for immobilization of biomolecules.

INTRODUCTION

The development of new fabrication methods of organized nanoparticles on surfaces is important for electronic, optoelectronic, biological, and sensing applications. Usually chemical modification of SiO₂ substrates with silanization techniques are used for potential biosensor and electronic applications where the targeted biological components are assembled onto modified substrates. By combining silanization methods with microfabrication technology, surfaces can be patterned with functional groups, making it possible to attach cellular structures such as microtubules or cells in specific locations [1]. However, there is great interest in alternative techniques, which enables the immobilization of such components onto silicon substrates for future device applications. More importantly, the method chosen should lead to well organized nanoparticles with full coverage. In order to use nanoparticles for device technologies, uniform distribution of nanoparticles on patterned surfaces with precise control of their organization and density is crucial.

In such components, there is a great need to increase the sensitivity level of the fabricated device, and one way to achieve this can be done via further assembly of nanoparticles on the modified substrates which show promising characteristics for the immobilization of the targeted chemical/biological compounds. Zeolite nanoparticles were shown to display good interactions with biological molecules [2]. They have a remarkably large surface area that is available for the immobilization of different molecules, tunable surface properties for controlled variation of surface charge and hydrophilic/hydrophobic characteristics. The zeolite monolayers are suggested to be used as ideal media for organizing semiconductor quantum dots and nonlinear optical molecules in uniform orientations [3]. Therefore, in the current study, zeolite nanoparticles were organized into functional entities on silanized SiO_2 substrates and microfabricated using the electron beam lithography (EBL). The effect of application of different techniques for zeolite assembly on the silanized surfaces was investigated.

EXPERIMENTAL DETAILS

Zeolite A was synthesized according to the published procedures using sodium metasilicate pentahydrate ($Na_2O.SiO_2.5H_2O$, 29 wt. % Na_2O, 28 wt. % SiO_2, 43 wt. % H_2O, Fluka AG) and sodium aluminate as silicate and aluminate sources, respectively. These solutions were put into polypropylene bottles and kept statically at the designated temperatures (Table 1). Crystals were obtained following vacuum-filtering or ultracentrifugation, washing and drying steps. The FE-SEM image of the obtained nano zeolite particles is shown in Figure 1.

Figure 1. SEM image of the synthesized zeolite nanoparticles.

Phase identification of the synthesized zeolites was achieved by powder X-Ray diffraction analysis (XRD) using Ni filtered Cu-Kα radiation (Philips PW 1729). Morphological properties of the zeolites as well as zeolite-PU composites were examined by scanning electron microscopy (FE-SEM) (QUANTA 400, FEI). Energy dispersive X-ray spectroscopy analysis was done by using EDAX X-ray detector attached to Quanta 400 FE-SEM.

For the preparation of silanized zeolite particles, 3-aminopropyltriethoxysilane was used as silanization agent. Typically, 200 mg zeolite particles were refluxed in boiling toluene (20 ml) and silanization agent (0.2 ml) mixture for 1 hour. The silanized zeolite particles were obtained after filtering for micron-sized zeolites and ultracentrifugation for nano-sized zeolites.

Thermally oxidized Si substrates were used throughout the study. SiO_2 grown substrates were coated with PMMA (950000 g/mol) and pattern through Electron beam lithography (EBL) process. Exposed substrates were dipped into methyl isobutyl ketone: isopropyl alcohol [1:2] mixture for 1 minute for development of patterns.

Three different methods are used for zeolite assembly process. The first method involves the spin coating of the silanized zeolites onto patterned and bare substrate surfaces. Zeolite micron and nanoparticles dissolved in IPA (3 wt%) was spun onto surfaces at 3000 rpm. For the ultrasound aided sonication method, bare substrates are dipped into silanized zeolite solution in 3 wt% in toluene and agitated for 5 minutes in an ultrasonic bath. This method is not applied to the patterned substrates because of the chemical vulnerability of PMMA. Lastly, direct attachment method was tried with non-modified zeolite nanoparticles on the patterned and bare SiO_2 substrates. For this method nano-zeolite powder was simply rubbed onto the substrate as described in the work of Yoon et al. [4].

DISCUSSION

In the current study, different experimental procedures were investigated to efficiently assemble zeolite crystals on SiO_2 substrates for future biosensor applications. The efficiency criteria for the zeolite assembly are full degree of coverage, strong binding, and organized assembly of zeolite particles with the ability to control zeolite patterns in the nanometer range. Degree of coverage was simply judged from SEM results. Strong binding is the remaining zeolite particles on the SiO_2 substrates after strong 15 min agitation. Organized assembly of zeolite particles refers to having all zeolites attached in a monolayer form and/or avoiding the agglomeration of zeolite nanoparticles.

Three different methodologies were investigated in order to achieve the best assembly of zeolite nanoparticles on the SiO_2 substrates. These methods are spin-coating, sonication [5], and direct attachment methods [4] of zeolite nanoparticles. The first two methods were tried by using silanized and bare zeolite nanoparticles, whereas direct attachment method was performed by using only bare zeolite nanoparticles. It was important not only to achieve the aimed criteria but also to use these methods successfully for lithography purposes.

Spin coating method was chosen as the first method to be investigated because of its compatibility and ease in use with patterning processes. This method usually does not necessitate the use of any chemical modification of substrates or zeolite particles to obtain a decent coverage; however, the spin coated bare zeolite particles were observed to fall off as soon as the substrates were sonicated. Enhanced binding properties were attained upon treating the zeolites with the chemical linker. Nevertheless, agglomeration of both bare and chemically treated zeolite particles was observed (Figure 2-A). Thus with the spin coating of bare zeolite particles, strong binding and organization of zeolite nanoparticles were not obtained.

As an alternative to spin coating method, sonication experiments were carried out to see if better organization of zeolite particles was going to be attained with respect to the spin coating method. The zeolite particles were observed to have an improved organization on the SiO_2 substrates as shown in Figure 2-B. The agglomeration of zeolites was also less in comparison with the spin coating method. Furthermore, most of the 3-aminopropyl attached zeolite particles still remained after 30 second sonication with only about 20-25 % loss.

93

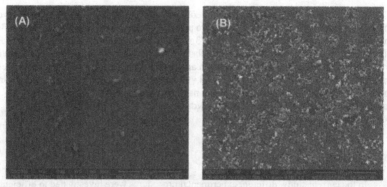

Figure 2. SEM images of SiO₂ wafers coated with zeolite microparticles by spin coating (A) and by sonication methods (B).

Lastly, direct attachment method was investigated. The advantage of this method is that there is no need to use any chemicals or reactors, which was the case for the spin coating and sonication methods. As seen in Figure 3, the desired criteria, which were stated as "full coverage" and "well organized monolayer" of zeolite nanoparticles on SiO₂ substrates were achieved by the direct attachment method. To check the binding strength of the particles, the substrates were sonicated in ultrasonic bath for about 5 minutes. It was seen that there was no significant change in the coverage and all zeolite nanoparticles remained on the substrate.

Figure 3. SEM images of SiO₂ wafers coated with zeolite nanoparticles by direct attachment method.

Microfabrication of SiO₂ surfaces with organized zeolite nanoparticles

The most efficient method among the ones used to organize zeolite nanoparticles on the whole SiO₂ substrates was investigated for the microfabrication purposes using EBL technique.

The discussed sonication method (Figure 3-A) was not suitable at all for EBL, since it necessitates the use of toluene as a solvent and it dissolves the mask. The spin coating method was tested by using bare (Figure 4-A) and 3-aminopropyl attached (Figure 4-B) zeolites. As shown in Figure 4-B, spin coating of 3-aminopropyl attached zeolites lead to a significantly denser zeolite patterns with better coverage with respect to the spinned bare zeolites on the substrate. However, the zeolite nanoparticles were again in agglomerates and were not fully covered in the obtained patterns using spin coating method of 3-aminopropyl attached zeolites. To check the binding strength of zeolites, the substrates were again sonicated for over 15 minutes and zeolites remained in their original positions.

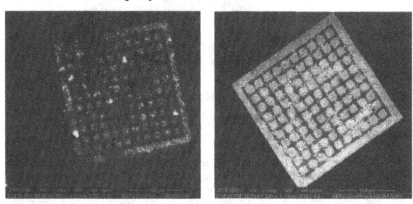

Figure 4. Zeolite patterns on SiO$_2$ wafer upon spinning bare (A) and 3-aminopropyl attached zeolites (B).

The desired criteria for zeolite nanoparticle organization on SiO$_2$ substrates, which are full coverage, well organized zeolite nanoparticles, and strong binding, is achieved with the direct attachment method as shown in Figures 5. More importantly, this method allowed full control during patterning of SiO$_2$ substrates with zeolite nanoparticles. If the final goal is only to increase the surface area for further immobilizations of biological compounds, the spin coating can also be used. The discussed methodologies are believed to open new gates for the future biosensor applications.

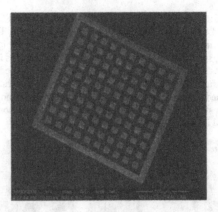

Figure 5. Zeolite patterns on SiO_2 wafer obtained after direct attachment method.

CONCLUSIONS

Different techniques for making a zeolite thin film on SiO_2 substrates were investigated. Three different methodologies were studied in order to obtain the fullest coverage of zeolite nanoparticles with improved organization and strong binding of nanoparticles. Spin coating of silanized zeolite particles showed improved coverage with enhanced binding; however the organization of zeolites were poor with agglomerates of zeolites. Sonication lead to better organized zeolite particles, but this method was not found to be compatible for patterning purposes. Direct attachment method showed a great potential for making zeolite thin films with organized nanoparticles composed of a single monolayer with strong binding properties. It is believed that this method opens new alternatives for the attachment of nanoscale components onto surfaces to fabricate functional nanostructured systems for advanced applications, such as biosensors.

REFERENCES

1. T. C. Turner, C. Chang, K. Fang, S. L. Brandow, D. B. Murphy, *Biophys. J.* **69**, 2782 (1995).
2. X. Zhou, T. Yu, Y. Zhang, J. Kong, Y. Tang, J-L, Marty and B. Liu, *Electrochem.Comm.* **9**, 1525 (2007).
3. H. S. Kim, M. H. Lee, N. C. Jeong, S. M. Lee, B. K. Rhee and K. B. Yoon, *J. Am. Chem. Soc.* **128**, 15070 (2006).
4. J. S. Lee, J. H. Kim, Y. J. Lee, N. C. Jeong and K. B. Yoon, *Angew. Chem. Int. Ed.* **46**, 3087 (2007).
5. J. S. Lee, K. Ha, Y-J. Lee and K. B. Yoon, *Adv. Mater.* **17**, 837 (2005).

Mater. Res. Soc. Symp. Proc. Vol. 1129 © 2009 Materials Research Society   1129-V08-04

6 Watt Segmented Power Generator Modules using Bi_2Te_3 and $(InGaAs)_{1-x}(InAlAs)_x$ Elements Embedded with ErAs Nanoparticles.

Gehong Zeng[1], Je-Hyeong Bahk[1], Ashok T. Ramu[1], John E. Bowers[1], Hong Lu[2], Arthur C. Gossard[2] Zhixi Bian[3], Mona Zebarjadi[3] and Ali Shakouri[3]

[1]Department of Electrical and Computer Engineering, University of California, Santa Barbara, CA 93106, U.S.A.
[2]Materials Department, University of California, Santa Barbara, CA 93106, U.S.A.
[3]Electrical Engineering Department, University of California, Santa Cruz, CA 95064, U.S.A.

ABSTRACT

We report the fabrication and characterization of segmented element power generator modules of 16 x 16 thermoelectric elements consisting of 0.8 mm thick Bi_2Te_3 and 50 μm thick ErAs:$(InGaAs)_{1-x}(InAlAs)_x$ with 0.6% ErAs by volume. Erbium Arsenide metallic nanoparticles are incorporated to create scattering centers for middle and long wavelength phonons, and to form local potential barriers for electron filtering. The thermoelectric properties of ErAs:$(InGaAs)_{1-x}(InAlAs)_x$ were characterized in terms of electrical conductivity and Seebeck coefficient from 300 K up to 830 K. Generator modules of Bi_2Te_3 and ErAs:$(InGaAs)_{1-x}(InAlAs)_x$ segmented elements were fabricated and an output power of 6.3 W was measured. 3D finite modeling shows that the performance of thermoelectric generator modules can further be enhanced by the improvement of the thermoelectric properties of the element materials, and reducing the electrical and thermal parasitic losses.

INTRODUCTION

Solid state thermoelectric generator modules composed of n and p semiconductor element couples can be used for directly thermal to electrical energy conversion. Their great potential in providing cleaner form of energy and reducing environmental contamination has been recognized. The power conversion performance of a thermoelectric generator module depends on the semiconductor's thermoelectric properties, in terms of the figure of merit, $Z = \alpha^2 \cdot \sigma / \kappa$, where α is the Seebeck coefficient, σ is the electrical conductivity and κ is the thermal conductivity. Thermal conductivity can be reduced due to the increase of phonon scattering by abundant surfaces and interfaces in nanostructured materials, and the Seebeck coefficient can be increased through thermionic emission across heterointerfaces,[1] and/or electron scattering by nanostructures.[2, 3] Thermal conductivity reduction using superlattice heterostructures or incorporation of nanoparticles has been demonstrated [4, 5]. When ErAs nanoparticles are incorporated into $(InGaAs)_{1-x}(InAlAs)_x$, potential barriers are formed at the interface between the particle and semiconductor. The Seebeck coefficient can therefore be enhanced through the electron filtering effects of these potential barriers.[6] The performance of a solid state generator also depends on the Carnot efficiency, which can be expressed as $\Delta T/T_h$., where ΔT is the temperature difference across the elements, and T_h the hot side temperature of the elements. Large ΔT is desirable for large output power and high efficiency. The performance of a thermoelectric generator module can be effectively enhanced by using segmented element structures with materials whose thermoelectric properties are optimized in successive temperature ranges,

In this paper, we report the fabrication, characterization and measurement of generator modules using segmented elements of 50 μm $(InGaAs)_{1-x}(InAlAs)_x$ and 0.8 mm Bi_2T_3. The generator modules were fabricated via the pick-up and place method, and flip-chip bonding technique. An output power of 6.3 W was measured with a heat source temperature at 610 K.

MATERIAL CHARACTERIZATION

Two 50 μm 0.6% $ErAs$:$(InGaAs)_{1-x}(InAlAs)_x$ samples for segmented generator modules were grown lattice-matched on InP(100) substrates of 520 μm thick using MBE. The growth rate was about 2 μm per hour and the growth temperature was maintained at 490 °C. The n-type $ErAs$:$(InGaAs)_{1-x}(InAlAs)_x$ consists of 80% InGaAs and 20%InAlAs, while the p-type sample is $ErAs$:$InGaAs$.

Characterization samples of $(InGaAs)_{0.8}(InAlAs)_{0.2}$ of 2 μm thick with 0.6% Er with the same material structure as that of the 50 μm material were grown on semi-insulating InP substrates of about 520 μm thick via MBE for the Seebeck coefficient and electrical conductivity measurements. To avoid side effects from InP substrate in electrical conductivity and Seebeck coefficient measurements at high temperatures, the semi-insulating InP substrate was removed in our material characterization. The epitaxial sample was bonded onto sapphire substrate using a SiO_2-SiO_2 oxide bonding technique. Then, the semi-insulating InP was removed by wet etching leaving just the 2 μm epitaxial layer of 0.6%Er $(InGaAs)_{0.8}(InAlAs)_{0.2}$ bonded on a 500 μm sapphire substrate. A Van der Pauw device pattern was formed on the 2 μm epitaxial layer by reactive ion etching. Metallization of TiWN was used as metal diffusion barrier, which showed very good Au barrier property and thermal stability up to 700 C. [7] The measurements were carried out in a vacuum chamber with the pressure pumped below 1 mTorr. The measurement results in figure 1 show that the Seebeck coefficient increases with temperature.

The electrical conductivity was measured using the same Van der Pauw device pattern as the

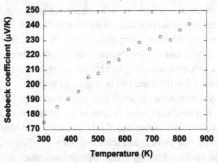

Figure 1. Measurement results of Seebeck coefficients of $(InGaAs)_{0.8}(InAlAs)_{0.2}$ with 0.6% ErAs nanoparticles from 300 K to 830 K.

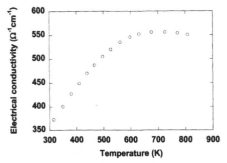

Figure 2. Measurement results of electrical conductivity of $(InGaAs)_{0.8}(InAlAs)_{0.2}$ with 0.6% ErAs nanoparticles from 300 K up to 800 K.

one used for Seebeck coefficient measurements. In electrical conductivity measurements, the device was placed in the center of one copper heater bar in the vacuum chamber. Two current and two voltage electrodes were connected to the four metal pads of the device, and data acquisition were done via a control computer. The Van der Pauw pattern is a cloverleaf shape, and four electrodes were placed at the far corners of the pattern, so the measurement errors from pattern geometry were almost negligible. Fig. 2 shows the measurement results with temperatures from 300 K up to 800 K.

DEVICE FABRICATION

The segmented generator modules were fabricated via pick-up and place approach method and flip-chip bonding techniques. Processing techniques are similar to those of standard large scale integrated circuits for the segmented elements of ErAs:InGaAlAs and Bi_2Te_3. The thin film element fabrication started with the front side metallization of the epitaxial layer: Ni/GeAu/Ni/Au contact metals were used for n-type ErAs:InGaAlAs, and Pt/Ti/Pt/Au were used for p-type ErAs:InGaAs, respectively. The InP substrate was removed through wet etching solution to expose the backside of the 50 μm epitaxial layer. The backside metallization was also the same Ni/GeAu/Ni/Au and Pt/Ti/Pt/Au were used for n-type and p-type, respectively. Then the n and p type thin film wafers were diced into 1.4 mm × 1.4 mm square chips ready for bonding. Ni was used as contact metallization for Bi_2Te_3 of both n and p type. The bulk Bi_2Te_3 was cut into square chips of 1.4 mm × 1.4 mm in area. All the Bi_2Te_3 elements were bonded on a lower ceramic plate; while the ErAs:InGaAlAs elements were bonded on an upper ceramic plate. Finally, the lower Bi_2Te_3 bonded plate and the upper ErAs:InGaAlAs bonded plate were bonded together using flip-chip bonding to form a 16 x 16 element generator module.

MEASUREMENT RESULTS AND DISCUSSIONS

The measurement setup consists of a heat sink with circulating cooling water, a heat source of aluminum block with two built-in electrical cartridge heaters, two thermocouples for temperature monitoring, and electrical probes for the output power measurements. One of the thermocouples was fixed in the aluminum heat source block, 1 mm away from the interface of the heat block and

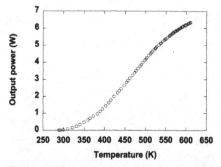

Figure 3. The output power measurement results for the 16 x 16 segment element power generator of 50 μm ErAs:(InGaAs)$_{1-x}$(InAlAs)$_x$ and 0.8 mm Bi$_2$Te$_3$. The data were obtained when the heat source temperature was increased from 290 K up to 610 K.

generator ceramic plate and used as the heat source temperature sensor; the other was placed on the heat sink surface at the interface of the heat sink and the generator as heat sink temperature sensor. The cooling water temperature was set at 285 K. The measurement results of output power are shown in Figure 3. In the low temperature range when heat source temperature is below 450 K, the output power shows quadratic increase with temperature, but when the heat source temperature rises above 500 K, the output power begins to saturate, which indicates that the thermoelectric properties of Bi$_2$Te$_3$ degrade when the temperature is above 500 K. To get a better understanding of the performance of the segmented generator modules, a 3D finite element model (FEM) was set up for theoretical analyses. The thermoelectric effects of the Peltier, Seebeck and Thomson are all taken into account in our finite analyses, and the material property values, metal to semiconductor electrical contact resistances were from experimental results. The water circulating heat sink was modeled using a constant heat transfer coefficient at the cold side of the module. The external electrical load was a constant resistor which was made to be an impedance match to the module. An output voltage to the external electrical load resistor due to the thermoelectric properties of the segmented elements was obtained at each heat source temperature, therefore the output current and power to the load resistor at each heat source temperature becomes known. Figure 4 shows the 3D modeling result of the electric potential distribution across the generator module's element couple. As the n and p elements are electrically connected in series, and thermally connected in parallel, the highest potential is at the cold side of p-type element, (red area); lowest electric potential is at the cold side of n-type element (blue area).

A comparison of the 3D FEM values with measurement results is shown in Fig. 5. When temperature is below 500 K, the two results fit well. This indicates that in the low temperature range from 300 K to 500 K, the 3D model is close to the real generator module and its working conditions. When the temperature is above 500 K, measurements and finite element module results begin to show their discrepancy, which comes from the thermal resistance increasing at

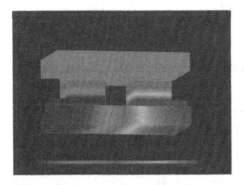

Figure 4. 3D simulation result shows the potential distribution across the module n and p elements when loaded with external electrical resistor R_o. The lowest potential is at cold side of the n-type segmented element leg; while the highest potential occurs at the cold side of the p-type segmented element.

the interface of heat source block and module as temperature rises, so the real temperature drop across the elements does not increase linearly with the rise of the heat source temperature. In the 3D model, the heat block on the top of the generator module was modeled as an ideal heat source at a constant temperature without any thermal interfaces at the interface between the heat source and the module. In real measurement setup, the interface of heat source and generator module was connected using thermal paste, which can quickly become dry at high temperatures and present larger thermal resistance, and therefore produce significant temperature drop at the interface. One solution to it is to use liquid metals instead of thermal paste for the hot side interface connection, such as indium or stannum, the thermal conductivity of which can be very high when melted at high temperatures. The optimization results of our finite element model also show that the output power can be improved by improving the thermoelectric properties of the

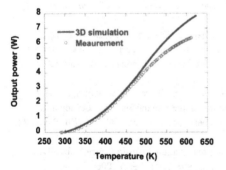

Figure 5. A comparison of 3D simulation values with real generator module measurements.

elements, reducing the thermal and electrical parasitic loss, and increasing heat transfer coefficient of the heat sink.

CONCLUSIONS

The incorporation of ErAs nanoparticles into $(InGaAs)_{1-x}(InAlAs)_x$ alloy results in significant improvement in the material's thermoelectric properties. Variable temperature measurements of 0.6% $ErAs:(InGaAs)_{1-x}(InAlAs)_x$ show that the Seebeck coefficient increases with temperature from 173 μV/K at 300 K up to 240 μV/K at 830 K, and electrical conductivity increases with temperature from 370 $\Omega^{-1}\cdot cm^{-1}$ at 300 K up to 550 $\Omega^{-1}\cdot cm^{-1}$ at 700 K. A 16 × 16 generator module was fabricated using segmented elements of 50 μm 0.6%$ErAs:(InGaAs)_{1-x}(InAlAs)_x$ and 0.8 mm Bi_2Te_3. An output power of 6.3 W was measured with heat source temperature at around 620 K. The 3D finite element modeling shows that the performance of thermoelectric generator modules can be further improved by improving material thermoelectric properties, reducing electrical and thermal parasitic resistance loss, and improving the heat transfer coefficient of the heat sink.

ACKNOWLEDGMENTS

The authors acknowledge useful discussions with Dr. Mihal Gross. This work is supported by the Office of Naval Research through contract N00014-05-1-0611.

REFERENCES

[1]   A. Shakouri and J. E. Bowers, "Heterostructure integrated thermionic coolers," *Applied Physics Letters,* vol. 71, pp. 1234-1236, SEP 1 1997.

[2]   S. V. Faleev and F. Leonard, "Theory of enhancement of thermoelectric properties of materials with nanoinclusions," *Physical Review B,* vol. 77, pp. -, Jun 2008.

[3]   M. Zebarjadi, K. Esfarjani, A. Shakouri, J.-H. Bahk, Z. Bian, G. Zeng, J. E. Bowers, H. Lu, J. M. O. Zide, and A. Gossard, *submitted to Applied Physics Letters* 2008.

[4]   R. Venkatasubramanian, E. Siivola, T. Colpitts, and B. O'Quinn, "Thin-film thermoelectric devices with high room-temperature figures of merit," *Nature,* vol. 413, pp. 597-602, OCT 11 2001.

[5]   W. Kim, S. L. Singer, A. Majumdar, D. Vashaee, Z. Bian, A. Shakouri, G. Zeng, J. E. Bowers, J. M. O. Zide, and A. C. Gossard, "Cross-plane lattice and electronic thermal conductivities of ErAs:InGaAs/InGaAlAs superlattices," *Applied Physics Letters,* vol. 88, p. 242107, 2006.

[6]   J. M. O. Zide, D. Vashaee, Z. X. Bian, G. Zeng, J. E. Bowers, A. Shakouri, and A. C. Gossard, "Demonstration of electron filtering to increase the Seebeck coefficient in In0.53Ga0.47As/ In0.53Ga0.28Al0.19As superlattices," *PHYSICAL REVIEW B,* vol. 74, p. 205335, 2006.

[7]   S. Bhagat, H. Han, and T. L. Alford, "Tungsten-titanium diffusion barriers for silver metallization," *Thin Solid Films,* vol. 515, pp. 1998-2002, Dec 5 2006.

Mater. Res. Soc. Symp. Proc. Vol. 1129 © 2009 Materials Research Society   1129-V03-08

Multiple Duplication of Electroformed Nano-Ni Stamps from Si Mother Mold

Si-Hyeong Cho[1], Jung-Ki Lee[1], Jung-Ho Seo[1], Hyun-Woo Lim[2], Jin-Goo Park[3]

[1]Hanyang University - Bionano Technology, Ansan, Republic of Korea

[2]Micro Biochip Center - R&D Ansan, Republic of Korea

[3]Hanyang University - Materials Engineering, Ansan, Republic of Korea

ABSTRACT

Nanoimprint lithography (NIL) is an alternative lithographic method that offers a sub-10 nm feature size, high throughput, and low cost. It requires a mold which has a low fabrication cost and long life time. A mold is the material with various patterns to transfer on the plastic substrate. Materials such as Si, quartz, plastic and Ni have been widely used for micron or sub micron sized mold fabrication depending on its pattern size and application. Si and quartz are very easy to be broken, and plastic can be deformed easily during process. However, Ni has enough hardness and life time as a mold material and can be fabricated from on Si micro to nano sized molds with a seed layer by a simple Ni electro-deposition. After electro-deposition, the sample is usually dipped in KOH solution to remove Si from Ni. The consumption of Si mold is necessary step to produce a Ni mold.

In this study, a method was developed to fabricate a Ni mold without the consumption of Si mold. Vapor SAM (self assembled monolayer) method was used to deposit hydrophobic layer on Si mold. A low surface energy release layer on stamp surfaces not only helps to improve imprint qualities, but it also increases the stamp lifetime significantly by preventing surface contamination. Hydrophobic layer, which has a low surface energy, makes possible to separate Ni from Si substrate without causing any damages on Si mold. The characteristics of deposited hydrophobic layer were analyzed by measurements of the contact angle, its hysteresis, surface energy, thickness and lateral friction force. The stiction of Ni on Si mold was observed when the separation of Ni from Si was tried without the SAM deposition.

The multiple duplication of Ni molds has been successfully developed without disposing costly Si mother mold. Duplicated patterns on Ni mold showed the same patterns as on Si mother mold when they were observed with optical microscope, 3D profiler, FE-SEM, and AFM (atomic force microscope).

INTRODUCTION

Nanoimprint lithography has made great progress in a relatively short period of time. One of the important issues still to be resolved is the useful lifetime of the mold. Presently, nano imprint molds require replacement after ~50 consecutive imprints [1]. Because of this reason, Ni which has enough strength and lifetime has been conventionally used for mold material.

For Ni mold manufacture, it needs Si master mold. After Ni is electrodeposited on Si master mold, it can be fabricated though a removal of Si mold in Si etchant. However, the conventional master mold fabrication technique is unsatisfactory in terms of time and costs [2].

In other words, the consumption of Si master mold is cost ineffective and inefficient. In order to prevent stiction between the master mold and replica during the demolding process, a hydrophobic surface treatment was required. An efficient way to avoid or reduce in use stiction is to deposit a thin layer of low surface energy material on the surfaces of the microstructure. Among various surface coatings and modification techniques, self assembled monolayers (SAMs) that are grown from organosilanes are promising candidates for antistiction coatings due to their good bonding strength, low surface energy, low friction energy, and good thermal stability [2, 3].

Recently vapor phases coating process has been developed. Vapor phase processing eliminates the use of organic solvents and greatly simplifies handling of the samples. Moreover, the stoichiometry of the precursor molecules can be more precisely controlled. The advantages of vapor self assembly monolayer (VSAM) process are easy setup and operation, easy scale-up at low cost, portability, repeatability and reproducibility [4].

In this study, we investigate a surface coating of Si mold by FOTS deposition with vapor self assembled monolayer (Vapor SAM) method to reduce surface energy and making it easy to release the Ni mold from the Si master mold. As a result, it was accomplished the multiple duplication of Ni molds without consumption of Si master mold.

EXPERIMENT

In our experiments to get a Si mother mold, 6inch Si wafer was used as a substrate. Through the conventional photolithography, micro and nano sized patterns were formed on Si wafer. Then the mother mold was cut into 2×2 cm^2 rectangular shapes and dipped in piranha solution for 15 min to remove organic contamination and form OH$^-$ on surface for the experiments. In the case of Ni substrate, O$_2$ plasma treatment was used. OH$^-$ is important reaction factor for vapor SAM method. After surface treatment, the surface of mother mold showed the static contact angles lower than 10°.

Trichloro[1H, 1H, 2H, 2H-perfluoroctyl] silane (FOTS, C$_8$H$_4$C$_3$F$_{13}$Si, Sigma-Aldrich, USA) was used as a precursor. Calculated molecular length and boiling point were 1.1 nm, 192°C at 760 Torr and density was 1.65 g/ml. To FOTS deposition, VP-SAM equipment (VAC-150, Sorona, Korea) was used. It was deposited under vacuum ~10^{-2} Torr. And it was kept 150 $^\circ$C for 10 min. Then, a seed layer for electrodeposion was deposited by sputtering (SRC-130, Sorona, Korea). To reproduction of Ni molds from Si mother mold, Ni electrodeposition was done at room temperature for 4 hours. Ni sulphuric (Hojin Polartech, Korea) electrolyte which has pH 4.7 was used. And Current density was applied 40 mA/cm^2. Finally, duplicated Ni molds were manufactured from Si mother mold through the physical separation. Figure 1 shows the experimental procedure for multiple duplication of Ni mold.

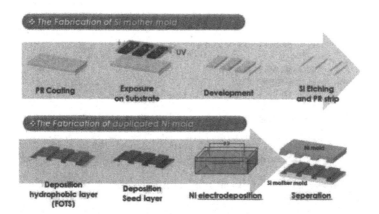

Figure. 1 Experiment procedure of the fabrication of Si mother mold and Ni mold for multiple duplication

Contact angle analyzer (Phoenix 300, SEO, Korea) was used to measure contact angle, hysteresis and surface energy of FOTS film with vapor SAM (self assembled monolayer) method. And the duplicated patterns on Ni mold were observed with that of Si mother mold by using optical microscope (L-150A, Nikon, Japan), 3D profiler (μsurf, Nano focus, Germany), FE-SEM (S4800, Hitachi, Japan), and AFM (XE-100, PSIA, Korea).

RESULTS AND DISCUSSION

Figure 2 (a) shows changes of contact angle and hysteresis of FOTS thin film on Si mother molds as a function of temperature. Contact angles around 100° and hysteresis around 20° were measured in all samples. Figure 2 (b) shows changes of surface energy of FOTS thin film on Si mother molds as a function of temperature. Surface energy around 15 mN/m was measured in all samples. It was lower than well-known antistiction layer Teflon of which surface energy is 18 mN/m.

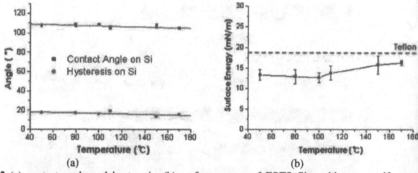

(a)   (b)

Figure.2 (a) contact angle and hysteresis, (b) surface energy of FOTS film with vapor self assembled monolayer method.

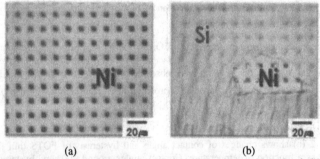

(a)   (b)

Figure 3. The optical microscope images of duplicated Ni molds (a) with hydrophobic layer (FOTS film) and (b) without hydrophobic layer

First, we investigated the effect of FOTS deposition on Si mother mold. Figure 3 shows the surface of duplicated Ni molds with FOTS coating and without that. The surface of duplicated Ni mold using FOTS coating was very clear without any residues or defects, on the other hand, Si residues were formed on surface of duplicated Ni mold without FOTS coating.

We fabricated a Si master mold which had micro sized patterns, and using vapor SAM method it was deposited FOTS film as an antistiction layer. Figure 4 contains the optical microscope images of patterns on Si mother mold and duplicated Ni mold. It was confirmed that patterns of Si and Ni mold were almost similar. And 3D profiler was used for measure more accurate dimension of patterns. Figure 5 shows width and depth of the patterns between Si mother mold and duplicated Ni mold.

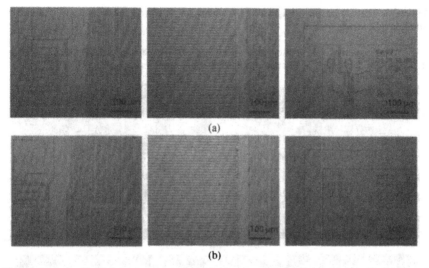

(a)

(b)

Figure 4. The optical microscope images of (a) patterns on Si mother mold and (b) duplicated Ni mold

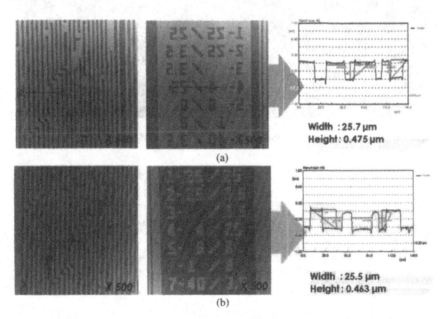

Width : 25.7 μm
Height : 0.475 μm

(a)

Width : 25.5 μm
Height : 0.463 μm

(b)

Figure 5. The 3D profiler images of (a) patterns on Si mother mold and (b) patterns on duplicated Ni mold

We fabricated a Si master mold which had array pattern of nano sized tips. Electrodeposition for duplication of Ni mold was done total two steps. In order to get Ni master mold,

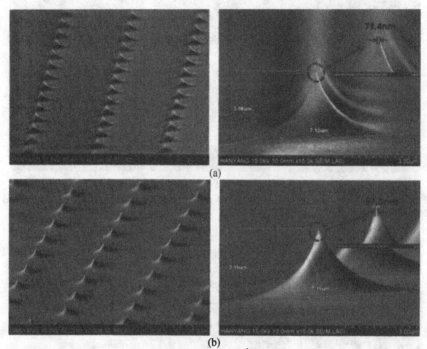

(a)

(b)

Figure 6. SEM images of (a) Si mother mold and (b) 2nd duplicated Ni mold. It waŝ used FOTS film as a hydrophobic layer.

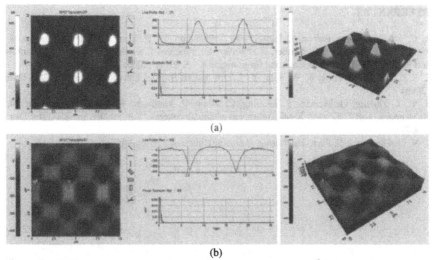

(a)

(b)

Figure 7. AFM images and topography of (a) Si mother mold and (b) 1ˢᵗ duplicated Ni mold.

1ˢᵗ Ni molds were fabricated by 1ˢᵗ electrodeposition on Si master mold. 2ⁿᵈ duplicated Ni molds were fabricated from 1ˢᵗ duplicated Ni mold. In figure 6 show SEM images of Si mother mold and 2ⁿᵈ duplicated Ni mold. They were observed almost same patterns, and it was confirmed duplicated Ni mold had high precision and high reproducibility. Somewhat there were about 10 nm difference. It seems that it is caused because seed layer which had 10 nm thickness was deposited on Si mother mold. Figure 7 shows AFM images and topography of Si mother mold and 1ˢᵗ duplicated Ni mold. Nano and micro sized patterns were reproduced, but it was not possible to observe exactingly.

CONCLUSIONS

Nanoimprint lithography (NIL) has a potential to be a method allowing low-cost, high-throughput production of nano-patterns over large areas employing a single lithographic step. The mold problems were occurred because of high cost and long time waste for master mold manufacture.

To solve the problems, in this study, a method was developed to fabricate a Ni mold without the consumption of Si mold. FOTS film was coated on mother mold by vapor SAM(self assembled monolayer) method. The FOTS coating on Si mother mold shows low surface energy and high precision of duplicated Ni molds.

ACKNOWLEDGMENTS

This research was supported by the small and medium business administration through the fostering project of the industry university partnership and BK21 program.

REFERENCES

1. C. M. Sotomayor Torres, S. Zankovych, J. Seekamp, A. P. Kam, C. Vlavijo Cedeno, T. Hoffmann, J. Ahopelto, F. Reuther, K. Pfeiffer, G. Bleidiessel, G. Gruetzner, M. V. Maximov and B. Heidari, Materials Science and Engineering C 23, 23-31 (2003)
2. S. J. Kim, H. S. Yang, K. W. Kim, Y. T. Lim and H. B. Pyo, Electrophoresis 27, 3284-3296 (2006)
3. Y. X. Zhuang, O. Hansen, T. Knieling, C. Wang, P. Rombach, W. Lang, W. Benecke, M. Kehlenbeck, and J. Koblitz, J. Microelectromechanical systems, Vol.16, No. 6, 1451-1460 Dec. (2007)
4. N. G. Cha and J. G. Park, PhD thesis, Hanyang University, Feb. (2007)

Shape Memory Alloys and
Magneto-Materials

Mater. Res. Soc. Symp. Proc. Vol. 1129 © 2009 Materials Research Society   1129-V12-01

Microstructures and Enhanced Properties of SPD-processed TiNi Shape Memory Alloy

Koichi Tsuchiya[1,3] Masato Ohnuma[1], Kiyomi Nakajima[1], Tadahiro Koike[2], Yasufumi Hada[3],
Yoshikazu Todaka[3] and Minoru Umemoto[3]
[1]National Institute for Materials Science, Sengen 1-2-1,
Tsukuba, Ibaraki 305-0047, Japan
[2]Asahi Intecc, Co., Ltd.,
Izumi, Osaka 594-1157, Japan
[3]Toyohashi University of Technology,
Tempaku-cho Hibarigaoka 1-1,
Toyohashi, Aichi 441-8580, Japan

ABSTRACT

Crystalline-to-amorphous transformation and nanostructure formation by severe plastic deformation was investigated for TiNi shape memory alloy by high pressure torsion deformation and cold drawing. Phase transformation and mechanical properties of partially amorphous materials were also investigated. The amorphization and nanocrystallization can be an effective mean to produce materials with high young's modulus and large recovery strain. The amorphous/nanocrystalline TiNi wires after aging exhibit high tensile strength (~2 GPa), high apparent elastic modulus (~71 GPa) and large pseudoelastic recovery (> 5%). Such properties are useful for novel medical devices.

INTRODUCTION

TiNi-based shape memory alloy is the most popular shape memory alloy and have been used in various areas of applications, such as house-hold appliances, airplanes, automobiles, arch-wires and eyeglasses flames. Recently there has been a growing market for various medical devices for minimally invasive surgery using TiNi, such as guide wires for catheter, stents and filters. One of the recent trend in the development of the medical devices is miniaturization, represented by neurovascular stents and guide wires. These new devices call for materials with high elastic modulus and large elastic recovery.

Both shape recovery force and recovery stress or superelastic temperature windows are limited by the yield stress of the austenite phase (B2 phase). Therefore, in the history of development of TiNi-SMA, the strengthening of the materials have been achieved mainly through the precipitation hardening by Ti_3Ni_4 phase, and work-hardening by cold working.

Apart from the shape memory effect and superelasticity the TiNi has another peculiar aspect. TiNi is susceptible to undergo solid-state crystalline-to-amorphous transformation (CTAT) by various methods, such as the particle beam irradiation (electrons, neutrons and ions) and severe plastic deformation (SPD) of bulk materials, such as cold rolling [1-3], shot peening [4] and high pressure torsion (HPT) [5]. Recently we have found out that amorphization also occurs by cold-drawing [6] which may have a large impact on application since most of the TiNi are provided and used in the form of wires or tubing.

The present paper describes the process and mechanism of CTAT in TiNi by SPD process and nanocrystallization by post-deformation aging as well as their impact on phase transformation and mechanical properties.

EXPERIMENT

Hot rolled sheets of Ti-50.2mol%Ni (Daido Steel., KIOKALLOY-R) were used in the present study. They were heat-treated at 1173 K for 3.6 ks under an argon atmosphere and quenched into water. The transformation temperatures for Ti-50.2mol%Ni are M_s = 298 K, M_f = 282 K, A_s = 310 K, and A_f = 332 K. High pressure torsion experiment was done using a disk sample (10 mmϕ and 0.85 mm thick). The sample was put between two anvils with a depression of 0.25 mm depth and 10 mm diameter. Torsion deformation was given up to 10 turns under an applied compressive stress of 5 GPa at room temperature. In HPT deformation, the extent of deformation is a function of radial position of a sample, r. In this method, shear strain, γ, is given by $\gamma = 2\pi r N/t$, where N is the number of turns and t is the thickness of a sample. Equivalent von Mises strain is given by $\varepsilon_{eq} = 3^{-1/2}\gamma$. Post deformation aging was done for HPT (N = 10) samples at various temperatures ranging from 573 K to 773 K for 3.6 ks.

Characterizations were done by X-ray diffractometry (XRD) with a Cu-K$_\alpha$ radiation and transmission electron microscopy (TEM) operated at 200 kV. TEM samples were electropolished on Tenupol-3 with an electrolyte of H_2SO_4/methanol at 253 K. Phase transformation behavior was characterized by differential scanning calorimetry (DSC) with a heating/cooling rate of 0.17 K s^{-1}. Micro-Vickers hardness measurements were done with an applied load of 2.94 N for 15 s. For XRD and hardness measurements a whole 10 mm disks were examined after removing the rough surface by mechanical polishing. For TEM and DSC measurements, a 3 mmϕ disk with its center at r = 2 mm were used.

For cold-drawing, annealed wires of Ti-50.9mol%Ni (0.5mmϕ) were cold-drawn to the areal reduction of 50~70% using diamond dies without intermittent annealing. Drawing speed was ~50 m/min. Structures of drawn wires were characterized by X-ray diffractometry with a Mo-K$_\alpha$ radiation. Microstructures of drawn wires were observed by TEM. Samples for transversal sections were prepared by FIB.

RESULTS and DISCUSSION

Changes in Microstructures by HPT

Figure 1 shows the evolution of XRD patterns for Ti-50.2mol%Ni by HPT deformation. Before deformation, the sample was in the B2 phase state with a small amount of the B19' martensite phase. By applying 5 GPa compressive stress, a large portion of the B2 phase changed to the B19' phase by stress-induced transformation. By applying the torsion deformation a significant decrease in peak intensity and peak broadening occurred. Crystallite size, d, was estimated from 110(B2) diffraction peaks and plotted as a function of N in Fig. 2. It is apparent that the decrease in crystallite size is significant in the first turn of deformation followed by a slow decrease. After N = 10, d reaches to about 4 nm.

Figure 3 is TEM bright field (BF) images showing microstructural evolution during HPT. Fig. 3(a) shows the microstructure of a sample after applying 5 GPa compressive stress. The sample is in B19' phase as indicated in the selected area diffraction pattern (SADP) shown as the inset. Numerous striations, which are most likely the deformation twins are also seen. By applying N = 0.125(ε_{eq} = 1.6), numerous gray band-like structures can be seen in the BF image. In the SADP the diffracting spots for the B19' are very faint, indicating that the B19' martensite reverse-transformed to B2 phase by deformation. Also the SADP exhibits hallo ring with its

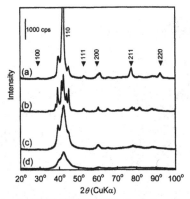

Figure 1 Change in XRD pattern by HPT deformation. (a) before deformation. (b) after compression. (c) N = 0.125. (d) N = 10.

radius corresponding to $|g_{110}|$ of B2. This is the indication that a small but appreciable amount of amorphous phase exists and the gray band structure may correspond to amorphous region. With increasing the extent of deformation the area of gray contrast increases and the hallo ring become more intense (Fig. 3(c)). After HPT deformation of N = 10 (ε_{eq} = 128) the sample is in a mixture of amorphous and nanocrystalline B2 phase. Volume fraction of amorphous phase was estimated to be about 0.96 from the recrystallization peak on a DSC profile. Figure. 4 shows the changes in crystallite size of the samples subjected to HPT deformation (N = 10) and post - deformation aging. (The data for 773 K aging is the grain size estimated from TEM observations.) Corresponding TEM micrographs are also shown. The crystallite size determined by XRD and grain size seen in TEM images seem to correspond well. After aging for 3.6 ks, grain size was about 20 nm (573 K) and 40 nm (673 K). In the case of 573 K aging, the amorphous region still existed even after 360 ks. Aging at 773 K led to a significant grain coarsening.

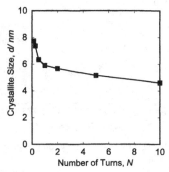

Figure 2 Change in crystallite size by HPT.

115

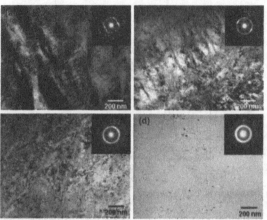

Figure 3 TEM micrographs of HPT deformed samples. (a) 5 GPa. (b) $N = 0.125$ ($\varepsilon_{eq}=1.6$). (c) $N = 5(\varepsilon_{eq}=64)$. (d) $N = 10(\varepsilon_{eq}=128)$.

    Figure. 4 shows the changes in crystallite size of the samples subjected to HPT deformation ($N = 10$) and post -deformation aging. (The data for 773 K aging is the grain size estimated from TEM observations.) Corresponding TEM micrographs are also shown. The crystallite size determined by XRD and grain size seen in TEM images seem to correspond well. After aging for 3.6 ks, grain size was about 20 nm (573 K) and 40 nm (673 K). In the case of 573 K aging, the amorphous region still existed even after 360 ks. Aging at 773 K led to a

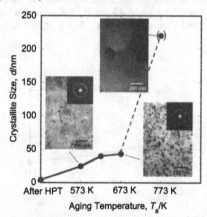

Figure 4 Changes in crystallite size by post-deformation aging. 773 K data is the grain size estimated from TEM micrograph.

significant grain coarsening.

Martensitic Transformation in TiNi after Post-deformation Aging

DSC profiles of HPT deformed (N = 10) and aged samples are shown in Fig. 5. As-deformed sample and one aged at 573 K for 3.6 ks do not exhibit any sign of transformation in DSC curves. For the sample aged at 623 K, single peak on the cooling run and two peaks on the heating run are seen. A peak on cooling and the peak at 328 K on heating can be attributed to B2-to-R phase transformation on the basis of the small temperature hysteresis ($\sim$ 7 K). The peak at 311 K on heating is peculiar since there is no corresponding forward transformation peak on the cooling curve. Similar situation can be found in the sample aged at 673 K. For the sample aged at 723 K. Two peaks can be seen on both cooling and heating runs. The peaks at 236 K on cooling and 309 K on heating can be attributed to forward and reverse transformation of B19' phase, respectively. This kind of multi-step transformation has been often observed in aged TiNi with Ni-rich compositions, in which semi-coherent precipitation of Ti$_3$Ni$_4$ phase occurs. They are attributed to the stress-field or composition gradient around the precipitates. In the present case, however, Ti$_3$Ni$_4$ precipitates are not involved and the multiple step transformation can be

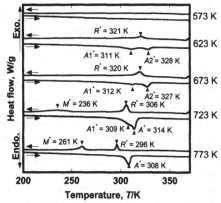

Figure 5 DSC profiles after aging of HPT deformed (N = 10) Ti-50.2mol%Ni.

attributed to the effect of small grain size.

Recently, Waitz et al. studied the effect of grain size on martensitic transformation in HPT deformed Ti-50.2mol%Ni[7]. Phase constitution at room temperature was found to depend on the grain size (D): (1) D > 150 nm: B19' was dominant. (2) 60nm < D <150 nm: R and B19' phases coexist. (3) 15 nm< D <60 nm: R and B2 phases coexist. (4) D < 15 nm: B2 phase. Thus the extremely small grain size in HPT sample stabilizes the B2 and R phases with respect to the B19' phase. The results of present investigation are in accordance with these observations.

The effect of internal stress on thermoelastic martensitic transformation was discussed by Tong and Wayman[8]. Although their model consider the strain energy caused only by the formation of martensite plates in the parent phase, it can be extended to take the strain energy caused the elastic interaction of martensite plates with grain boundaries and dislocations. If one assumes the both the irreversible energy E_{irr} and elastic energy E_{el} are monotonically increasing functions of volume fraction of martensite (f_M) and defect density (a function of strain, ε) , martensitic transformation temperature $T_{\mp}(f_M,\varepsilon)$ is expressed as

$$T_{\mp}(f_M,\varepsilon) = \frac{\Delta H + E_{el}(f_M,\varepsilon) \pm E_{ir}(f_M,\varepsilon)}{\Delta S} \qquad (1),$$

where, ΔH (<0) and ΔS (<0) are the difference in enthalpy and entropy between the austenite and martensite phase, respectively. T_{-} and T_{+} corresponds to the forward and reverse transformation temperature. It is apparent from (1) that the large elastic energy stabilizes the austenite phase, and that large irreversible energy stabilizes the austenite on cooling but stabilizes the martensite on heating. Transformation strain for B19' phase ($\sim$0.08) is much larger than that for R phase ($\sim$0.01) thus the strain energy effect and suppression of transformation temperature is much more pronounced in B19' transformation. It should be noted that the R phase transformation temperature in the 623 K aged sample is about 25 K higher than in the 773 K aged sample. Thus R phase is even stabilized by the small grain sizes. Elastic energy and irreversible energy also have marked effect of transformation intervals $\Delta T_{-} = M_s - M_f$ and $\Delta T_{+} = A_f - A_s$. In our previous investigation in situ cooling X-ray diffraction for cold rolled TiNi revealed that very small amount of the B19' phase forms on cooling even when no transformation peak is seen on DSC curve[9]. Therefore absence of transformation peaks on DSC curves implies not only the absence of transformation but also "smearing out" of transformation.

Transformation intervals for the forward and reverse transformation can be written as

$$\Delta T_{\mp}(\varepsilon) = -\frac{\Delta E_{el}(\varepsilon) \pm \Delta E_{ir}(\varepsilon)}{\Delta S} \qquad (2),$$

where $\Delta E_{el} = E_{el}(1,\varepsilon) - E_{el}(0,\varepsilon)$ and $\Delta E_{ir} = E_{ir}(1,\varepsilon) - E_{ir}(0,\varepsilon)$. An increase in elastic energy and irreversible energy during transformation causes the increase in transformation interval. One peculiar aspect of the transformation is the difference in transformation intervals between the cooling and heating runs. As seen in DSC profiles for the 723 K aged sample in Fig. 5, the interval is larger in forward transformation than in reverse transformation. Elastic energy cannot account for this asymmetry. It was reported that in the nanocrystalline conditions the twinning mode for self-accommodation changes from <011>type-II or (11-1) type-I twins to a high density of (001) compound twins[7]. The (001) compound twins are not the one giving the invariant plane but known as the deformation twins in B19' [10]. The formation of this twin may make large contribution in irreversible energy in forward transformation leading to a large interval.

Mechanical properties of HPT and aged TiNi

Vickers hardness of HPT deformed and aged samples are shown in Fig. 6. By HPT deformation of $N = 10$, $H_V = 532$ was achieved. Aging decreases the hardness but the value remains above 400Hv even after aging at 673 K, which corresponds to nanocrystalline states and undergoes the R and B19' transformation. It indicates that the combination of severe plastic deformation and post-deformation aging is useful way to achieve the high yield strength. Since

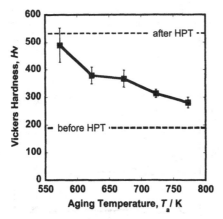

Figure 6 Hardness as a function of aging temperature.

the shape recovery stress and temperature window for superelasticity are usually limited by the yield stress of the B2 phase, the strengthening by the grain refinement can be useful to improve shape memory and superelastic properties of TiNi.

Microstructures of Severely Cold Drawn Ti-50.9mol%Ni wires

Figure 7 is the X-ray diffraction patterns of Ti-50.9mol% wires before and after cold-drawing. The wire before the drawing is in the B2 state. The absence of 200,310 and 420 reflection are due to the strong <111> fiber texture. As the area reduction increases, peak broadening and intensity decrease become more significant. After 70% reduction, only two broad peaks, 110 and 211, are apparent.

The crystallite size was determined from the FWHM of 110 diffraction peaks. After 70% reduction, the crystallite size of the wire was about 3 nm.

In order to study the microstructures of the wire in more details, TEM observation was made for as-drawn wires. Figure 8(a)-(c) shows the microstructures of the 70% drawn wire. Figure8 (a) and (b) are the bright and dark field images, respectively. SAD pattern shown as the inset exhibits hallo ring with some diffuse diffraction spot superimposed, suggesting that the area is the mixture of amorphous and nanocrystalline B2 phase. The dark field image (b) was obtained from the part of the hallo ring shown by the circle in the SAD pattern. The layers with gray uniform contrast in Fig. 8(a) appear as the brighter layers in Fig. 8(b). This is an indication that the layers are predominantly amorphous phase. The layers between the amorphous composed of the dark, strongly diffracting region and brighter area which is out of diffraction condition. These layers are nanocrystalline B2 phase. It is apparent that the wire is composed of the layers of nanocrystalline (denoted as *c*) and amorphous (*a*). The layers are parallel to the drawing direction shown by the arrows (DD). Fig.8 (c) is the high resolution TEM image of the same sample. Fig. 8(c) is the high resolution TEM image of the different area of the sample. The amorphous layer are separated by the debris crystalline areas. The interfaces between the crystalline and the amorphous, delineated by the broken lines, are very irregular and not well-defined. The amorphous area contained some indication of the medium range order (MRO) clusters as seen in the amorphous phase processed by HPT[11], which may correspond to the

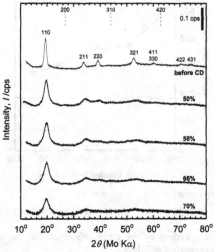

Figure 7 X-ray diffraction patterns of Ti-50.9mol%Ni wire before and after cold drawing.

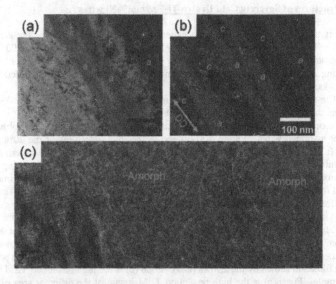

Figure 8 TEM micrographs of 70% drawn Ti-50.9mol%Ni wire. (a) bright field image. (b) dark field image. (c) high resolution image.

bright speckle contrast in Fig. 8(b).

Mechanical Properties of Severely Cold Drawn Ti-50.9mol%Ni wires

Fig. 9 shows the tensile stress-strain curves of as-drawn Ti-50.9mol%Ni wires. The curve corresponding to the commercial superelastic wire is also indicated for comparison. It is apparent that the cold-drawn wires possess very high tensile strength exceeding 2 GPa. The tensile elongation decreases with increasing reduction in area. It is about 8% for 50% drawn wires and about 4% in 70% drawn one, which is remarkably large for amorphous containing material. Also, it should be emphasized that the initial slope of the stress-strain curves depends

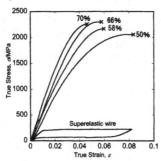

Figure 9 Tensile stress-strain curves of as-drawn Ti-50.9mol%Ni wires.

strongly on the reduction. Young's modulus, E, was evaluated from the slope of the initial part of the curves and is plotted in Fig. 10. The values change from 43 GPa in the 50% drawn wire to 71 GPa in the 70% drawn one. In a coarse grained TiNi, Young's modulus was reported to be 30 GPa for the B2 phase and 18 GPa for the B19' martensite phase [12]. For amorphous thin film of Ti-47.5mol %Ni and -52.7mol%Ni, the value of about 93 GPa is reported [13]. Thus the Young's modulus of the amorphous/nanocrystalline wires lies between those for the crystalline

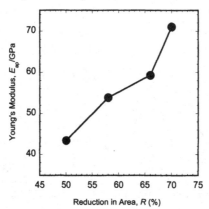

Figure 10 Young's modulus of as-drawn Ti-50.9mol%Ni wires.

TiNi and amorphous TiNi, although the tensile elongation of the wires is much better than that of the amorphous films (~ 1%).

Assuming that the as-drawn wire is the mixture of B2 phase and amorphous, the amount of amorphous phase was estimated on the basis of the rule of mixtures (Voigt average),

$$E = V_f^{B2} E_{B2} + V_f^{am} E_{am} \qquad (3),$$

where V_f^{B2}, V_f^{am} are the volume fractions of the B2 and amorphous phases, and E_{B2} and E_{am} are Young's moduli of the B2 and amorphous, respectively. V_f^{am} was estimated to be 0.65 for the 70% drawn wire.

Cyclic tensile stress-strain curves of the wire after 50% and 70% drawing are shown in Figs. 11(a) and (b), respectively. For the 50% drawn wire, the stress hysteresis between the loading and unloading curves develops with increasing imposed strain, which suggests the occurrence of stress-induced martensitic transformation. Meanwhile for the 70% drawn wire, the hysteresis is much less than the one in the 50% drawn wire. For the 70% drawn wire, permanent

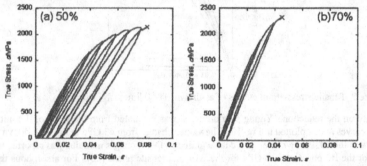

Figure 11 Cyclic tensile stress-strain curves of (a) 50% and (b) 70% drawn Ti-50.9mol%Ni wires.

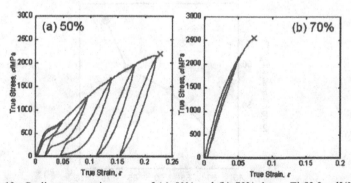

Figure 12 Cyclic stress-strain curves of (a) 50% and (b) 70% drawn Ti-50.9mol%Ni wires after aging at 573 K for 3.6 ks.

122

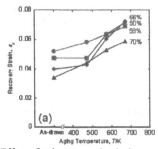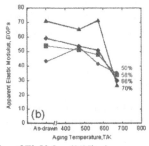

Figure 13 Effect of aging on mechanical properties of Ti-50.9mol%Ni wires. (a) Recovery strain. (b) Young's moduls.

plastic strain (ε_{pl}) was almost zero for the first 2 cycles and the ε_{pl} developed markedly after the 3rd cycle. Yet the wire exhibits a pseudo-elastic recovery strain (ε_r) of 3.4%. This value of ε_r exceeds the elastic strain of metallic glasses, which is about 2% under compression. The mechanism of the peculiar pseudoelastic recovery is not clear at this point. It cannot be the microtwinning of the (001) deformation twins [14] since both the X-ray diffraction and TEM observations support that the dominant phase is the amorphous or B2 nanocrystalline phase. It could be due to combined effect of superelastic deformation of nano-B2 grains and elastic deformation of amorphous phase. *In situ* XRD study is needed to clarify the mechanism.

Effect of aging on the mechanical properties was also investigated. Figure 12 shows the cyclic stress-strain curves after aging at 573 K for 3.6 ks. For the 50% drawn wire, stress-plateau becomes apparent and stress hysteresis is larger than in as-drawn wires as shown in Fig. 12(a). For the 70% drawn wires, by aging at 573 K, tensile strength, elongation and ε_r significantly increased to 2255 MPa, 71.4 GPa and 5.8 %, respectively, as shown in Fig. 12(b). The reason for the simultaneous increase in the tensile strength and elongation is not clear at this point. An increase in strength by aging was reported for SPD aluminum and interpreted as the results of the annihilation of dislocation sources during aging [18]. It can be speculated that an increase in elongation can be related to crystallization of amorphous phase and/or structural relaxation in amorphous phase. More detailed investigation is needed to clarify this aging behavior. The effect of aging on the mechanical properties are summarized in Figure 13. Although there are some scatter in the data, aging at temperature lower than 673 K lead to good balance between Young's modulus and recovery strain without the loss in much of the fracture stress.

CONCLUSIONS

In the present paper the effect severe plastic deformation and post-deformation aging on the microstructures, martensitic transformation and mechanical properties were described. Post-deformation aging resulted in the formation of nanocrystalline B2 phase. Nano-scale grain size stabilizes the B2 phase and suppress the formation of B19' martensite while it stabilize the R phase to some extent. The nanocrystalline TiNi exhibits very high value of microhardness which can be beneficial to increase shape recovery stress and a wider temperature window for superelasticity. Thin wires of Ti-50.9mol%Ni were processed by cold drawing of 50~70 % reduction (~0.3 mmϕ). XRD and TEM investigations revealed that amorphous is the dominant phase in as-drawn wires with the coexisting nanocrystalline B2 phase. The amorphous/nanocrystalline TiNi wires after aging exhibit high tensile strength (~2 GPa), high apparent elastic modulus (~71 GPa) and large pseudoelastic recovery (> 5%). Therefore, this material has very high potential as new medical device materials.

ACKNOWLEDGMENTS

A part of this work was supported by "Nanotechnology Network Project" of the Ministry of Education, Culture, Sports, Science and Technology (MEXT), Japan.

REFERENCES

[1]   J. Koike and D. M. Parkin, J. Mater. Res., **5** (1990) 1414.
[2]   H. Nakayama, K. Tsuchiya, Z.-G. Liu, M. Umemoto, K. Morii and K. Shimizu, Mater. Trans., **42** (2001) 1987.
[3]   H. Nakayama, K. Tsuchiya and M. Umemoto, Scripta Mater., **44** (2001) 1781.
[4]   D. M. Grant, S. M. Green and J. V. Wood, Acta Metall. Mater., **43** (1995) 1045.
[5]   Y. V. Tat'yanin, V. G. Kurduymov and V. B. Fedorov, Phys. Metall. Metallography. **62** (1986) 133.
[6]   K. Tsuchiya, Y. Hada, T. Koyano, K. Nakajima, M. Ohnuma, Y. Todaka and M. Umemoto, Scripta Mater., to be published.
[7]   T. Waitz, V. Kazykhanov and H. P. Karnthaler, Acta Mater., **52** (2004) 137.
[8]   H. C. Tong and C. M. Wayman, Acta Metall., **22** (1974) 887.
[9]   K. Tsuchiya, M. Inuzuka, D. Tomus, A. Hosokawa, H. Nakayama, K. Morii, Y. Todaka and M. Umemoto, Materials Science and Engneering A. **438-440** (2006) 643.
[10] M. Nishida, S. Ii, T. Furukawa, A. Chiba, T. Hara and K. Hiraga, Scripta Mater., **39** (1998) 1749.
[11] M. Peterlechner, T. Waitz and H. P. Karnthaler, Scripta Mater., **59** (2008) 566.
[12] Y. Liu and H. Xiang, J. Alloys. Comp., **270** (1998) 154.
[13] A. Gyobu, Y. Kawamura, T. Saburi and H. Horikawa, Mechanical Properties of Sputter-Deposited Amorphous Ti-Ni Films, PRICM 3, 1998, M. A. Imam,R. Denale,S. Hanada,Z. Zhong andD. N. Lee (Eds.) (TMS, pp. 2719.
[14] Y. F. Zheng, B. M. Huang, J. X. Zhang and L. C. Zhao, Mater. Sci. Eng., **A279** (2000) 25.

Mater. Res. Soc. Symp. Proc. Vol. 1129 © 2009 Materials Research Society   1129-V13-01

Sensing Shape Recovery Using Conductivity Noise in Thin Films of NiTi Shape Memory Alloys

U. Chandni[1], M. V. Manjula[1], Arindam Ghosh[1], H. S. Vijaya[2] and S. Mohan[2]

[1]Department of Physics, Indian Institute of Science, Bangalore 560 012, India.

[2]Department of Instrumentation, Indian Institute of Science, Bangalore 560 012, India.

ABSTRACT

Low frequency fluctuations in the electrical resistivity, or noise, have been used as a sensitive tool to probe into the temperature driven martensite transition in dc magnetron sputtered thin films of nickel titanium shape-memory alloys. Even in the equilibrium or static case, the noise magnitude was more than nine orders of magnitude larger than conventional metallic thin films and had a characteristic dependence on temperature. We observe that the noise while the temperature is being ramped is far larger as compared to the equilibrium noise indicating the sensitivity of electrical resistivity to the nucleation and propagation of domains during the shape recovery. Further, the higher order statistics suggests the existence of long range correlations during the transition. This new characterization is based on the kinetics of disorder in the system and separate from existing techniques and can be integrated to many device applications of shape memory alloys for in-situ shape recovery sensing.

INTRODUCTION

Near equiatomic nickel titanium (NiTi) shape memory alloys have gained considerable industrial focus owing to its varied applications including micro-actuators, positioners, micro-valves and so on [1]. In the thin film form, the displacive structural transformation from the high temperature cubic (B2:CsCl) austenite phase to the low temperature monoclinic (B19/19') martensite phase is often influenced by external parameters such as the substrate, finite grain size, surface defects etc. The information content from the standard characterization schemes such as x-ray diffraction (XRD), differential scanning calorimetry (DSC), resistivity measurements, acoustic emission etc. is often limited and there is a need to have microscopically intuitive estimates of the temperature scales of martensite transformation that may be relevant to growing applications of NiTi thin films. Here, we report two different aspects of the conductivity noise measurements on dc-magnetron sputtered NiTi thin films [2]: (a) equilibrium resistance noise that arises from the coupling of electron transport to slow variations in the structural disorder, which was found to be sensitive to instantaneous bi-crystalline topology during the thermoelastic martensite transition; (b) the dynamic noise and higher order avalanche statistics while the system is being temperature-driven which serves as an excellent probe to the resistivity avalanches that correspond to the athermal martensitic phase transition.

EXPERIMENTAL DETAILS

The samples were prepared by dc magnetron sputtering of a mosaic target, which consists of a patterned titanium disk of 76mm diameter and 0.8mm thickness laminated over a circular nickel disk of 76mm diameter and 1.6mm thickness. Si(1,0,0) cleaned with dilute HF and methanol were used as substrates. The samples were deposited under 2×10^{-3} mbar Argon pressure and were of thickness $\approx$ 1mm. Both length ($\approx$ 5mm) and width ($\approx$ 2mm) of the film were kept macroscopically large. The samples were characterized using XRD, scanning tunneling microscopy and DSC. The resistivity measurements were carried out using van der Pauw method.

For the noise measurements, the sample was current biased by a low noise source and the fluctuations in the voltage developed across the sample $\delta V(t) = V_0(t) - \langle V \rangle$ (where $\langle V \rangle$ is the average voltage) is measured. Special care has been taken during experiment as well as during the subsequent digital signal processing, to minimize extraneous noise from the electronics, sample contacts, electromagnetic interferences, ground loops, impedance mismatch and temperature fluctuations of the sample and hence a noise floor lower than $S_V(f) \leq 10^{-20} V^2 Hz^{-1}$ over a frequency range from 10Hz to 1mHz was obtained. The technique is primarily based on the idea put forward by Scofield [3] and uses an ac bridge configuration to measure conductivity fluctuations, which allows a direct determination of the background noise as well as the resistivity noise from the sample. The background noise primarily consists of Johnson's noise and is at least two orders of magnitude lower than device noise at all temperatures. For the equilibrium noise measurements the noise was measured at fixed temperatures (T) stabilized to an accuracy of 3ppm after a waiting period of ~1800-2000 seconds at each T. For the dynamic noise measurements, data was acquired in successive windows of 510 seconds using a16 bit digitizer, while ramping the T at several rates across the transition, and the statistics of the deviation of the resistivity from its mean value, $\Delta\rho(t)$, were evaluated for these windows.

DISCUSSION

The sample was characterized using resistivity, XRD as well as scanning tunneling microscopy, which indicated a co-existence of martensite and austenite phases at room temperature with a grain size of ~160nm. Resistivity measurements have long been used as an empirical detector of martensite phase transformation temperature scales [1]. The alloys in general exhibit interesting changes and peaks in the resistivity by up to 20% over the transformation temperature range; however, correlating these changes with the measured phase changes or mechanical properties has not always been very successful. Also, there are often large changes in the resistivity curves after cycling samples through the transformation a number of times. Nevertheless, resistivity is still used for the basic characterization of shape memory alloys. Figure 1 shows the T-dependence of $\langle\rho\rangle$ of a typical 0.9μm thick NiTi film where each point of $\langle\rho\rangle$ was obtained on averaging over $\approx$ 5 sec. Before measuring noise the system was subjected to several tens of thermal cycles till the $\langle\rho\rangle - T$ traces for two successive cycles were identical in the transition zone. No external stress was applied for any of the measurements.

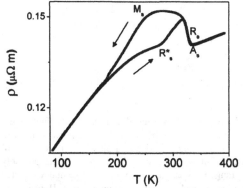

Figure 1: Resistivity vs Temperature plot. The transition temperatures are indicated. M, R, R* and A correspond to the martensite, intermediate R-phase, intermediate R*phase and austenite phases respectively.

Equilibrium (or Static) Noise Measurements

Equilibrium resistance noise arises from the coupling of electronic transport to slow variations in the configuration of structural disorder, change in the structural and scattering properties or the density of states at the Ti sites due to the decrease in the energy of the B19/B19' structure. Such a study would also probe into the influence of grains and grain boundaries on the thermoelastic properties of the NiTi films.

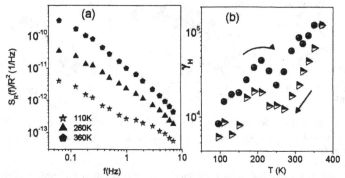

Figure 2: (a) Normalized Power Spectral Density as a function of frequency for three different temperatures. (b) Hooge parameter vs T. Note that the regions at the two ends are linear where as the middle portion is flattened out.

The power spectral densities corresponding to three different temperatures are shown in Figure 2(a), which represents the conventional $1/f^{\alpha}$-noise ($\alpha{\sim}1$-1.2) observed in a wide range of disordered metallic films [4]. The T-variation of equilibrium noise is shown in Figure 2(b). The measured power spectral density $S_{\rho}(f)$ has been normalized to the conventional Hooge parameter γ_H defined as

$$S_{\rho}(f) = \frac{\gamma_H \rho^2}{nVf^{\alpha}} \qquad (1)$$

where n, V and ρ are the number density of atoms, sample volume and the resistivity of the sample respectively.

γ_H is ~ 9 orders of magnitude higher than for typical diffusive metallic films [5]. We find a clear difference in the T-dependence of γ_H in the region ~200-300K *i.e.* between martensite start and finish temperatures M_s and M_f respectively, which might be due to a plausible co-existence of phases during the transformation. The weak T-dependence of noise in this regime probably indicates defect relaxation caused by long range (in space and time) elastic potential associated with the martensite transformation, dominating thermal fluctuations in the process. However, γ_H shows a nearly activated behavior deep inside the martensite and austenite regimes with activation energies of ~ 0.17eV and 0.28eV, respectively, implying thermal fluctuations as the source of noise in these regimes.

Dynamic Noise Measurements

In order to probe the statistics of avalanches we have also carried out dynamic noise experiments, while ramping the temperature at several rates across the transition region. The statistics of $\Delta\rho(t)$, was evaluated within each of these windows. Here, we present results for three different ramp rates: 0.3K/min, 0.5K/min and 1K/min, and also compare the results with that obtained in the static case.

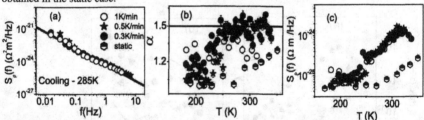

Figure 3: (a) Power spectral densities corresponding to three different cooling rates at 285K. (b) Frequency exponent α and (c) power spectral density as a function of temperature for the three rates as well as the static case.

Figure 3(a) shows the power spectral density corresponding to different ramp rates at 285K during the cooling cycle. The absolute value of $S_{\rho}(f)$ is nearly independent of ramp rate over three decades of f, as well as temperature (Figure 3(c)). This can be attributed to the athermal nature of the transition, so that once the disorder is quenched, the transformation proceeds through the same set of metastable states in every cycle. Further, the power spectrum during

128

ramping is far larger compared to the static (Figure 3(c)), confirming noise to mainly arise from the avalanche dynamics [6] in the dynamic mode. The frequency exponent α (Figure 3(b)) is ~ 1.5±0.1 for slow ramps (0.3K/min and 0.5K/min) over a broad range of T (340-240K), but reduces to ~1.3-1.35 for 1K/min. This can also be identified with the universality in the behavior of avalanche dynamics close to various non-equilibrium phase transitions [7].

We have also investigated correlation effects close to the structural transition by estimating the higher order statistics of fluctuations [8]. To achieve this we evaluated the 'second spectrum' $S^{(2)}(f)$ which is the Fourier transform of the four point correlation function and is given as:

$$S^{(2)}(f) = \int_0^\infty \langle \Delta \rho^2(t) \Delta \rho^2(t+\tau) \rangle_t \cos(2\pi f \tau) \, d\tau \tag{2}$$

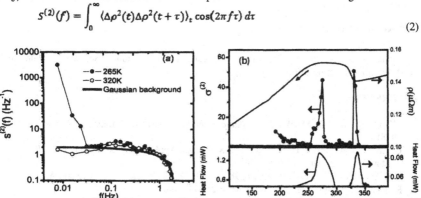

Figure 4: (a) Normalized second spectra for the cooling cycle (0.3K/min ramp rate). The dark line shows the gaussian expectation. (b) Integrated second spectra showing divergence of the correlation length at the transition temperatures. The lower panel shows the heat flow obtained from the DSC measurements.

Figure 4(a) clearly shows the non-gaussianity involved in the second spectra at 265K during cooling at 0.3K/min, as compared to the gaussian expectation. Figure 4(b) shows the normalized second spectra integrated over the entire bandwidth (1Hz-3Hz) as a function of temperature for cooling at 0.3K/min. Clear peaks are obtained at the phase boundaries namely between A→R and R→ M during the cooling cycle. This is essentially due to the long range correlations prevalent in the system at the onset of transformation. Thus, at the critical temperature, the system is correlated and the fluctuations are maximally non-gaussian. The peaks can be interpreted as direct manifestations of non-gaussianity associated with the divergence of the correlation length [2] during the two-stage structural phase transition. It is clear that at the transition temperatures defined by the DSC, which represents the latent heat released in the first order phase transition, the correlation length diverges as well.

129

CONCLUSION

In conclusion, we have realized a new sensing mechanism of shape recovery which clearly distinguishes the transformation temperature scales. The equilibrium noise magnitude was found to be unusually large and displays the signature of shape recovery and martensite transformation on the dynamics of structural defects. The avalanche dynamics which is characteristic of athermal phase transitions is also directly probed into. The higher order statistics have been used to analyze the long range correlations prevalent in the system and constitutes a new non-invasive technique of sensing the transformation temperature scales that is of utmost importance to many device applications as well.

ACKNOWLEDGMENTS

We thank the Central Facility, Department of Physics, Indian Institute of Science, Bangalore for the differential scanning calorimetry measurements.

REFERENCES

1. K. Otsuka and C. M. Wayman, *Shape memory Materials*, (Cambridge University Press, 1998)
2. U. Chandni, A. Ghosh, H. S. Vijaya and S. Mohan, Appl. Phys. Lett. **92**, 112110 (2008); e-print arXiv:cond-mat/0811.0102 (2008).
3. J. H. Scofield, Rev. Sci. Instrum. **58**, 985 (1987); A. Ghosh, S. Kar, A. Bid and A. K. Raychaudhuri, e-print arXiv:cond-mat/0402130 (2004).
4. D. M. Fleetwood and N. Giordano, Phys. Rev. B **31**, 1157 (1985); M. B. Weissman, Rev. Mod. Phys. **60**, 537 (1988).
5. P. Dutta and P. M. Horn, Rev. Mod. Phys. **53**, 497 (1981); J. H. Scofield and J. V. Mantese, Phys. Rev. B. **32**, 736 (1985).
6. F. J. Perez-Reche, E. Vives, L. Manosa and A. Planes, Phys. Rev. Lett. **87**, 197501 (2001).
7. L. Laurson and M. J. Alava, Phys. Rev. E. **74**, 066106 (2006).
8. G. T. Seidler and S. A. Solin, Phys. Rev. B. **53**, 9753 (1996).

Mater. Res. Soc. Symp. Proc. Vol. 1129 © 2009 Materials Research Society 1129-V02-04

Magnetoelastic Material as a Biosensor for the Detection of Salmonella Typhimurium

Ramji S Lakshmanan[1*], Rajesh Guntupalli[4,1], Shichu Huang[1], Michael L Johnson[1], Leslie C Mathison[1], I-Hsuan Chen[3], Valery A Petrenko[2], Zhong-Yang Cheng[1] and Bryan A Chin[1];
[1]Materials Engineering, Auburn University, Auburn, Alabama;
[2]Department of Pathobiology, Auburn University, Auburn, Alabama;
[3]Department of Biological Sciences, Auburn Unviersity, Auburn, Alabama;
[4]Department of Anatomy, Physiology and Pharmacology, Auburn University, Auburn, Alabama.

ABSTRACT

Magnetoelastic materials are amorphous, ferromagnetic alloys that usually include a combination of iron, nickel, molybdenum and boron. Magnetoelastic biosensors are mass sensitive devices comprised of a magnetoelastic material that serves as the transducer and bacteriophage as the bio-recognition element. By applying a time varying magnetic field, the magnetoelastic sensor thin films can be made to oscillate, with the fundamental resonant frequency of oscillations depends on the physical dimensions and properties of the material. The change in the resonance frequency of these mass based sensors can be used to evaluate the amount of analyte attached on the sensor surface. Filamentous bacteriophage specific to S. typhimurium was used as a bio-recognition element in order to ensure specific and selective binding of bacteria onto the sensor surface. The sensitivity of magnetoelastic materials is known to be dependent on the physical dimensions of the material. An increase in sensitivity from 159Hz/decade for a 2mm sensor to 770Hz/decade for a 1mm sensor and 1100Hz/decade for a 500micron sensor was observed. The sensors were characterized by scanning electron microscopy (SEM) analysis assayed biosensors to provide visual verification of frequency responses and an insight into the characteristics of the distribution of phage on the sensor surface. The magnetoelastic sensors immobilized with filamentous phage are suitable for specific and selective detection of target analyte in different media. Certain modifications to the measurement circuit resulted in better signal to noise ratios for sensors with smaller dimensions (L<1mm). This was achieved by tuning the circuit resonance close to that of the sensor. According to models and preliminary tests, this method was anticipated in about a 5 times increase in signals for a 200×40×6microns. This technique and further studies into the design and modification of the measurement circuits could yield better, sensitive responses for sensors with smaller dimensions. The magnetoelastic materials offer further advantages of potential miniaturization, contact-less nature and ease of operation.

INTRODUCTION

Each year approximately 76 million cases, 325,000 hospitalizations and 5000 deaths are reported in the United States alone [1] caused by food-borne illnesses. Food-borne illnesses are primarily caused by viruses, bacteria and parasites. These illnesses cause mild diarrhea to severe life-threatening neurological ailments. About 90% of food-borne illnesses caused can be attributed to bacterial contamination of food. *Salmonella* species are the most frequent and commonly occurring bacterial food-borne pathogens worldwide [2]. The ever growing need for rapid detection of pathogenic microorganisms present in food, environmental and clinical samples has resulted in an increased interest in the research efforts towards the development of novel diagnostic methodologies. In this article we report the results of the characterization of a phage-based magnetoelastic biosensor developed towards the detection of *Salmonella typhimurium*.

THEORY AND EXPERIMENTAL PROCEDURES

Magnetostriction is a property observed in ferromagnetic materials. A magnetostrictive material experiences a change in shape when exposed to a magnetic field. The opposite effect is known as the Villari effect or inverse magnetostriction. In this inverse effect, the application of a mechanical stress on a magnetostrictive material results in a change in the magnetization of the material. In a typical sensor application both these effects are utilized. Magnetoelastic sensors are actuated by the application of an AC magnetic field. The applied varying magnetic field causes the sensors to oscillate mechanically. The characteristic fundamental resonance frequency of these oscillations is dependent on the physical dimensions of the material. For a thin, planar, ribbon shaped sensor of length L, vibrating in its basal plane, the fundamental resonant frequency of longitudinal oscillations is given by [34]

$$f = \sqrt{\frac{E}{\rho(1-\sigma^2)}} \frac{1}{2L} \quad ---------- (1)$$

where, E is Young's modulus of elasticity, ρ is the density of the sensor material, σ is the Poisson's ratio, and L is the long dimension of the sensor.

The resonance frequency of the sensor can be read from the frequency spectrum and can be tracked for any changes. These changes can then be related to the magnitude of non-magnetoelastic mass attached to the sensor surface. Addition of a small mass of Δm (much smaller than the original mass of the sensor ($\Delta m << M$)) on the sensor surface results in a mechanical damping of the frequency and hence would result in the change of natural resonance frequency (f_0) by an amount of Δf which can be related to the mass added as:

$$\Delta f = -\frac{f}{2} \cdot \frac{\Delta m}{M} \quad -------- (2)$$

where f is the initial resonance frequency, M is the initial mass, Δm is the mass change and Δf is the shift in the resonant frequency of the sensor. Recently, the use of magnetoelastic materials as transducer platforms for the remote monitoring of food-borne pathogens [5-9] and other applications in chemical detection and environmental monitoring [3,10,11] have attracted considerable interest. The wireless nature of magnetostrictive sensors allows these sensors to be used *in-vivo* in closed containers.

In this work, METGLAS 2826MB was chosen as the platform for the biosensor. Rectangular strips were diced and then sputtered with thin layers of chromium and gold in that order. The details of the procedures have been explained elsewhere [6,7]. The filamentous phage E2, used in this work, was provided by Dr. James M. Barbaree's lab in the Department of Biological Sciences at Auburn University. The magnetoelastic sensor platforms were placed in vials containing 100 µL of phage E2 (5×10^{11} virions/mL). The immobilization of phage on the sensor surface was done by placing these vials on a rotor (running at 8 rpm) for 1 hour. The use of a rotor for the immobilization step resulted in a uniform distribution of physically adsorbed phage on the sensor surface. These sensors were then washed three times with Tris-Buffered Saline (TBS) solution and two times with sterile distilled water in order to remove any unbound or loosely bound phage. Dose response was studied as the frequency response of the sensor to different concentrations of *S. typhimurium*, while monitoring the frequency while the binding was occurring. The different concentrations (5×10^1 cfu/ml through 5×10^8 cfu/ml) were prepared by successive dilutions of the as-received *S. typhimurium* suspensions (5×10^8 cfu/ml).

The phage immobilized sensors are placed in the center of a solenoid coil for resonance frequency measurements. A schematic of the experimental setup is shown in Figure 1.

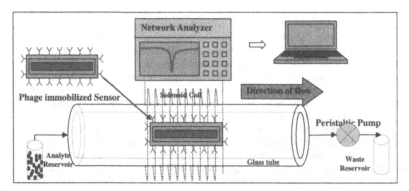

Figure 1: Schematic of experimental setup for dose response studies

The analyte reservoir was then successively filled with suspensions containing *S. typhimurium* (5 × 10¹ cfu/ml through 5 × 10⁸ cfu/ml). These solutions were allowed to flow (50 μl/min) over the sensor with the help of an Ismatec Reglo Digital peristaltic pump. As the target analyte attached to the sensor surface the changes in the resonance frequency of the sensors were recorded and monitored. Each aliquot of *S. typhimurium* was allowed to flow for 20 minutes allowing exactly 1 ml of each concentration pass over the sensor. To improve sensitivity of the sensors we studied the dose response of the magnetoelastic sensors with two different dimensions (1×0.2×0.015 mm and 0.5×0.1×0.015 mm).

DISCUSSION
Dose response curves and electron microscopy of sensors with different dimensions

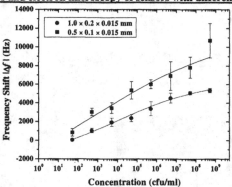

Figure 2: Comparison of magnetoelastic biosensor's dose responses, when exposed to increasing concentrations (5×10¹ to 5×10⁸ cfu/mL) of *S. typhimurium* suspensions on two different sizes of sensors (0.5×

0.1×0.015 mm ($\bullet$- χ^2=0.048, R^2=0.99) and 1×0.2×0.015 mm ($\bullet$- χ^2=0.7231, R^2=0.91) The curves represent the sigmoidal fit of signals obtained.

The dose response curves obtained for the sensors of the two sizes has been shown in Figure 2. Upon exposure to a higher concentration of S. typhimurium it was observed that the frequency shift responses were higher. The sensitivity of the response in the linear region was measured to be 770 Hz/decade and 1152 Hz/decade for a 1 mm and a 500 μm sensor respectively. Also the detection limits were observed to have improved from the order of 10^5 cfu/ml to 10^3 cfu/ml. The assayed sensors were viewed in a scanning electron microscope to visually confirm the obtained frequency shifts. Figure 3(a)-3(c) shows the SEM images of entire sensors (L=0.5 mm) exposed to different concentrations of the target analyte. Figure 3(d) and 3(e) show high magnification images of sensor surfaces before and after phage immobilization.

A clear decrease in the number of bacteria bound to the sensor can be observed for the biosensors exposed to the three different concentrations (5×10^2 cfu/mL, 5×10^4 cfu/mL and 5×10^8 cfu/mL). The number of bacteria attached to the entire biosensor surface (one side) was counted manually. The total number of attached S. typhimurium cells were 14278, 3286 and 157 for the three different biosensors exposed to concentrations of 5×10^8, 5×10^4 and 5×10^2 cfu/mL, respectively. An attachment of 14278 cells on the sensor surface theoretically corresponds to a frequency shift of 9.5 kHz. This number is very similar to the measured resonance frequency shift of 10 kHz for the biosensor. From Figure 3(d) and 3(e) it can clearly be seen that the phage immobilized on the sensor surface can be viewed with the help of SEM. Uniformly distributed phage filaments oriented in random directions were observed to saturate the entire surface.

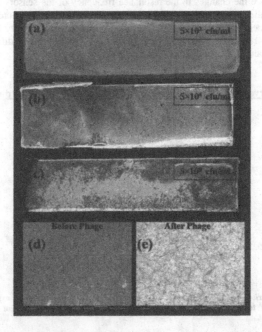

Figure 3:Typical SEM images of the entire surfaces of assayed 500μm sensors at three different concentrations ((a) 5×10² cfu/mL, (b) 5×10⁴ cfu/mL and (c) 5×10⁸ cfu/mL). High magnification images of sensor surface (d) before and (e) after phage immobilization.

Enhanced sensor resposes due to modified circuit

A modification to the measurement setup is described that enhances the signal amplitude of the resonance frequency response (enhanced oscillation displacements) of magnetoelastic sensors. A solenoid coil measures changes in magnetic flux caused by changes in the magnetization of the oscillating magnetoelastic sensor platform. A sensor with smaller dimensions will produce smaller changes in the magnetic flux. During the development of the equivalent circuit representing the measurement setup, it was observed that the addition of a capacitor in series with the solenoid coil resulted in enhanced signal amplitudes for the sensor's frequency spectrum. The addition of the tuned capacitor in the circuit resulted in an increase in the signal amplitudes. The signal amplitudes increased by 6 times for a 200 μm length sensor and by 20 times for a 500 μm length sensor. One such example of enhanced signal amplitudes is shown in Figure 4, that is a comparison of the S_{11} frequency responses of a tuned and an untuned circuit for a 200 μm length sensor.

A clear increase in the amplitudes of the signal can be observed for the frequency spectrum of the sensors. Our current work focuses on understanding the increase in the signal responses and also to utilize this circuit modification to improve the detection capabilities for smaller sized magnetoelastic sensors.

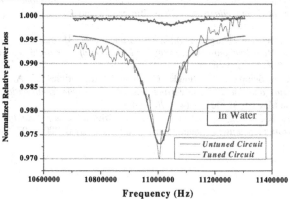

Figure 4: Comparison of frequency responses of a 200 μm sensor in water with a tuned and an untuned circuit.

CONCLUSIONS

It was demonstrated that the sensitivity of the magnetoelastic biosensors could be increased by decreasing the dimensions of the sensor. Also, the use of a tuned circuit was shown to enhance the sensor's resonance peak amplitude. Phage can be used as a highly specific and sensitive bio-recognition element for biosensor applications

REFERENCES

1. Mead, P.S., Slutsker, L., Dietz, V., McCaig, L.F., Bresee, J.S., Shapiro, C., Griffin, P.M., and Tauxe, R.V., *Food-Related Illness and Death in the United States*. Emerging Infectious Diseases, 1999. 5(5): 607-625.
2. Todd, E.C., *Costs of acute bacterial foodborne disease in Canada and the United States*. International Journal of Food Microbiology, 1989. 9(313-326).
3. Stoyanov, P.G. and Grimes, C.A., *A Remote Query Magnetostrictive Viscosity Sensor*. Sensors and Actuators A, 2000. 80: 8-14.
4. Landau, L.D. and Lifshitz, E.M., *Theory of Elasticity*. 1986: Pergamon
5. Lakshmanan, R.S., Guntupalli, R., Hu, J., Kim, D.-J., Petrenko, V.A., Barbaree, J.M., and Chin, B.A., *Phage immobilized magnetoelastic sensor for the detection of Salmonella typhimurium*. Journal of Microbiological Methods, 2007. 71(1): 55-59.
6. Lakshmanan, R.S., Guntupalli, R., Hu, J., Petrenko, V.A., Barbaree, J.M., and Chin, B.A., *Detection of Salmonella typhimurium in fat free milk using a phage immobilized magnetoelastic sensor*. Sensors and Actuators B: Chemical, 2007. 126(2): 544-549.
7. Guntupalli, R., Lakshmanan, R.S., Johnson, M.L., Hu, J., Huang, T.S., Barbaree, J.M., Vodyanoy, V.J., and Chin, B.A., *Magnetoelastic biosensor for the detection of Salmonella typhimurium in food products*. Sensing and Instrumentation for Food Quality and Safety, 2007. 1 (1): 3-10.
8. Guntupalli, R., Hu, J., Lakshmanan, R.S., Huang, T.S., Barbaree, J.M., and Chin, B.A., *A magnetoelastic resonance biosensor immobilized with polyclonal antibody for the detection of Salmonella typhimurium*. Biosensors and Bioelectronics, 2007. 22: 1474-1479.
9. Ruan, C., Zeng, K., Varghese, O.K., and Grimes, C.A., *Magnetoelastic Immunosensors: Amplified Mass Immunosorbent Assay for Detection of Escherichia coli O157:H7*. Anal. Chem., 2003. 75: 6494-6498.
10. Jain, M.K., Schmidt, S., Mungle, C., Loiselle, K., Grimes, C. A., *Measurement of temperature and liquid viscosity using magneto-acoustic/magneto-optical sensors*. IEEE. Trans. on Magnetics, 2001. 37(4): 2767-2769.
11. Shankar, K., Zeng, K., Ruan, C., and Grimes, C.A., *Quantification of ricin concentrations in aqueous media*. Sensors and Actuators B, 2005. 107: 640-648.

Mater. Res. Soc. Symp. Proc. Vol. 1129 © 2009 Materials Research Society　　　　1129-V04-16

Multiple Phage-Based Magnetoelastic Biosensors System for the Detection of *Salmonella typhimurium* and *Bacillus anthracis* spores

S. Huang[1], S. Li[1], H. Yang[1], M.L. Johnson[1], R.S. Lakshmanan[1], I.-H. Chen[2], V.A. Petrenko[3], J.M. Barbaree[2], and B.A. Chin[1]
[1]Materials Engineering, 275 Wilmore Labs, Auburn, AL 36830, USA
[2]Dept. of Botany & Microbiology, Auburn University, Auburn, AL 36849, USA
[3]Dept. of Pathobiology, Auburn University, Auburn, AL 36849, USA

ABSTRACT

This paper presents a multiple magnetoelastic (ME) biosensor system for in-situ detection of *S. typhimurium* and *B. anthracis* spores in a flowing bacterial/spore suspension ($5 \times 10^1 - 5 \times 10^8$ cfu/ml). The ME biosensor was formed by immobilizing filamentous phage (specific to each detection target) on the ME platforms. An alternating magnetic field was used to resonate the ME biosensor to determine its resonance frequency. When cells/spores are bound to a ME biosensor surface, the additional mass of the cells/spores causes a decrease in the resonance frequency of the biosensor. The detection system was composed of a control sensor, an E2 phage-based biosensor (specific to *S. typhimurium)* and a JRB7 phage-based biosensor (specific to *B. anthracis* spores). The frequency response curves of the ME biosensors as a function of exposure time were then measured and the detection limit of the ME biosensor was observed to be 5×10^3 cfu/ml. The results show that phage-based ME biosensors can detect multiple pathogens simultaneously and offer good performance, including good sensitivity and rapid detection.

INTRODUCTION

Every year, there are over 76 million Americans that suffer from food-borne illnesses, out of which there are about 325,000 hospitalizations and 5,000 deaths annually [1]. Bacterial pathogens account for more than 50% of these food-borne illnesses. In the United States, human illness due to *Salmonella* infection is most commonly caused by *S. typhimurium*, *S. enteritidis*, or *S. heidelberg* serotypes [2]. This year (2008), *Salmonella* outbreaks seem to be increasing in frequency. In mid-April, the *Salmonella St. Paul* outbreak on tomatoes became one of the largest *Salmonella* outbreaks in history and was far worse than the *E. coli* outbreak on spinach in 2006. This outbreak affected at least 869 people, 257 of whom were reported to be hospitalized. Additionally, *B. anthracis* is a possible bioterrorism agent that could be used to deliberately contaminate our food supply. Recently, fears of deliberate contamination of our food supply have become a concern, *S. typhimurium* and *B. anthracis* (anthrax) being primary pathogenic targets of interest. There is an increasing need to develop rapid and cost-effective methods to monitor pathogens and spores. So in this paper we present a multiple phage-based magnetoelastic biosensor system which is able to detect *S. typhimurium* and *B. anthracis* spores simultaneously.

THEORY AND EXPERIMENT

Working principle of ME platform

In this research, Metglas 2826 MB ($Fe_{45}Ni_{45}Mo_7B_3$) ribbon, acquired from Honeywell International, was selected as the biosensor platform due to its excellent magnetoelastic (ME) properties. The fabrication of the ME sensor is based on dicing and sputtering and has been described in detail elsewhere [3, 4]. Briefly, after dicing the ME platform into the desired size, a Cr layer and then an Au layer are deposited onto both sides of platforms sequentially. A time-varying AC external magnetic field is used to resonate the ME platform. The large magnetostriction and soft magnetic properties of the material allows the ME platform to exhibit a pronounced magnetoelastic resonance, which results in the emission of a magnetic signal that can be remotely detected using a pickup coil. Since no physical connection is needed between the ME platform and the device, the ME platform is used in a wireless (remote) manner.

For a thin, ribbon-shaped ME platform of length L excited by an AC magnetic field, and vibrating in its basal plane, the fundamental resonant frequency f_0 of the longitudinal vibrations is given by [5]:

$$f_0 = \sqrt{\frac{E}{\rho(1-v)}} \frac{1}{2L} \quad (1)$$

Where E is Young's modulus of elasticity, v is the Poisson ratio, ρ is the density of the sensor material, L is the long dimension of the ME platform.

If the mass increase (Δm) is much smaller compared to the mass of the sensor (M), and it is uniformly applied on the sensor surface, then the shift in resonant frequency (Δf) is given by [6]:

$$\Delta f = -\frac{f_0}{2} \frac{\Delta m}{M} \quad (2)$$

Eq. (2) shows that the resonance frequency shifts linearly and decreases with increasing mass on the sensor surface. Hence, with the capture of the target organism to the sensor surface, the resonance frequency will shift to a lower value.

ME biosensor

Affinity selected filamentous phage clones E2 and JRB7 used for this work (with an initial concentration of 1.06×10^{13} vir/ml) as well as S. typhimurium (ATCC13311) and B. anthracis spores (5×10^8 spores/ml) were provided by Dr. Barbaree of the Department of Biological Sciences at Auburn University. These phages have been shown to be highly specific and selective to S. typhimurium (E2 phage clone) and B. anthracis spores (JRB7 phage clone) by Dr. Petrenko and his coworkers [7, 8]. All phage suspensions were diluted to 5×10^{11} vir/ml in 1 × TBS (25 mM Tris, 3 mM KCl, and 420 mM NaCl at a pH of 7.4). The bacterial and spore suspensions were serially diluted with sterile distilled water to prepare bacterial suspensions ranging from 5×10^1 to 5×10^8 cfu/ml.

To form the ME biosensor, the two different sizes of sputtered ME platforms were immersed into two genetically engineered phage solutions (E2 and JRB7 phage) for 1 hour, allowing the phage to attach to the sensors' surface by physical adsorption. After rinsing the platforms in distilled water, BSA solution (1% (w/v)) was then coated onto the sensor for 40 min., followed by another distilled water rinse. The purpose of BSA is to serve as a blocking agent to prevent nonspecific binding during exposure to the analyte solutions. At this point, the ME biosensors are ready for the testing of S. typhimurium bacteria and anthrax spores. In this research, the reason for using two sensors of different lengths is to obtain two separated characteristic resonance frequencies that can be used to distinguish each target, which allows the detection to be conducted simultaneously. A control sensor (no phage immobilization, but blocked with BSA) was used as a reference to eliminate the effects of nonspecific binding with possible contaminants in the analyte, varying flow conditions, and variable temperature conditions.

Multiple-sensor detection system

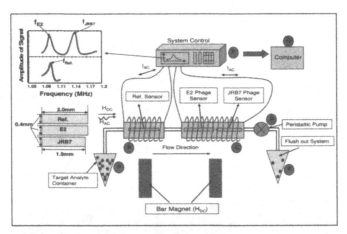

Fig. 1. Scheme of multiple ME biosensors measurement approach

A schematic of the simultaneous detection system is shown in Fig. 1. Two sensors, an E2 phage-based biosensor (2 × 0.4 × 0.15mm) and a JRB7 phage-based biosensor (1.9 × 0.4 × 0.15mm), were placed together in the primary measurement chamber (C). The control sensor (2 × 0.4 × 0.15mm) was placed in the second measurement chamber (B). Each chamber was connected to the network analyzer (F) via its own separate channel. The analyzer is able to measure and display the signals from both channels simultaneously, enabling the simultaneous measurement, display and recording of the three sensors' resonance frequencies in real time. The experiment was conducted by using the peristaltic pump (D) to drive an analyte containing only *S. typhimurium* or *B. anthracis* spores at a concentration of 5 × 10^8 cfu/ml from container (A), through the detection chambers (B&C), and finally into the flush out container (E) for collection and disposal of waste solution. When the target bacteria/cells bind to their corresponding ME biosensor, a frequency shift was observed to confirm the interaction, while the frequencies of the other two biosensors remained relatively stable. A flow rate of 50 µl/min was used in order to ensure laminar flow over the ME sensors. Resonance frequencies were continuously monitored and recorded by the analyzer (F) and computer system (G).

Microscopic Analysis

A JEOL-7000F scanning electron microscope (SEM) was used to confirm and quantify the binding of target antigens to the phage-coated ME biosensors. In preparation for SEM observations, the tested biosensors were washed with distilled water and then exposed to osmium tetroxide (OsO$_4$) vapor for 40 min.

DISCUSSION

Response with *S. typhimurium* / *B. anthracis* spores introduced sequentially

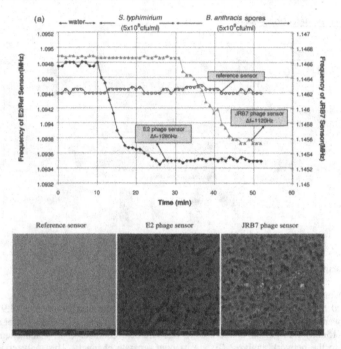

Fig.2. Response curves and SEM pictures for three diced ME sensors tested simultaneously when exposed to the *S. typhimurium* / *B. anthracis* spores suspension with a concentration of 5×10^8 cfu/ml.

Fig. 2 shows the simultaneous response of the multiple phage-based ME biosensors to water and the highest concentration (5×10^8 cfu/ml) of *S. typhimurium* and *B. anthracis* spore suspensions. The experiment was controlled by pumping filtered water through the two chambers continuously until the frequencies for the three ME biosensors stabilized, which usually took about 10 min. Then the concentrated *S. typhimurium* suspension was introduced into the system and passed across the sensors for 20 min under the same flowing conditions. After 20 minutes, a concentrated *B. anthracis* spore suspension was introduced continuously and allowed to flow for another 20 min. In Fig. 2, steady state response of all the sensors in water is observed during the first 10 minutes of the test. Then, within one minute after the introduction of *S. typhimurium*, the E2 phage-based biosensor showed a smooth decrease in resonance frequency due to the binding of these bacteria onto the biosensor surface. Steady state was achieved after about 15 minutes for the E2 phage-based biosensor due to the saturation of the *S.*

140

typhimurium-phage conjugates on the sensor surface. During this time, the other two biosensors showed a negligible frequency change, indicating that no appreciable binding had occurred to the control or JRB7 phage-based biosensors. As soon as the 5×10^8 cfu/ml *B. anthracis* spore suspension was introduced, a similar response was observed for the JRB7 phage biosensor in that a sudden drop of the resonance frequency was observed. This time, no significant change in the resonance frequency of the control sensor or the E2 phage-based biosensor was observed, which again indicated negligible non-specific binding to the control and E2 phage-based biosensors. In either case, the decrease in frequency is only present when bacteria cells or spores bind to the specific phage on the sensor's surface. During the entire test, the frequency of the control biosensor exhibited no appreciable change regardless of the composition of analyte introduced. This shows that non-specific binding was effectively blocked during testing by the 1 mg/ml BSA pre-treatment. Also, the sensor showed good environmental stability in the flowing liquid system, as evidenced by a lack of corrosion. This interaction is confirmed by the SEM photographs of the control and treated sensors. It is visually clear that the frequency shifts were due to the spores/bacteria cells attached to the corresponding sensors.

Dose response of ME biosensors within a multi-sensor system

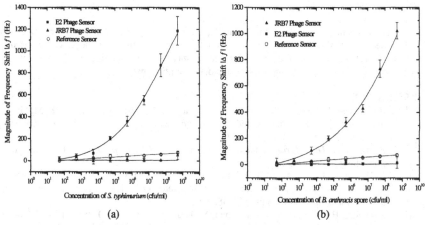

(a)   (b)

Fig.3. Dose response curve of ME biosensors within a multi-sensor detection system exposed to increasing concentrations (5×10^1-5×10^8 cfu/ml) of *S. typhimurium* and *B. anthracis* spores suspensions, sequentially. (a) E2 phage-based biosensor represents the specific response of ME biosensor coated with phage specific to *S. typhimurium* (($\blacksquare$) (R^2=0.987)). Control sensor represents the uncoated (devoid of phage) sensor's response ((o) (R^2=0.99)). JRB7 phage-based biosensor represents the non-specific response of ME biosensor coated with phage only specific to *B. anthracis* spores (($\blacktriangle$) (R^2=0.984)); (b) E2 phage-based biosensor (($\blacksquare$) (R^2=0.997)), JRB7 phage-based biosensor (($\blacktriangle$) (R^2=0.988)), control sensor ((o) (R^2=0.996)). The curve represents the sigmoidal fit of signals obtained.

141

The magnitude dose response of the three sensors simultaneously exposed first to *S. typhimurium* suspensions in the range of 5×10^1 to 5×10^8 cfu/ml is shown in Fig. 3 (a). As concentration increases, the frequency shifts for the E2 phage-based biosensor increase dramatically, which is because of the specific interaction between *S. typhimurium* cells with the E2 phage coated on the sensor surface. Both the control sensor and JRB7 phage-based biosensor show a slight frequency shift with increasing analyte concentration, with a maximum shift of less than 70 Hz at the largest concentration. This demonstrates that non-specific binding has been blocked by the BSA coating. The average Δf_{max} for the E2 phage-based biosensor was 1185Hz, and the detection limit for the sensor is estimated to be 5×10^3 cfu/ml. Saturation of the sensor occurred at greater than 5×10^8 cfu/ml. A linear response is found between the concentrations of 5×10^4 to 5×10^8 cfu/ml. A similar trend was found for the simultaneous exposure of the three sensors to *B. anthracis* spores in the range of 5×10^1 to 5×10^8 cfu/ml, and is shown in Fig. 3(b), immediately following the exposure to *S. typhimurium* suspension. The detection limit estimated to be 5×10^3 cfu/ml. The average Δf_{max} for the JRB7 phage-based biosensor was 1000Hz. Saturation of biosensor occurred at greater than 5×10^8 cfu/ml. Overall, the sensitivity for E2 phage-based biosensor with a length of 2mm is 238.5 Hz/Decade, while the sensitivity for JRB7 phage-based biosensor with a length of 1.9mm is 279.5 Hz/Decade.

CONCLUSIONS

These results demonstrate that multiple, targeted, phage-based ME biosensors can be simultaneously monitored to detect and discriminate between different pathogens. Immobilized phage on the sensor's surface provides the biomolecular recognition ability that insures sensitivity, specificity, and a simple method to detect multiple pathogens.

REFERENCES

[1]   P. S. Mead, L. Slutsker, V. Dietz, L. F. McCaig, J. S. Bresee, C. Sharpiro, P. M. Griffin, and R. V. Tauxe, "Food-related illness and death in the United States," *Emerging Infectious Diseases*, vol. 5, pp. 607-625, 1999.

[2]   R. Berkow, *The Merck Manual of Diagnosis and Therapy 1992*, 16th ed.: Merck Publishing Group, 1992.

[3]   S. Huang, J. Hu, J. Wan, M. L. Johnson, H. Shu, and B. A. Chin, "The effect of annealing and gold deposition on the performance of magnetoelastic biosensors," *Materials Science and Engineering C*, vol. 28 pp. 380-386, 2008.

[4]   S. Huang, H. Yang, R. S. Lakshmanan, M. L. Johnson, I. Chen, J. Wan, H. C. Wikle, V. A. Petrenko, J. M. Barbaree, Z. Y. Cheng, and B. A. Chin, "The Effect of Salt and Phage Concentrations on the Binding Sensitivity of Magnetoelastic Biosensors for B. anthracis Detection," *Biotechnology and Bioengineering, in press,* 2008.

[5]   C. Liang, S. Morshed, and B. C. Prorok, "Correction for longitudinal mode vibration in thin slender beams," *Applied Physics letter. ,* vol. 90, p. 221912, 2007.

[6]   P. G. Stoyanov and C. A. Grimes, "A remote query magnetostrictive viscosity sensor," *Sens. Actuators,* vol. 80, pp. 8-14, 2000.

[7]   V. A. Petrenko and G. P.Smith, "Phages from landscape libraries as substitue antibodies," *Protein Engineering,* vol. 13, pp. 589-592, 2000.

[8]   V. A. Petrenko, G. P. Smith, X. Gong, and T. Quinn, "A library of organic landscapes on filamentous phage," *Protein Engineering,* vol. 9, pp. 797-801, 1996.

Mater. Res. Soc. Symp. Proc. Vol. 1129 © 2009 Materials Research Society   1129-V01-02

Magnetostrictive Fe-Ga Wires with <100> Fiber Texture

S.P. Farrell[a,*], P.E. Quigley[a], K.J. Avery[a], T.D. Hatchard[b,c], S.E. Flynn[b], R.A. Dunlap[b,c]

[a] Defence R&D Canada - Atlantic, 9 Grove St., Dartmouth, NS, Canada B2Y 3Z7
[b] Dept of Physics and Atmospheric Science, Dalhousie University, Halifax, NS, Canada B3H 3J5
[c] Institute for Research in Materials, Dalhousie University, Halifax, NS, Canada B3H 3J5
[*] author for correspondence - email: shannon.farrell@drdc-rddc.gc.ca

ABSTRACT

Recently, low-cost processing approaches that produce textured thin bodies have engendered interest as cost-effective approaches for fabrication of magnetostrictive Fe-Ga alloys. In particular, wire-forming methods that strictly control the solidification direction could lead to some measure of crystallographic texture control. This is critical for development of large magnetostriction in polycrystals and for use of the alloys in actuators, sensors, energy harvesters and other systems. Magnetostrictive Fe-Ga wires have been prepared using an innovative cost-effective approach – based on the Taylor wire method – that combines rapid solidification and deformation processes. The procedure for making magnetostrictive wires is discussed and the wires are evaluated in terms of microstructure, crystallographic texture and magnetostriction. Results show that the Taylor-based approach is an effective and versatile means to draw 1-3 mm diameter textured Fe-Ga wire. Experimentation on the influence of drawing technique and quench conditions on texture development resulted with production of a strong <100> fiber texture in the Fe-Ga wire. Magnetostriction measurements, in the absence of prestress, indicated a maximum magnetostriction of ~165 ppm in a saturation field of less than 200 mTesla. This is considered a significant strain for bulk polycrystalline Fe-Ga alloys without a pre-stress or a stress-annealing treatment. The unique properties of wires made with the Taylor-based approach coupled with the low intrinsic cost make this an attractive approach for production of textured magnetostrictive wire for a variety of applications.

INTRODUCTION

There has been a proliferation of research into cost-effective non rare-earth element based magnetostrictive materials for actuators, sensors, motors, adaptive structures, energy harvesters and other applications. Although Fe-Ga alloys exhibit approximately 20 % of the strain of traditional giant magnetostrictive alloys (ex., Terfenol-D), they are less expensive, more robust and combine advantages of toughness, formability, machinability and mechanical strength with large magnetostriction (and little magnetic hysteresis) at low saturation fields [1]. While Fe-Ga single crystals offer better magnetoelastic properties over their polycrystalline counterparts, fabrication of single crystals is expensive. Recently, low-cost processing approaches that produce highly textured (near <100>) thin bodies of Fe-Ga alloys have engendered interest (for example, see the work of [2]). The strong <100> texture is desired due to the anisotropic nature of magnetic field-induced strains that produce a maximum magnetostriction along the <100> in these alloys. Although textured polycrystalline materials tend to achieve lower magnetically-induced strains than single crystals, they are less expensive and easier to fabricate. These advantageous features, as discussed above, when coupled with the thin-bodied alloy being less susceptible to eddy current-related losses, make thin bodies desirable for many applications.

The objective of this study is to investigate the use of the Taylor wire method for the production of <100> textured magnetostrictive Fe-Ga alloys. The Taylor wire method (also known as the "Taylor-Ulitovski or "microwire" process) was originally developed by Taylor in 1924 [3]. This method is similar to other non-equilibrium techniques (such as melt spinning) that combine rapid solidification and deformation processes. Although traditionally used for the production of microwire (micron diameter glass-encapsulated metal filaments) [4], the method has shown promise for the production of large millimeter diameter wire [3,5]. To date there has been very little evidence in the open literature to relate the properties of Taylor drawn Fe-based wires to their detailed crystallographic structure. This paper reports the results of employing this innovative thermomechanical processing approach for production of textured Fe-Ga alloys. Magnetostrictive Fe-Ga wires will be discussed in terms of the resultant microstructure, crystallographic texture and magnetostriction.

EXPERIMENT

Fe-Ga precursory alloys with a target composition of 19 at% Ga were prepared from high purity Fe (99.98 %) and Ga (99.99 %) elements. A suction extraction technique, similar to that developed by Srinivas and Dunlap [6], was employed to fabricate uniform 2 to 3 mm diameter rapidly solidified Fe-Ga rod precursors. Fe-Ga wires were then prepared from suction cast rods using a variation of the Taylor wire method [3,5]. In this approach, the Fe-Ga rod precursor was inserted into fused quartz glass tubing that was sealed at one end, evacuated and flushed with argon several times, backfilled to 1/3 atmosphere argon and sealed. The quartz vessel was heated to near 1500°C using an oxy-acetylene torch. This temperature was well above the composition dependent melting point of the Fe-Ga alloy (near 1400°C) [1] and above the softening temperature of the quartz. This made the quartz vessel pliable and, when axially loaded, required little force to initiate plastic deformation of the quartz. The quartz was then hand-drawn to produce glass-sheathed Fe-Ga wire and allowed to cool slightly. The glass-encapsulated wires were then re-heated to a few hundred degrees below the melting temperature (near 1000°C) and then quenched in cold water. The glass provided an effective medium to prevent oxidation of the Fe-Ga wires. Uniform wires from 1-3 mm diameter were routinely made with this approach. The diameter and microstructure of the final wires were found to be dependent on drawing (loading magnitude and rate) and quench conditions.

Grain size and structure were investigated using optical microscopy. The concentration of Fe and Ga in the Fe-Ga alloys was determined via energy dispersive x-ray spectrometry (EDXS) on a Leo 1455VP scanning electron microscope (SEM). Room temperature structure analysis was investigated using Cu $K_{\alpha 1}$ radiation operating at 40 kV and 55 mA with a Philips X'Pert Pro X-ray diffraction (XRD) system. The room temperature crystallographic texture of the wires was determined on the plane normal to the wire axis using Laue Back-Reflection XRD (Laue XRD) and an Electron Back-Scatter Detector (EBSD) on a SEM. Laue XRD patterns were typically collected on Polaroid films using W radiation and a Philips X-ray generator operating at 20 kV and 30 mA. EBSD patterns were determined using a Channel 5 EBSD from HKL Technology on a Cold Field Emission Hitachi S-4700 FEG SEM that operated at 20 kV and 30 mA. Axial magnetic field-induced strains and magnetization were measured as a function of magnetic field applied parallel to the wire axis. Magnetization measurements were determined using a PAR-155 vibrating sample magnetometer (VSM). Two strain gauges were employed to identify occurrence of inhomogenous strain.

RESULTS

Compositional and Microstructural Analysis

Compositional analysis showed that the wires contained approximately 18 at% Ga. A consistent loss of ~1 at% Ga was found between the Fe-Ga nominal starting compositions and the final wire products. The radial and axial optical microstructures for a typical Fe-Ga wire are shown in Figures 1 and 2, respectively. Typically, the grains had a bimodal distribution with 1 or a few large columnar grains in the center and smaller grains near the edge. Average grain diameters on the cross-sectioned surface (Figure 1) were 1.5 and 0.5 mm for the larger and smaller grains, respectively. This is in contrast to rod precursors that showed equiaxed grains from 100 - 300 microns in diameter. The axial section showed a few large grains with lengths beyond the polished area of the sectioned wire (~6 mm).

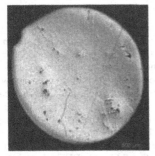

Figure 1. Cross-sectional microstructure of Fe-Ga wires from optical microscopy.

Figure 2. Microstructure of Fe-Ga wires along the axial direction using optical microscopy.

Crystallography

XRD powder patterns of the crushed wire indicated the disordered body-centered-cubic (bcc) A2 phase [5]. Laue XRD patterns were determined normal to the cross-sectional surface (parallel to the axial direction) of the Fe-Ga wires. The Laue pattern of a typical quenched Fe-Ga wire is shown in Figure 3a. The cubic symmetry of the Laue spots is apparent due to the position of one of the <100> axes near the centre of the film. Figure 3b shows the diffraction pattern with dashed lines that represents sets of diffracting planes characteristic of the <100> orientation of the bcc alloy. This analysis indicated that the plane normal to the axial direction of the wire was within ~ 3° of <100>.

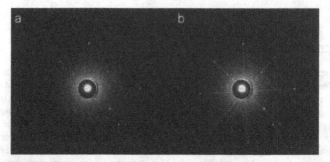

Figure 3. Laue XRD patterns showing (a) the cubic symmetry of the bcc structure of α-Fe and (b) with dashed lines representing sets of diffraction planes typical of the <100> orientation.

Laue analysis was supported by EBSD analysis normal to the surface of the quenched Fe-Ga wire. A typical EBSD color contrast orientation map is shown in Figure 4. This shows the typical distribution of plane normals (using results from ~13,000 analysis locations) parallel to the axial direction of the quenched Fe-Ga wire. The preponderance of the red color suggests a near <100> fiber texture is present in the grains of the quenched wire. Equal area projections (pole figures) and orientation distribution graphs (ODGs) were determined for the {111}, {100} and {110} planes (Figure 5). A strong <100> fiber texture is revealed by the large density of crystallographic planes surrounding the pole normal to {100}. In this case, the number of measurements that exhibit the {100} planes normal to the specimen surface (parallel to wire axis) is proportionately larger that the number of grains with other orientations; when comparing the multiples of uniform distribution (M.U.D.). ODGs show that nearly all grains exhibit a fiber texture within 2° of <100>.

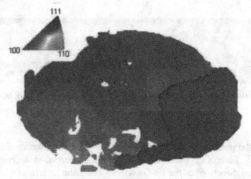

Figure 4. EBSD color contrast orientation map showing the typical distribution of plane normals parallel to the axial direction of the quenched Fe-Ga wire. Color variation is indicative of the distribution of measurement of ~13,000 plane normals.

a

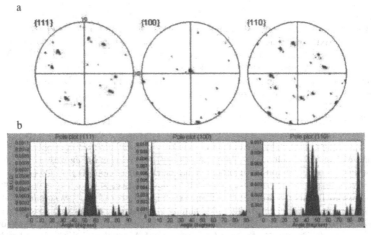

b

Figure 5. (a) Inverse pole figures and (b) ODGs showing distribution of {111}, {100}, and {110} plane normals parallel to the axial direction of Fe-Ga wire. The center of the pole figures correspond to the wire axis. ODGs show the M.U.D. as a function of angle for each orientation.

<u>Magnetic Characterization</u>

Room temperature axial magnetization and magnetostriction measurements as a function of applied field were determined for the quenched wires. Magnetization curves show that the Fe-Ga alloys are magnetically soft with magnetization saturation near 210 emu/g in a saturation field of 200 mTesla. Magnetostriction curves indicate magnetostriction saturation of ~165 ppm in saturation magnetic field of ~ 200mTesla with little hysteresis. This may have important implications for design and development of systems that incorporate magnetostrictive wires.

DISCUSSION

It is believed that the interplay of high temperature and constraint conditions imparted by the drawing process during the Taylor process drives the wire morphology and causes individual crystals to adopt a preferred orientation. Texture evolution of Taylor-drawn Fe-Ga alloys is discussed in more detail in Farrell *et al.* [7] (in preparation). The paper by Sinha *et al.* [8] indicated a correlation between the wire axis and easy axis of magnetization for the inner core of amorphous magnetic materials. This is believed to indicate at least a partial crystallographic texture in the axial direction.

Both Laue XRD and EBSD indicated that the axial direction of the Fe-Ga wires had a fiber texture within 3° of <100> and that magnetic field-induced strains up to 165 ppm strain were achievable along the <wire axis. The microstructure and magnetic properties may be compared with those produced by other non-equilibrium techniques. A paper on solidification of undercooled $Fe_{80}Ga_{20}$ rods by Ma *et al.* [9] had shown pole figure plots that indicated the <100> preferred orientation was ~10 ° from the axial direction. Similarly, Srisukhumbowornchai and

Guruswamy [10] had fabricated polycrystalline $Fe_{100-x}Ga_x$ rods by directional growth in which the <100> orientation was 14° from the axial direction. In that study [10], a rod of similar composition ($Fe_{80}Ga_{20}$) to the wires produced in this study exhibited a magnetostriction of 228 ppm along the axial direction, in the absence of a prestress. Although magnetostriction values from wires are lower than those achieved for polycrystals of similar composition [10], 165 ppm is still considered a significant strain for bulk polycrystalline alloys without pre-stress or stress-annealing treatments. It is anticipated that with careful refinement of the Taylor-based approach, magnetostriction values may improve considerably.

CONCLUSIONS

The present results demonstrated the viability of the Taylor-based approach for the production of small diameter <100> fiber-textured magnetostrictive Fe-Ga wire. Strict control of rapid solidification and deformation processes has enabled the introduction of the desired <100> texture in the final alloy product that is required for development of large magnetostriction. Taylor-drawn wires with a strong <100> fiber texture showed a maximum magnetostriction of ~165 ppm in a saturation field of ~200 mT – similar to oriented Fe-Ga bulk polycrystalline alloys of similar composition. The unique geometry, texture and large magnetostriction found in Fe-Ga wires made with this Taylor-based approach coupled with the low intrinsic cost make this an attractive approach for production of textured wire and glass-coated Fe-Ga filaments. Availability of textured magnetostrictive Fe-Ga wires broadens possibilities for design and development of novel magnetostrictive-based systems.

Studies are underway to adapt the Taylor-based approach to produce higher quality continuous <100> textured Fe-Ga wires. This includes alloy additions to improve drawability for conventional wire-drawing methods, and automation of the method for better control of solidification (temperature, time), drawing (load, drawing speed) and cooling conditions.

ACKNOWLEDGMENTS

The authors are grateful to D. Bligh, and A. George of Dalhousie University and A. Nolting of DRDC – Atlantic for their contributions.

REFERENCES

1. A.E. Clark and M. Wun-Fogle, *Proc. of SPIE*, **4699**, 421 (2002)
2. N. Srisukhumbowornchai and S. Guruswamy, *Met. Mat. Trans A*, **35**, 2963 (2004).
3. G.F. Taylor, *Phys. Rev.* **23**, 655 (1924).
4. K. Han, J.D.Embury, J. Petrovic and G.C.Weatherly, *Acta Mat.* **46**, 4691 (1998)
5. S.P. Farrell, R.A. Dunlap, N.C. Deschamps, J.D. McGraw, J. Lawless and R. Ham-Su, *Proc. of CanSmart 2006*, October 2006, Toronto, Ontario.
6. V. Srinivas and R.A. Dunlap, *Mat. Res. Bull.* **30**, 581 (1995).
7. S.P. Farrell, P.E. Quigley, K.J. Avery, T.D. Hatchard, S.E. Flynn, R.A. Dunlap, to be submitted to *J Phys: Cond. Mat.*
8. S. Sinha, K. Mandal and B. Das, *J. Phys. D: Appl. Phys.* **40**, 2710 (2007).
9. W. Ma, H. Zheng, M. Xia and J. Li, *J. Alloys Compd.* **379**, 188 (2004).
10. N. Srisukhumbowornchai and S. Guruswamy, *J. Appl. Phys.* **90**, 5680 (2001).

Mater. Res. Soc. Symp. Proc. Vol. 1129 © 2009 Materials Research Society   1129-V13-03

Hysteresis in Temperature- and Magnetic Field-Induced Martensitic Phase Transitions in Ni-Mn-Sn Heusler Alloys

Patrick J. Shamberger[1] and Fumio S. Ohuchi[1]

[1]Materials Science & Engineering Dept., University of Washington, Box 352120, Seattle, WA.

ABSTRACT

We have investigated the hysteresis of the temperature- and field-induced phase transition of $Ni_{50.5}Mn_{34.0}Sn_{15.6}$. Specifically, we have characterized the fraction of martensite and parent present along a variety of complex temperature and magnetic field paths, using magnetization as a proxy for phase fractions. We demonstrate that both temperature and magnetic field are thermodynamically equivalent driving forces for the phase transition and result in complementary hysteresis behavior. Additionally, we report that the effective magnetocaloric effect is limited by both large hysteresis losses, as well as the limited extents of phase transformation occurring upon application and removal of a magnetic field. Our results suggest that the hysteresis behavior of a material is critical in evaluating its potential as a functional magnetic refrigerant.

INTRODUCTION

Magnetic field-induced first-order phase transitions result in a number of functional material properties, including the metamagnetic shape memory effect [1] and the "giant" magnetocaloric effect (MCE) [2]. These properties can be used in high-performance magnetic field controlled actuators and magnetic refrigerators, respectively. Ni-Mn-X (X=Ga, In, Sn...) Heusler alloys exhibit this type of behavior, having a martensitic phase transition at a critical temperature, T_{cr}, which is controllable by alloying (e.g., [3]). Applying a magnetic field while at temperatures near T_{cr} induces a phase change in these materials, resulting in a strain or a magnetic entropy change. However, the martensitic transition in Ni-Mn-X alloys is typically associated with hysteresis, which both reduces the fraction of phase transformed for a given field and diminishes the amount of useful work that a material can perform (hysteretic loss).

Further advancement of first-order phase change materials towards the ultimate goal of practical magnetic refrigeration requires an improved understanding of their hysteretic behavior around the phase transition. It is not well understood how the hysteresis of first-order phase change materials behaves under the complex temperature and magnetic field path that a working magnetic refrigerant is likely to follow, nor is it well characterized how the hysteresis of most materials affects their potential use as magnetic refrigerants. To these ends, we report here on the hysteresis of temperature- and field-induced martensitic transformations of a representative polycrystalline Heusler alloy, $Ni_{50.5}Mn_{34}Sn_{15.6}$.

EXPERIMENTAL

A polycrystalline Ni-Mn-Sn alloy was prepared by melting ~4g of >99.95% purity starting metals in an arc-melting furnace under a gettered Ar atmosphere. After melting, the alloy was annealed at 1200K for 24 hours under vacuum, then was cooled slowly (~10K/min). The alloy composition, determined by wavelength dispersive spectrometry is: Ni = 50.5 ± 0.1, Mn = 34.0 ± 0.3, Sn = 15.6 ± 0.2 (in atomic %). The measured post-melting composition is within 1 at% of the pre-melting composition. Powder x-ray diffraction patterns at room temperature indicate only the Heusler phase (L2$_1$ structure) was present. Magnetic characterization was performed with a vibrating sample magnetometer in a Quantum Design Physical Properties Measurement System on a small (~5mg) equant alloy chip. Heating and cooling rates were limited to 1°C /min, while magnetic field was changed at a rate of 0.002 T /sec. These rates were chosen to reduce instrumental lag caused by thermal heterogeneity in the sample, while still allowing for the practical conduction of the experiment.

RESULTS AND DISCUSSION

Alloy magnetization was measured during repeated thermal cycling at a constant magnetic field (5 T) to record the temperature-induced hysteresis behavior. Heating cycles were initiated well below the phase transition (160 K), and warmed the sample through various degrees of completion of the phase transformation, before cooling the sample back to the initial temperature (figure 1a). Both heating cycles at low field (0.005 T, not shown) and cooling cycles (5 T, not shown) behave similarly to those illustrated in figure 1a. Magnetization data were transformed to $f(T)$, the mass fraction of parent present, by assuming that total magnetization is proportional to the mass fraction of parent and martensite phases. This allows for a linear mapping from magnetization to fraction of parent, as illustrated by figure 1.

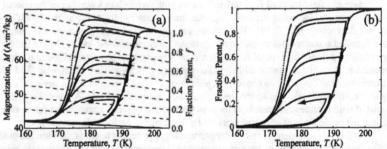

Figure 1. Thermal hysteresis at 5 T in (a) magnetization, and (b) mass fraction parent, calculated assuming magnetization is proportional to $f(T)$, as illustrated by the red dashed lines.

Magnetization of the alloy was measured at a constant temperature while applying and removing an external magnetic field (magnetic hysteresis curves). Magnetic hysteresis curves measured at temperatures above (210 K) and below (160 K) the transition, indicate that both martensite and Heusler phases are magnetically soft. Magnetic hysteresis curves were measured at 190 K while the field was increased to a local maximum (1, 3, 5, 7, or 9 T, sequentially) and then decreased to zero field (figure 2a). The measurement from 0 to 9 T was repeated three times to determine reproducibility. Magnetization data were transformed to fraction of parent present, $f(T)$, following a similar approach as used with thermal hysteresis data (see dashed lines in figure 2a). The fraction of parent present was calculated assuming that the total magnetization is proportional to the mass fraction of parent and martensite phases figure 2b.

Application of a magnetic field readily converts martensite to the parent phase at 190 K, but the removal of the magnetic field leads to only a small degree of recovery (figure 2b). Decreasing the applied field from a local maximum, H', to zero field, and increasing it again to H' defines a cyclic transformation that is highly reproducible, as evidenced by the repeated application and removal of a 9 T field.

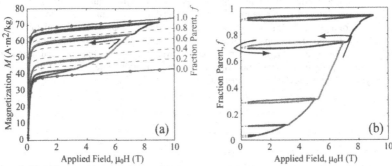

Figure 2. Field-induced hysteresis at 190 K in (a) magnetization, and (b) mass fraction parent, calculated assuming magnetization is proportional to $f(T)$, as illustrated by the red dashed lines.

Equivalence of thermal and magnetic hysteresis

Our results demonstrate that both temperature and magnetic field can induce a martensitic phase change in the Ni-Mn-Sn alloy system. These results agree with direct observation of the field-induced martensitic transition by high-field x-ray diffraction experiments [4]. Temperature and field dictate the free energy of martensite and parent Heusler phases in the Ni-Mn-Sn alloy system through the magnetic Gibbs free energy:

$$dG = -\mu_o MdH - SdT .\qquad(1)$$

At the martensite/parent phase boundary, phase transformations are induced by temperature or field changes when the free energy of one phase is decreased relative to the free energy of the other. For any arbitrary change in temperature, there is an equivalent change in field that causes the same extent of phase transformation (i.e., the same relative difference in free energies):

$$\left(dG^M - dG^P\right)_{isothermal} = \left(dG^M - dG^P\right)_{isofield} .\qquad(2)$$

Equation (2) assumes only that the two phases have different magnetizations and entropies. Combining equations (1) and (2) and integrating over a range of temperature and field (assuming that ΔM_{tr} and ΔS_{tr} are constant over this region) leads to:

151

$$\Delta T \Delta S_{tr} = \Delta(\mu_o H) \Delta M_{tr},$$ (3)

where ΔM_{tr} and ΔS_{tr} are the differences in magnetization and entropy between the two phases at the transition.

To test this equivalence, we compared field-induced hysteresis observations with temperature-induced hysteresis observations (figure 3). The field-induced data were translated to an "effective temperature," θ, by using equation (3), with the observed average difference in magnetization ($\Delta M_{tr} = 29.4$ Am^2/kg) and the entropy of transition ($\Delta S_{tr} = 1.33$ J /mole /K), calculated from magnetic data via the magnetic Clausius-Clapeyron equation. Both temperature- and field-induced phase transformations 1) have identical minimum fractions of phase transformed for a given θ, 2) follow similar interior hysteresis loops upon removal and re-application of θ, and 3) encompass nearly identical hysteresis loss over the full θ cycle. The identical behavior of field- and temperature-induced hysteresis implies that the hysteresis is a fundamental property of the phase transition itself, and is therefore independent of the thermodynamic driving force (i.e., temperature or magnetic field).

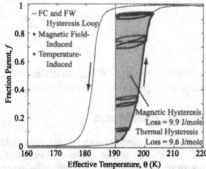

Figure 3. Temperature-induced (red) and Field-induced (blue) hysteresis, converted to an effective temperature scale, using equation (3).

Effect of hysteresis on magnetocaloric properties

Hysteresis of the martensitic phase transformation limits the useful work that can be performed by a MCE material through two distinct mechanisms: irreversible dissipation of energy (hysteresis loss), and limitation of the extent of phase transformation. The hysteresis loss caused by a complete forward and reverse martensitic transformation can be estimated as [5]:

$$E_{irr} = \Delta S_{tr} \Delta T_{irr},$$ (5)

where ΔT_{irr} is the average thermal hysteresis of the material. In Ni$_{50.5}$Mn$_{34}$Sn$_{15.6}$, hysteresis loss at zero applied field is 21.3 ± 2.2 J /mole, as calculated from the Clausius-Clapeyron derived ΔS_{tr} and the observed ΔT_{irr}. Hysteresis loss for partial phase transitions may be calculated from the relative area of hysteresis loops in temperature-induced phase transitions [6] or directly from magnetic hysteresis data, and is illustrated in figure 4. For comparison, figure 4 also includes representative relative cooling powers (RCPs) for $Ni_{50}Mn_{35}Sn_{15}$ found in the literature [7]. The RCP is a measure of the useful work a MCE material can perform over a refrigeration cycle. The hysteresis loss caused by the complete martensitic transformation presented in this study is nearly three times the relative cooling power of Ni-Mn-Sn alloys with similar compositions.

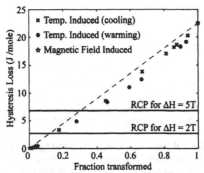

Figure 4. Calculated hysteresis loss for partial thermal and magnetic hysteresis as a function of fraction transformed. The dashed line illustrates a linear relationship.

In first order phase change materials, the magnetic entropy change, ΔS_M, is due primarily to the difference in the entropy between the two phases at the transition temperature, ΔS_{tr} [8]. If a magnetic field change only induces a fraction of a given material to transform from one phase to another, the resulting ΔS_M of the material will decrease proportionally. Large unidirectional extents of phase transformation may be induced in Ni-Mn-Sn Heusler alloys by applying magnetic fields up to 9 T (figure 2). However, the repeated cycling of a magnetic field leads to much smaller extents of phase transformation, as the transition is limited by both hysteresis and elastic strain effects. The extents of reaction caused by removing and reapplying a magnetic field are measured at a single temperature, 190K (figure 5). In the alloy reported here, application and removal of magnetic fields up to 9 T resulted in cyclic phase transformations of at most 5% of the alloy. Therefore, the effective magnetic entropy change at 190 K is at most ~5% of the entropy of transition, or $(\Delta S_M)_{eff} \approx 0.066$ J /mole /K. This is over an order of magnitude lower than the reported thermodynamically reversible magnetic entropy change for $Ni_{50}Mn_{35}Sn_{15}$, $(\Delta S_M)_{rev} \approx 0.965$ J /mole /K [7]. Effective magnetic entropy changes as a function of magnetic field change are shown in the inset of figure 5.

CONCLUSIONS

While Ni-Mn-Sn alloys and other ferromagnetic materials with first-order phase transitions have promising intrinsic magnetocaloric properties, hysteresis effects commonly limit these materials from practical use. Our study draws a number of important conclusions regarding the hysteresis in these materials: 1) Both temperature and applied magnetic field are

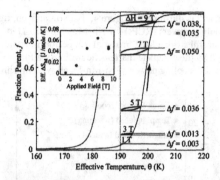

Figure 5. Extent of repeatable phase transformations at 190 K as a function of applied magnetic field change. Inset illustrates the "effective" magnetic entropy change.

thermodynamically equivalent driving forces and are related by a simple expression (given by eq. 3). 2) Hysteresis loss from the complete forward and reverse transformation is 21.3 J /mole, which is ~3 times greater than cooling capacities measured for similar NiMnSn alloys at 5 T. 3) Hysteresis limits the extent of the repeatable phase transformation induced by a magnetic field to only ~5% of the alloy. Therefore, only ~5% of the entropy of the transition, ΔS_{tr}, contributes to the magnetic entropy change of the material. Our results suggest that the hysteresis behavior of a material is critical in evaluating its potential as a functional magnetic refrigerant.

ACKNOWLEDGMENTS

We would like to thank the M.J. Murdock Charitable Trust and Micron Foundation for their generous funding of facilities utilized in this study. PJS acknowledges support from DoD.

REFERENCES

1. O. Tegus, E Bruck, K.H.J. Buschow, and F.R. De Boer, *Nature*, **415**, 150 (2002).
2. R. Kainuma, Y. Imano, W. Ito, Y. Sutou, H. Morito, S. Okamoto, O. Kitakami, K. Oikawa, A. Fujita, T. Kanomata, K. Ishida, *Nature*, **439**, 957 (2006).
3. Y. Sutou, Y. Imano, N. Koeda, T. Omori, R. Kainuma, K. Ishida, K. Oikawa, *Appl. Phys. Lett.*, **85**, 4358 (2004).
4. K. Koyama,T. Kanomata, R. Kainuma, K. Oikawa, K. Ishida, and K. Watanabe, *Physica B* **383**, 24 (2006).
5. J. Ortín and A. Planes, *Acta Metall.* **36**, 1873 (1988).
6. Y. Deng and G. S. Ansell, *Acta Metall. Mater.* **39**, 195 (1991).
7. T. Krenke, E. Duman, M. Acet, E. Wassermann, X. Moya, L. Mañosa, and A. Planes, *Nat. Mater.* **4**, 450 (2005).
8. V. K. Pecharsky, K. A. Gschneidner, Jr., A. O. Pecharsky, and A. M. Tishin, *Phys. Rev. B* **64**, 144406 (2001).

Mater. Res. Soc. Symp. Proc. Vol. 1129 © 2009 Materials Research Society   1129-V12-05

Shape Memory Actuation by Resistive Heating in Polyurethane Composites of Carbonaceous Conductive Fillers

I. Sedat Gunes, Guillermo A. Jimenez, Sadhan C. Jana
Department of Polymer Engineering, The University of Akron, Akron, OH 44325-0301, U.S.A.

ABSTRACT

The dependence of electrical resistivity on specimen temperature and imposed tensile strains was determined for shape memory polyurethane (SMPU) composites of carbon nanofiber (CNF), oxidized carbon nanofiber (ox-CNF), and carbon black (CB). The SMPU composites with crystalline soft segments were synthesized from diphenylmethane diisocyanate, 1,4-butanediol, and poly(caprolactone)diol in a low-shear chaotic mixer and in an internal mixer. The materials synthesized in the chaotic mixer showed higher soft segment crystallinity and lower electrical percolation thresholds. The soft segment crystallinity reduced in the presence of CNF and ox-CNF; although the reduction was lower in the case of ox-CNF. The composites of CB showed pronounced positive temperature coefficient (PTC) effects which in turn showed a close relationship with non-linear thermal expansion behavior. The composites of CNF and ox-CNF did not exhibit PTC effects due to low levels of soft segment crystallinity. The resistivity of composites of CNF and ox-CNF showed weak dependence on strain, while that of composites of CB increased by several orders of magnitude with imposed tensile strain. A corollary of this study was that a high level of crystallinity may cause a PTC effect and prevent any actuation through resistive heating. However, a carefully tailored compound which has reduced crystallinity and which requires minimum amount of filler may prevent PTC phenomenon and could supply necessary electrical conductivity over the operating temperature range, while offering enough soft segment crystallinity and rubberlike properties for excellent shape memory function.

INTRODUCTION

Shape memory polymers (SMPs) are a class of stimuli responsive materials, which recover the original shapes from the states of large deformation when subjected to an external stimulus, such as heat, electrical voltage, magnetic field, etc. [1 -5]. Shape memory functions can be actuated by resistive heating [6-8], if appropriate electrically conductive fillers are combined with electrically non-conductive matrix polymers. A major challenge in material design and proper functioning of SMP composite actuators based on semi-crystalline SMP is the occurrence of positive temperature coefficient (PTC) of resistivity. An otherwise electrically conductive polymer composite transforms into insulator as the electrical resistance increases with temperature. Another critical aspect of shape memory actuation by resistive heating is the dependence of electrical conductivity on applied strain. Note that SMPs experience strains on the order of several hundred percents both during deformation and shape recovery [9-12]. In view of this, the knowledge of temperature and strain dependence of electrical resistivity is of central importance in the design of SMP composites.

In this study, the PTC effects and the relationship between applied strain and electrical resistivity were investigated by considering shape memory polyurethane (SMPU) with semi-crystalline reversible phase and its composites with three electrically conductive filler particles.

The relationships between soft segment crystallinity and the electrical percolation behavior of SMPU composites filled with carbon nanofiber (CNF), oxidized carbon nanofiber (ox-CNF), and CB are reported in this paper. The nature and origin of PTC effects, the effect of strain on electrical conductivity of the composites, and their importance on shape recovery triggered by resistive heating are discussed.

EXPERIMENTAL WORK

SMPU with 33% hard segment was synthesized from diphenylmethane diisocyanate (MDI, Bayer MaterialScience, Pittsburg, PA), 1,4-butanediol (BD, Avocado Organics, UK), and polycaprolactonediol (PCL diol, Solvay Chemical, UK) of molecular weight 4000 by mixing the ingredients in the stoichiometric ratio of 6/5/1 by moles respectively. Carbon nanofiber (CNF, Pyrograph III© PR-24-PS) and oxidized carbon nanofiber (ox-CNF, Pyrograph III© PR-24-PS-ox), both vapor grown grade carbon nanofibers with mean diameter 60-200 nm and mean length 30-100 µm, were obtained from Applied Sciences, Inc. (Cedarville, OH). High structure, conductive carbon black (CB, Ketjenblack® EC 300J) with pore volume of $0.310-0.345 \times 10^{-3}$ $m^3/100g$ [13] as determined by dibutyl phthalate absorption was obtained from Akzo Nobel (Norcross, GA).

Composites were prepared separately in a chaotic mixer and in a commercial internal mixer (Brabender Plasticorder internal mixer, model EPL 7752). The details on design and operation of the chaotic mixer can be found elsewhere [14]. Composites of poly(methyl methacrylate) (PMMA) and CNF prepared in chaotic mixer showed electrical percolation at much lower carbon nanofiber content than materials prepared in the internal mixer when operated under similar conditions of shear rate and temperature [15]. This was attributed to low shear nature of the chaotic mixer due to which fibers were much less damaged and produced electrically conductive networks at much lower volume fraction of CNF. Another study showed that ox-CNF dispersed well in polar polymers, such as PMMA and polyurethanes (PU) [16]. Consequently, its percolation threshold was higher than that of CNF, irrespective of the mixer used in preparation of the composites. In this work, all composites were prepared using a two-step bulk polymerization method. The prepolymer was synthesized from MDI and PCL diol and chain extended in the mixers with BD in the presence of tin catalyst. The fillers were first mixed with prepolymer, then BD and catalyst were introduced, and the polymerization was performed in the presence of fillers in the mixer preheated to a set temperature of 110°C for 10 minutes. The composites were compression molded at a pressure of 25 MPa and temperature of 220°C into specimens of 0.5 mm thickness. These molding conditions did not result in any degradation of molecular weight.

The electrical resistivity of the composites was measured using a four-probe Keithley 8009 resistivity tester and a Keithley 487 picoammeter/voltage source obtained from Keithley (Cleveland, OH). The dependence of electrical resistivity on temperature was determined by measuring the electrical resistivity while heating the composite specimens to desired temperatures in a compression molder at a pressure of 5 MPa. It was confirmed that a minimum of 5 MPa pressure was required for good mechanical contact between the flat probe and the sample specimen. The susceptibility of electrical resistivity to imposed tensile strain was determined by subjecting the composite specimens to desired tensile strain at 60°C in a tensile tester (Instron 4204, Norwood, MA) fitted with a heating chamber. The transmission electron microscopy (TEM) images were taken using TEM device JEM-1200EXII (JEOL, Japan) at 120 kV. The crystallinity of the soft segment phase was determined using a differential scanning calorimetry (DSC) device, TA instruments DSC-29210 (New Castle, DE) under nitrogen atmosphere at a heating rate of 10°C per minute in the range 25°C to 230°C which was representative of the actual shape memory testing cycle. The percentage of crystallinity was determined by comparing the heat of fusion obtained from DSC with the heat of fusion of 100 % crystalline PCL diol, 136 J/g [17].

RESULTS AND DISCUSSION

The single fibers of CNF and ox-CNF showed resemblance to the typical stacked morphology of multi-walled carbon nanotubes (figure 1). The ratio of carbon to oxygen atoms on the surfaces of filler particles was determined from the areas under the curves for $C1s$ and $O1s$ peaks in the x-ray photoelectron spectroscopy (XPS) spectrum. It is respectively 66.7, 7.7, and 19.8 for CNF, ox-CNF, and CB. These data indicate the presence of significant amounts of oxygen on the surfaces of ox-CNF and CB comprising primarily of polar functional groups, such as ester and carboxylic acid [16].

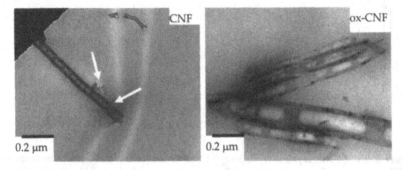

Figure 1. TEM images of CNF and ox-CNF. The stacked morphology typical for MWNT is evident. Note also the carbon nanotubes (shown by arrow) on the surface of CNF, purportedly grown from an occasional catalyst particle left on the CNF surface. A fracture site on CNF is also indicated by an arrow.

The soft segment crystallinity of SMPU and its composites was determined by DSC. It was found that the SMPU prepared in the chaotic mixer contained higher soft segment crystallinity than the materials prepared in Brabender internal mixer (figure 2a). The higher level of soft segment crystallinity in this case resulted from higher degree of phase separation in SMPU prepared in chaotic mixer [14]. Of the filler particles studied, CB was observed to exert weak influence on the soft segment crystallinity, while the presence of CNF and ox-CNF greatly reduced the soft segment crystallinity irrespective of the mixer used. The relatively higher soft segment crystallinity seen for composites of ox-CNF (figure 2a) can be attributed to polar-polar interactions between carbonyl groups of soft segment and polar functional groups on ox-CNF surface [16]. The polar-polar interactions between ox-CNF and SMPU may have promoted adsorption of polymer chains on the fiber surface, which in turn induced favorable chain conformations and hence higher rates of nucleation and crystallization [18]. A survey of literature on polymer crystallization in the presence of different fibers suggested that the rate for fast crystallizing polymers is not affected by the presence of fibers [19, 20]. On the other hand, crystallization of slowly crystallizing polymers is significantly affected in the presence of fibers [21, 22]; the crystallinity significantly increases if the polymer and the fiber surface are both polar or both non-polar. Note that SMPU formulation used in this work is a slow crystallizing polymer [23].

The volume electrical conductivity (σ) as function of filler weight fraction of SMPU composites is presented in figure 2b. It is seen that electrical percolation occurred at low filler content ~ 4 wt%, $\sigma \sim 10^{-5}$ S/cm, in composites of CNF prepared in the chaotic mixer. In comparison, the composites prepared in Brabender internal mixer remained insulators, $\sigma \sim 10^{-11}$ S/cm at up to 7 wt% CNF content. The low percolation threshold of materials prepared in the chaotic mixer can be attributed to much less attrition of high aspect ratio fibers and orientation of the fibers rendered by the chaotic flow [16].

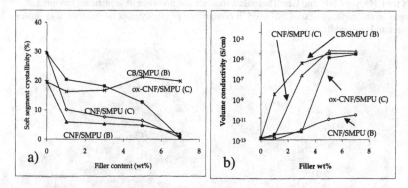

Figure 2. a) The soft segment crystallinity of SMPU composites. **b)** Electrical percolation in SMPU composites. Composites prepared in Brabender internal mixer and in chaotic mixer are represented respectively by (B) and (C).

The resistivity versus temperature data (figure 3a) indicate that CB composites showed pronounced PTC effects whereby the resistivity first increased up to approximately 45°C and then reduced. On the other hand, composites of CNF and ox-CNF did not show PTC effect. Composites with lower or higher filler contents than the ones reported in figure 3a did not show any significant change in resistivity upon heating and the data reported in figure 3a was reproducible during cyclic testing over a temperature interval of 25°C to 60°C. In order to interpret such trends, it is noted that the soft segment crystallinity of these composites were 7.6 wt%, 12.7 wt%, and 16.7 wt%, respectively for CNF, ox-CNF, and CB composites. The data on crystallinity and PTC effects in SMPU composites suggested that the PTC effects are associated with the melting of relatively large amounts of crystals in composites of CB.

The resistance of composites prepared in chaotic mixer and in Brabender internal mixer as function of uniaxial tensile strain is presented in figure 3b. The measurements were performed at 60°C, whereby the soft segment crystals were all melted. The resistances of composites with filler content approximately at the percolation threshold are presented in figure 3b. Composites with lower or higher filler contents did not show any significant change in resistance upon tensile deformation. It was observed that CB/SMPU composites with 5 wt% or higher CB content failed beyond 10% elongation and, therefore, was not considered. It is noted from figure 3b that the resistance in both axial and transverse directions of CB/SMPU composite containing 3 wt% CB increased linearly with strain up to 50% strain, although the changes in resistance in axial and transverse directions were different (figure 3b). The difference between resistance in axial and transverse directions was attributed to preferred orientation of the filler particles in the strain direction [24]. The data presented in figure 3 indicate that CB/SMPU composites cannot be actuated by the application of electrical voltage as the resistance increases, thus limiting the flow of current and hence resistive heating. CNF/SMPU composites on the other hand can be easily actuated by resistive heating.

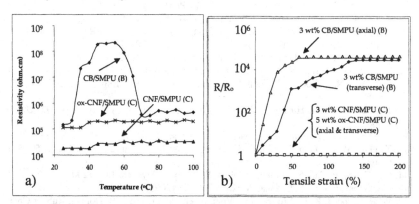

a)

b)

Figure 3. a) Temperature dependent volume resistivity of SMPU/3 wt% CNF, SMPU/ 3 wt% CB, and SMPU/ox-CNF 5 wt% composites. **b)** The ratio of resistance R/R_o of SMPU composites as function of tensile strain, where R_o is the original resistance and R is the resistance of the stretched specimen. The composites prepared in Brabender internal mixer and in chaotic mixer are represented respectively by (B) and (C).

Resistive heating induced shape recovery is demonstrated in CNF/SMPU composite with 5 wt% filler content (figure 4). The electrical conductivity of the composites was high and showed stability under thermal and mechanical strain gradients.

Figure 4. Resistive heating induced shape recovery of CNF/SMPU nanocomposite with 5 wt% filler content. The applied voltage is 300V.

159

CONCLUSIONS

The data showed that soft segment crystallinity of SMPU reduced in the presence of CNF and ox-CNF. The reduction was more severe in the presence of CNF. As a consequence of low level crystallinity, PTC effect was absent in composites of CNF and ox-CNF, although composites of CB with almost the same filler loading showed pronounced PTC effects. It was also observed that CNF/SMPU and ox-CNF/SMPU composites had highly stable electrical conductivity under uniaxial tensile strain, as compared to CB/SMPU composites.

REFERENCES

1. S. Hayashi, S. Kondo, P. Kapadia, E. Ushioda, Plast. Eng. 51, 29 (1995)
2. A. Lendlein, S. Kelch, Angew. Chem. Int. Ed. 41, 2034 (2002)
3. V. A. Beloshenko, V. N. Varyukhin, Yu. V. Voznyak, Russ. Chem. Rev. 74, 265 (2005)
4. C. Liu, H. Qin, P. T. Mather, J. Mater. Chem. 17, 1543 (2007)
5. I. S. Gunes and S. C. Jana, J. Nanosci. Nanotechnol. 8, 1616 (2008)
6. H. Koerner, G. Price, N. A. Pearce, M. Alexander, and R. Vaia, Nat. Mater. 3, 115 (2004)
7. I. H. Paik, N. S. Goo, K. J. Yoon, Y. C. Jung, J. W. Cho. Key Eng. Mater. 297-300, 1539 (2005)
8. J. S. Leng, W. M. Huang, X. Lan, Y. J. Liu, S. Y. Du. Appl. Phys. Lett. 92, 204101 (2008)
9. H. Tobushi, T. Hashimoto, S. Hayashi, E. Yamada. J. Intel. Mat. Syst. Str. 8, 711 (1997)
10. K. Gall, M. L. Dunn, Y. Liu, D. Finch, M. Lake, N. A. Munshi, Acta Mater. 50, 5115 (2002)
11. I. S. Gunes, F. Cao, S. C. Jana, J. Polym. Sci. Polym. Phys. 46, 1437 (2008)
12. I. S. Gunes, F. Cao, S. C. Jana, Polymer 49, 2223 (2008)
13. W. F. Verhelst, K. G. Wolthuis, A. Voet, P. Ehrburger, and J. B. Donnet, Rubber Chem. Technol. 50, 735 (1977)
14. C. Jung, I. S. Gunes, and S. C. Jana, Ind. Eng. Chem. Res., 46, 2413 (2007)
15. G. A. Jimenez and S. C. Jana, Compos. Part A.- Appl. S. 38, 983 (2007)
16. G. A. Jimenez and S. C. Jana, Carbon 45, 2079 (2007)
17. F. K. Li, J. N. Hou, W. Zhu, X. Zhang, M. Xu, X. L. Luo, D. Z. Ma, and B. K. Kim, J. Appl. Polym. Sci. 62; 631 (1996)
18. Yu. S. Lipatov and L. M. Sergeeva, Adsorption of polymers. (Wiley, New York, 1974) (chapter 4)
19. D. Fitchmun and S. Newman, J. Polm. Sci. Polym. Lett. Ed. 7, 301 (1969)
20. D. R. Fitchmun and S. Newman, J. Polym. Sci. Part A-2 8, 1545 (1970)
21. L. C. Lopez and G. L. Wilkes, J. Thermoplas. Compos. Mater. 4, 58 (1991)
22. G. P. Desio and L. Rebenfeld, J. Appl. Polym. Sci. 44, 1989 (1992)
23. P. Ping, W. Wang, X. Chen and X. Jing, J. Polym. Sci. Part B. Polym. Phys. 45, 557 (2007)
24. D. Bulgin, Rubber Chem. Technol. 19, 667 (1946)

Mater. Res. Soc. Symp. Proc. Vol. 1129 © 2009 Materials Research Society 1129-V12-02

Numerical Study on Fretting Fatigue Life of NiTi Shape Memory Alloys

Wang Xiao Xue[1] and Wang Rong Qiao[1]
[1] P.O. Box 405, School of Jet Propulsion, Beihang University, Beijing 100191, China

ABSTRACT

This paper presents a proposal by using NiTi Shape Memory Alloys (Smas) for improving the fretting fatigue life of mechanical components, and gives it a preliminary numerical investigation.

Firstly, with the Saeedvafa & Asaro thermo-mechanical constitutive model provided by finite element analysis (FEA) software Marc, the uniaxial tension numerical results of NiTi pseudoelastic (PE), which is considered as the main cause of its fretting fatigue resistant characteristics, is obtained and identical to that of the related test. This says that the model can be used in the following analysis.

The fretting experiment device, in which TC11 is the fretting bridge and GX8&NiTi for the test pieces, is adopted in the FEA numerical analysis. A complicated theory with material, geometry and contact interface nonlinear is used to establish the 2-D contact model whose friction type is Coulomb's contact.

Based on FEA results, a comprehensive parameter of fretting fatigue at low cycle fatigue (LCF) FFD was investigated. From the calculated results of GX8 and NiTi, we can find that the initial crack is stemmed from the same location on which the loading is the biggest in despite of the different properties of the two materials. However, under the same conditions, the FFD value of GX8 is higher than NiTi, which indicated GX8 crack appears easily than NiTi.

Finally, the GX8 and NiTi fretting fatigue lives were predicted by using the fourth strength theory and the Goodman curve considering size factors. The error of the theoretical method in this paper is 8.82% comparing with the experiments using GX8, which is within acceptable range. It is proved that the numerical method is valid and the prediction of NiTi fretting fatigue life is feasible.

The numerical results show that NiTi is of wearing resistance than other traditional materials because of its specific characteristics. The work in this paper set up the basis of the following investigation for the numerical and experimental study on the fretting fatigue resistant application of NiTi in an actual component.

RESEARCH PROPOSAL

The dovetail platform between the blade roots and disks in compressors and turbines of aero-engines has been identified as a critical area for the complicated coupling forces leading to fretting fatigue, which limits the life and reliability of engines and often brings catastrophic component failures[1]. It is assumed that the failure mode of dovetail is: The interaction of small-scale relative displacements or micro-slip at the contact surfaces and sharp gradients in the near-surface contact stress field drives a local area fretting, resulting in crack extension and fatigue fracture with the vibration stress [2].

At present, with the excellent corrosion resistance, high intensity and low density, titanium alloys are the main materials for compressor disk and blade of high-performance aero-engines. However, its poor wear-resisting performance causes fretting and fatigue fracture.

Two measures are adopted to improve the resistance to surface fatigue: Firstly, improve the structure to relieve the fretting fatigue from design stage [3]. Secondly, surface modification such as cloudbursttreatment, silver plating and dry film. It is not always effective since such surface modification technology needs to combine the operating condition of fretting fatigue components [4].

The NiTi Shape Memory Alloys (SMAs), a kind of new functional materials, could be an excellent candidate for various tribological applications benefiting from its excellent wear & fatigue resistance and flexibility due to its shape memory effects (SME) and pseudoelasticity (PE) [5].

Based on the overall development, a new method of fretting fatigue resistance is presented, which uses NiTi SMAs in dovetail components of aero-engines.This paper studies the fretting fatigue and life prediction method and compares with traditional materials to verify the high resistance, for the next coating complex numerical simulations and further experimentation.

VERIFICATION OF NiTi SMAs CONSTITUTIVE MODELS

The conventional models do not provide an adequate framework for representing the unusual macro-behavior of SMAs. The available literature shows different constitutive models of shape memory alloys, which are well described the characteristics from different theories [6].

The Saeedvafa & Asaro thermo-mechanical constitutive model is adopted provided by the finite element analysis (FEA) software MSC. Marc provides in this paper. Based on the small strain kinematics, the strain increment is the sum of the following:

$$\Delta \varepsilon = \Delta \varepsilon^{el} + \Delta \varepsilon^{th} + \Delta \varepsilon^{pl} + \Delta \varepsilon^{ph} \tag{1.1}$$

Where $\Delta \varepsilon^{el}$ is the elastic strain increment, $\Delta \varepsilon^{pl}$ is the visco-plasticity increment, $\Delta \varepsilon^{th}$ is the thermal strain increment and $\Delta \varepsilon^{ph}$ is the strain increment about thermo-elastic phase transition.

The relocation of stress and free thermodynamics direction forms the martensite, resulting direction formation leading from the phase transition to the strain. We express the strain $\Delta \varepsilon^{ph}$ as follows:

$$\Delta \varepsilon^{ph} = \Delta \varepsilon^{TRIP} + \Delta \varepsilon^{TWIN} \tag{1.2}$$

$$\Delta \varepsilon^{TRIP} = \Delta f^{(+)} g\left(\sigma_{eq}\right) \varepsilon_{eq}^{T} \frac{3}{2} \frac{\sigma'}{\sigma_{eq}} + \Delta f^{(+)} \varepsilon_{v}^{T} I + \Delta f^{(-)} \varepsilon^{ph} \tag{1.3}$$

$$\Delta \varepsilon^{TWIN} = f \Delta g\left(\sigma_{eq}\right) \varepsilon_{eq}^{T} \frac{3}{2} \frac{\sigma'}{\sigma_{eq}} \left\{\dot{\sigma}_{eq}\right\}\left\{\sigma_{eq} - \sigma_{eff}^{g}\right\} \tag{1.4}$$

Where f is the martensite volume rate:

$$\Delta f = \Delta f^{(+)} + \Delta f^{(-)}$$

$\{ \}$ is McCauley Parenthesis:

$$\{x\} = \frac{1}{2}\left(\frac{x + |x|}{|x|}\right), \quad x \neq 0$$

In Equation (1.3), $\Delta f^{(+)}$ and $\Delta f^{(-)}$ are the rate of martensite and austenite formation, ε_{eq}^{T} is the equivalent value of phase strain transition deviator and ε_{v}^{T} is the bulk strain of phase strain transition. Equation $g\left(\sigma_{eq}\right)$ is the relation of martensite phase strain transition and deviatoric stress tensor, where σ_{eq} is the equivalent stress. The last part of equation (1.3) is the phase strain transition recovery.

A table function established in MSC.Marc could data fitting most experiment as follows:

$$g\left(\sigma_{eq}\right) = 1 - \exp\left[g_a \left(\frac{\sigma_{eq}}{g_0}\right)^{g_b} + g_c \left(\frac{\sigma_{eq}}{g_0}\right)^{g_d} + g_e \left(\frac{\sigma_{eq}}{g_0}\right)^{g_f} \right] \tag{1.5}$$

This paper studies the pseudoelasticity of NiTi SMAs at the certain temperatures. Numerical simulations of typical uniaxial loading tests performed on NiTi SMA are presented and compared with available experimental data [8]. The uniaxial tension simulation, which accorded with the experiment data as well as the specimen specifications, was performed controlling the displacements at four given temperature. The stress-strain response of simulation and experiment data is reported in Fig.1.

In Fig.1, the simulated stress-strain curves match quite well the experimental data. The Saeedvafa & Asaro thermo-mechanical constitutive model is accuracy reproduces the pseudoelasticity features of NiTi SMAs, which can be use in the following study.

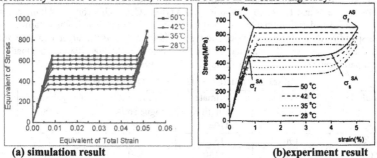

(a) simulation result   (b)experiment result

Fig.1 2-D uniaxial tension stress-strain curves

DISCUSSION NiTi SMAs FRETTING BRIDGE FEA MODELING AND SIMULATION

A bridge-shaped fretting fatigue experiment devices is adopted [9], and the setting of the test is idealized in Fig.2. The fretting bridge bears a load at the top, so that the bottom of bridge contacts with the specimen, forming normal pressure. Through imposing the variable load to the specimen, the local deformation between fretting bridge and specimen leads relative sliding and fretting wear, even the fatigue failure.

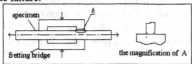

Fig.2 experiment device scheme

In the experiment, the bridge adopts TC11 and the specimen is a board whose material is GX8 and NiTi SMAs. GX8 and TC11 are the commonly used materials of high and low pressure rotors in aero-engines, which could contrast to available test results to verify the feasibility of the simulation and life prediction. Then compare the wearing resistance of GX8 and NiTi SMAs.

Due to geometric and loading symmetry conditions, in FEA only consider half of the contact face needs to be studied. For the numerical simulation with MSC.Marc, we use the mesh presented in Fig.3.

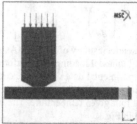

Fig.3 Bridge-Shaped Fretting FEA Model

The bridge-shaped problem belongs to a contact problem. A complicated contact theory with material, geometry and contact interface nonlinear is used to establish the 2-D contact model with 1187 elements and 1329 nodes, whose friction type is Coulomb's contact. All the contact elements are defined as deformable body and the friction coefficient is 0.3. The bottom of the bridges specified to the contact interface and the specimen surface is the target area. The bridge was performed under 200MPa load and the specimen bear the 100~500MPa variable load. Material parameters are shown in the Table 1 and Literature [8].

Table1 Parameters of GX8 , TC11 and NiTi [10]

Material Mark	ElasticModulus (GPa)	Ultimate strength (MPa)	Limiting fatigue stress (MPa)	Fatigue Life cycles
GX8	190	1340	647	3×10^6
TC11	120	1130	540	0.16×10^6
NiTi	72	1250	729.7	0.37×10^6

During the simulation, the convergence of the algorithm is given much consideration in two ways: Firstly, increase the loading steps properly. Impose 200 loading steps on fretting bridge and specimen totally. For the beginning 100 steps, exert linear loading to the bridge. Then at last 100 steps, simulate the fretting process so that the specimen bears the normal pressure and axial tension synchronously. Secondly, adjust the relative speed of Coulomb's contact as a result of 0.09mm/s.

Fig.4 and Fig.5 show the simulating mesh equivalent stress contour in the normal pressure (100 steps) and fretting maximum deformed configuration (200 steps) respectively. To take NiTi in 50℃ as an example. It could be illustrate from the contours that NiTi SMAs is better than GX8 in the stress distribution of both conditions, while the max stress is smaller than GX8.

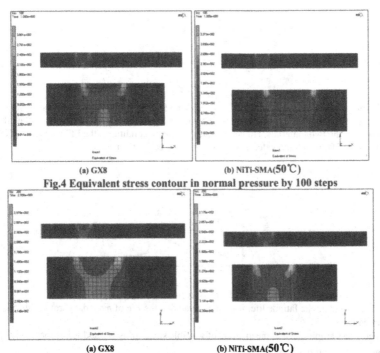

(a) GX8   (b) NiTi-SMA(50 ℃)

Fig.4 Equivalent stress contour in normal pressure by 100 steps

(a) GX8   (b) NiTi-SMA(50℃)

Fig.5 Equivalent stress contour in fretting progress by 200 steps

LOW CYCLE FRETTING FATIGUE LIFE STUDY

Based on FEA results, a comprehensive parameter of fretting fatigue at low cycle fatigue (LCF) FFD was investigated, whose max value means the dangerous point leading to fatigue crack of generation and development. Life prediction is studied using nominal method through stress and fretting conditions of the point.

According to Literature [1], FFD can be expressed as follows:

$$FFD = \mu\sigma_N\sigma_{tt}\Delta \tag{1.6}$$

Where σ_N is the normal stress in interface, Δ is the relative sliding amplitude, σ_{tt} is the tangential stress and μ is the fretting friction coefficient in interface.

Assign the nodes in the interface as shown in Fig.6, calculate the FFD of each node to confirm the dangerous points.

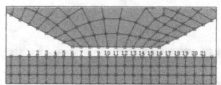

Fig.6 Partial FEA Model

From the calculated results in Fig.7, we can find that the initial crack is stemmed from the same location (identified in node 5) on which the loading is the biggest in despite of the different properties of the two materials. However, under the same conditions, the FFD value of GX8 is higher than NiTi, which indicated GX8 crack appears easily than NiTi.

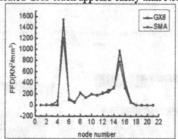

Fig.7 FFD Comparison of SMA and GX8

Then calculate the fatigue life. It is considered that the life of most dangerous node 5 is the specimen life.

The stress amplitude and mean stress of variable stress σ is σ_a and σ_m respectively, while the fatigue cyclic times is N.

The equation obtained from Goodman Curves can be written as

$$\sigma_{-1d} = \sigma_a + \left(\frac{\sigma_{-1N}}{\sigma_b}\right)\sigma_m \tag{1.7}$$

The equivalent stress defined in strength theory is σ_{-1DN}, which is

$$\sigma_{-1DN} = \sqrt{(\sigma_n)^2 + (\sigma_{tt})^2_{-1d} - (\sigma_n)_{-1d}(\sigma_{tt})_{-1d} + 3(\sigma_t)^2_{-1d}} \tag{1.8}$$

Considering the size factor ε, the fatigue notch factor K_σ and the surface machining factor β_1, the life equation is expressed as

$$\lg N = \lg N_0 - (\lg N_0 - 3)\frac{\lg \sigma_{-1DN} - \lg \sigma_{-1D}}{\lg(0.9\sigma_b) - \lg \sigma_{-1D}} \tag{1.9}$$

Through referring to the manual[11], we set β_1 =0.95, ε =0.95, K_σ =1.4. With Table 1 and the factors above, equivalent stress could be fixed to calculate the life using Equation (3.4).

The simulations in which GX8 as the specimen contrast to the test data demonstrate in Table 2. The error of the theoretical method in this paper is 8.82%, which is within acceptable range. It is proved that the numerical method is valid, which can be useful to predict and evaluate the NiTi SMAs fretting life.

Table 2 GX8 Life Comparison of Calculation and Experiment

Calculate result	1.623×10^7
Experiment result	1.78×10^7
error	8.82%

Table 3 displays the different stress and life parameters comparison. It could be seen that the life of NiTi SMAs is much higher than GX8. In deep analysis, it is clear that the normal stress in the interface of NiTi SMAs is smaller than GX8. As the same friction coefficient, the tangential stress and finally the equivalent stress is smaller than GX8, the advantages of NiTi SMAs is obvious.

Table 3 Stress and Life in Dangerous point

		GX8	NiTi SMAs (50℃)	NiTi SMAs (28℃)
Stress amplitude (MPa)	tangential stress σ_t	71.8492	45.2712	45.2937
	inner tangential stress σ_{tt}	14.795	16.9337	17.0064
	normal stress σ_n	239.4974	150.9041	150.9791
mean stress (MPa)	tangential stress σ_t	63.8957	55.0716	55.1163
	inner tangential stress σ_{tt}	107.6651	104.1454	104.8337
	normal stress σ_n	212.9856	183.572	183.721
equivalent symmetrical cyclic variable stress σ_{-1d} (MPa)	tangential stress σ_t	102.7003	77.4198	77.4684
	inner tangential stress σ_{tt}	66.7795	77.7062	78.204
	normal stress σ_n	342.3345	258.066	258.228
equivalent stress σ_{-1DN} (MPa)		361.1559	265.6397	265.72
$K_{\sigma D}$		1.5263	1.5263	1.5263
σ_{-1D} (MPa)		423.8966	353.7931	353.7931
Life N (cyclic times)		1.623×10^7	2.147×10^7	2.142×10^7

CONCLUSIONS

The paper proposes an initial assumption of NiTi SMAs to be used in dovetails of aero-engines to improve the fretting fatigue resistance. To reach this goal, a preliminary numerical investigation of NiTi SMAs is given and verified through comparison of available experiment data.

From simulation and life prediction, the force situation dominates the location of crack generation instead of materials properties. The NiTi SMAs fretting fatigue life has little effect on temperature between 28℃~50℃, which is much higher than GX8 commonly used material in aero-engines.

The numerical results show that NiTi is of wearing resistance than other traditional materials because of its specific characteristics. The work in this paper set up the basis of the following investigation for the numerical and experimental study on the fretting fatigue resistant application of NiTi SMAs in aero-engines.

REFERENCES

1. Cowles B. High cycle fatigue in aircraft gas turbines—an industry perspective[J]. International Journal of Fracture, 1996, 80: 147~163.
2. Liu Qingquan. Fracture Analysis for Rotor Blade Tenon of an Aeroengine Compressor[J]. Journal of Materials Engineering.1997,6:9~11.
3. Zhao Shaobian, Wang Zhongbao. Antifatigue Design[M]. China Machine Press, 1995.
4. Xu Guizhen, Liu Jiajun, Zhou Zhongrong. Application of Surface Modification Technology in Fretting Tribology[J]. Tribology. 1998,18(2): 185~190.
5. Jin Jianling, Wang Hongliang. The research of NiTi SMAs wear resistance[J]. Acta Metallurgica Sinica. 1988, 24(1):66~69.
6. Muller. A model for body with shape memory [J]. Arch Rat Mech Anal,1979,70(1):61~77.
7. H C Lin, S K Wu, M T Yeh. Damping Characteristics of TiNi Shape Memory Alloys [J].Metallurgical Transactions A, 1993, 24A: 2189~2194.
8. Wang Xinmei. Research on Fatigue and Rupture of NiTi Shape Memory Alloys [D]. Northwestern Polytechnical University. 2005.3.
9. Zhu Rupeng, Pan Shengcai, Gao Deping. A Numerical Study of Stress State Parameters and Fretting Wear Parameters in Fretting Fatigue. Engineering Mechanics. 1998,15(4):116~121.
10. Zhao Shaobian. Fatigue Resistance Design[M]. Beijing Machine Press, 1994: 23~28.
11. Aero-engine Design Data Book[Z]. National Defence Industry Press, 1990:532~538.

Mater. Res. Soc. Symp. Proc. Vol. 1129 © 2009 Materials Research Society   1129-V01-03

Rapid-solidified Magnetostrictive Polycrystalline Strong-Textured Galfenol (Fe-Ga) Alloy and its Applications for Micro Gas-Valve

Chihiro Saito [1], Masamune Tanaka[2], Teiko Okazaki [3] and Yasubumi Furuya [4],

[1]NJC Research Center, Namiki Precision Jewel Co.,Ltd, Shinden 3-chome, Adachi-ku, Tokyo, Japan 123-8511
[2] Graduate Student, Hirosaki University, 3 Bunkyo-cho,Hirosaki, Japan 036-8561,
[3]Physical Science, Hirosaki University, 3 Bunkyo-cho,Hirosaki, Japan 036-8561,
[4]Intelligent Machines and System Engineering, Hirosaki University, Hirosaki, Japan 036-8561.

ABSTRACT

The melt-spun, rapid solidified Galfenol (Fe-Gax, $(17 \leq x \leq 19)$) ribbon sample showed a clear angular dependency on magnetostriction and large magnetostriction (180-200ppm) and good ductility, higher strength etc..as compared with conventional bulk sample. This large magnetostriction is caused by non-precipitating of the ordered phases the release of considerable large internal stresses in as-spun ribbon as well as the remained [100] oriented strong textures after annealing. These are worthy for engineering application as a actuator/sensor devices, therefore, we will introduce the micro-valve which was designed as a device for micro gas-flow control. Bimorph-type FeGa/Ni with the opposite value (FeGa(+),Ni(-)) of magnetostriction coefficient. Large displacement of opposite type could be obtained and it degree and hysteresis curve changed depending the volume fraction of the composite actuator structure.

INTRODUCTION

Magnetostrictive materials are ideally required to have high strength, good ductility, large magnetostriction at low saturation magnetic field, high magnetomechanical coupling coefficients and low cost for engineering applications as shown in **Fig.1** .Rare-earth intermetallic compound, Terfenol-D single crystal exhibits giant magnetostriction ε=2000 ppm at room temperature. However, Terfenol-D single crystal is expensive and brittle to be worked. Recently, magnetic, magnetostrictive and elastic properties of so-called GALFENOL,, Fe100-xGax $(13 \leq x \leq 35)$ single crystals were studied by A. E. Clark *et al.*, and they found that the Fe100-xGax $(17 \leq x \leq 19)$ single crystal showed good durability, high ductility and large magnetostriction of 400 ppm (10-6) in low magnetic field [1]-[3]. But, this alloy's saturation magnetostriction decreased at higher composition of gallium (19<x). In the equilibrium phase diagram of Fe-Ga alloy, phase transformation from bcc disordered A2 phase to fcc ordered L12 as well as bcc ordered D03 phases appears at higher composition of gallium, they generally suppress microscopic magnetic domain movements and influence magnetic and magnetostrictive properties. As we know, the advantage of melt-spinning method is extension of solid solubility, grain refinement, reduction or elimination of microsegregation, and formation of metastable phase. If the disordered phase at high-temperature region can be frozen at room temperature without precipitating the ordered phases, larger magnetostriction can be expected. we focus our attention on the formation of non-equilibrium condition that might suppress the ordered phases.

Magetostrictive Actuator Material

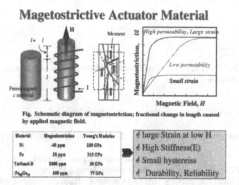

Fig. Schematic diagram of magnetostriction; fractional change in length caused by applied magnetic field.

Material	Magnetostriction	Young's Modulus
Ni	-40 ppm	190 GPa
Fe	10 ppm	210 GPa
Terfenol-D	2000 ppm	30 GPa
Fe₈₃Ga₁₇	300 ppm	77 GPa

- large Strain at low H
- High Stiffness(E)
- Small hystereiss
- Durability, Reliability

Fig.1 Good magnetostriction material condition and their material characteristics

The former part of this paper, the experimental results of the changes of magnetostriction in the rapid-solidified Galfenol(Fe-Ga system) ribbons are overviewed including their special unique microstructures and anisotropic tendency between external magnetic field and the plane direction of the ribbon sample. In fact, the melt-spun, rapid solidified Galfenol (Fe-Gax, ($17 \leq x \leq 19$)) ribbon sample showed a clear angular dependency on magnetostriction and large   magnetostriction (-180-200ppm) and good ductility, higher strength etc.. as compared with conventional bulk sample. This large magnetostriction is caused by non-precipitating of the ordered phases the   release of considerable large internal stresses in as-spun ribbon as well as the remained [100] oriented strong textures after annealing. These bring the materials improvement in strength toughness, hardness, wear resistance, heat resistance, and corrosion resistance and these are worthy for engineering application as a actuator/sensor devices.

Therefore, in the latter part of this paper, we will introduce the micro-valve which was designed as a device for micro gas-flow control. Bimorph-type FeGa/Ni with the opposite value (FeGa (+),Ni(-)) of magnetostriction coefficient. Large displacement of opposite type could be obtained and   it degree and hysteresis curve changed depending the volume fraction of the composite actuator structure,
The dynamic properties of the developed prototype magnetic micro-gas valve will be shown in more detail at the presentation.

EXPERIMENT

Ingots of Fe-17at%Ga were prepared by using of arc melting method from electrolysis iron (purity 99.999 %) and gallium (purity 99.9999 %) as shown in **Fig.2** They were homogenized by holding for 24 h at 1173 K. ribbon samples with 80 µm in thickness and 5 mm in width were produced by single-roll melt-spinning method in argon atmosphere from the ingot. Crystallographic structure of the ribbon samples was examined by X-ray diffraction with CuK   radiation. Magnetostriction was measured by the strain gauge attached on both surfaces. These . measurements were carried out under the tensile

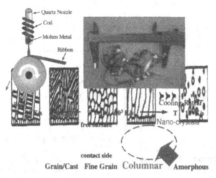

Fig.2 Metallic microstructure changes by controlling the rapid-solidification rate by changing rotating single-roll speed.

pre-stresses of 10MPa, where the pre-stresses were applied to the direction perpendicular to the magnetic field by using dead weight. Magnetostriction was determined by averaging the values obtained from both strain gauges. Two kinds of directions of measurement to were as follows,θ=0° at the minimum strain value and =80°at the maximum strain value, respectively. θ is defined as the angle between applied magnetic filed, H, and the ribbon surface. Applied magnetic field was up to 796kA/m. The strain along ribbon thickness was estimated from the strain of in-plane.

Next, in order to develop a micro-valve actuator in this study, we design the bimorph-type magnetostrictive thin film laminated composite where magnetostdctive positive magnetostriction substrate (araldite) negative magnetostriction thin films that is, one is a bi-morph type, FeGa ribbon (thickness, t=120μm adhesion (araldite) substrate Ni t=50μm and another is a three layered FeGa(t=30μm)/seam welding substrate Ni t=50μm composite thin plate as shown in Fig.3

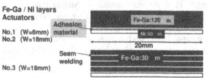

Fig.3 Bimorph type thin film laminated actuator made by FeGa and Ni magnetostrictive alloys with opposite magnetostriction coefficient.

RESULTS AND DISCUSSION

Rapid-solidified FeGa ribbon

Based on above-mentioned situation, we applied this melt-spinning method to develop rapidly solidified thin Fe-Ga ribbons. These ribbons had fine columnar grains with strong [001] oriented texture(see, **Fig.4(a)**), and maximum magnetostriction is obtained when magnetic field was applied to the direction nearly normal to the thickness direction. Longitudinal magn etostriction λ. (TD at θ　0°) reaches easily a saturation value of 60 ppm in rather weak applied field of 1 kOe. Otherλ's for θ=40°, 60 and 80° increase continuously with H. Among these　a maximum strain was obtained for θ　80°,andλfor θ=80°reaches 200 ppm at H=10KOe.　It can be judged from these results that the large

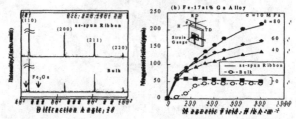

Fig.4 XRD profiles and magnetostriction of the rapid-solidified FeGa ribbon and bulk alloy.

magnetostriction originates in the strong [001]-oriented texture near θ　80°. It should be noted that this valueθ　80°at is about four times as large as that of the bulk material and it becomes effective　for thin plate-type actuator applications.

Bi-morph magnetostrictive gas-valve actuator

Giant magnetostrictive Fe-Ga alloy is expected as actuator/sensor material with high velocity and large stress responses.　In order to develop a micro-valve actuator in this study, we design three types of the bimorph-type magnetostrictive thin film laminates whose structure are magnetostdctive　positive magnetostriction FeGa　substrate (araldite)　negative magnetostriction Ni　thin films　composite. **Fig.5** shows the relationship between displacement D during one cycle of magnetization and applied external magnetic field strength, H in the three actuators. Among three types, actuator 1(W==6mm) showed the largest displacement and most sensitive as a actuator, and as the number of laminates increases, it is clear that the displacement decreases inversely. As a result, actuator force F decreased in inverse proportion to the displacement capability in each type actuator as shown in **Table 1**. Therefore, actuator type 3 was used for prototype micro-gas valve where large cyclic force becomes necessary to control the gas flow.

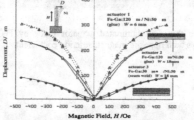

Fig.5　Relationship between displacement D during one cycle of magnetization and applied external magnetic field strength, H in the three types of actuators

172

Table 1 Comparison of the actuator capability of displacement D and Force F at 500Oe in the developed three composite laminates

Bi-morph	Displacement, $D/$ m	Actuated force, $F/$ mN
actuator 1 Fe-20at%Ga:120 m/Ni:50 m (glue) 20mmL × 6mmW	298	9.3
actuator 2 Fe-20at%Ga:120 m/Ni:50 m (glue) 20mmL × 18mmW	231	16
actuator 3 Fe-20at%Pd:30 m× 4/Ni:50 m (seam weld) 20mmL × 18mmW	87.5	19

The technical features of magnetostriuctive gas valve are summarized as follows, i.e. 1) rapid-response , 2) non-contact and remote controllable 3) easy to small size and 4) simpler structures with no wire than PZTs and SMAs etc. actuator. Opening and closing of gas valve are controlled by the magnetic field induced within the solenoid coil. The flow mass of Ar gas in SUS pipe (diameter=1mm) was measured by digital; flow micrometer (minimum scale :1cc/min.) **Fig.6** shows the structure and equipment for the flow test by this bimorph magnetostrictive composite actuator.

The performance of the prototype micro gas valve is shown in **Fig.7**. . Upper figure (a) shows the relationship between flow mass Q and pressure difference P of the gas valve. Lower (b) shows the relationship between gas flow mass Q and applied magnetic field strength. H. It is verified that the starting minimum gas pressure difference Pmin.is 9.9kPa in this laminated actuator type and the gas flow is controllable almost in proportional to applied magnetic field H from 0 to 16cc/min. These experimental results will show that the proposed ,agnetostrictive composite laminated-actuator has considerably good potential for a micro gas valve application. It is interesting in very small quantity intravenous drip pump in medical field and hydrogen supply type small fuel cell etc, in the near future.

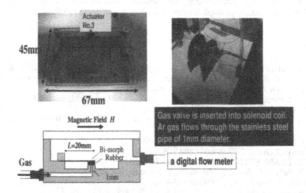

Fig.6 The structure and equipment for the flow test by this bimorph magnetostrictive composite actuator..

173

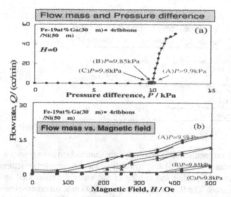

Fig.7 The performance of the prototype micro gas valve. Upper figure (a) shows the relationship between flow mass Q and pressure difference P of the gas valve. Lower (b) shows the relationship between gas flow mass Q and applied magnetic field strength. H.

CONCLUSIONS

Based on the technical advantage of the melt-spun, rapid solidified Galfenol (Fe-Ga) ribbon sample which showed a large magnetostriction (180-200ppm) and good ductility, higher strength etc..as compared with conventional bulk sample, we tried to make a prototype micro-valve for gas-flow control. By designing the bimorph-type FeGa/Ni composite laminates with the opposite value (FeGa(+),Ni(-)) of magnetostriction coefficient., large displacement of actuator device could be obtained.

REFERENCES

1. A. E. Clark, M. Wun-Fogle, J. B. Restorff, T. A. Lograsso, A. R. Ross, and D. L. Schlagel, "Magnetostrictive Galfenol/Alfenol single crystal alloys under large compressive stresses," in *Proc. 7th Int. Conf. on New Acutuators*, Bremen, Germany, 2000, pp.111.
2. A. E. Clark, J. B. Restorff, M. Wun-Fogle, T. A. Lograsso, and D. L. Schlagel. IEEE Trans Magn 2000; 36: 3238
3. A. E. Clark, M. Wun-Fogle, J. B. Restorff, T. A. Lograsso, and J. R. Cullen. IEEE Trans Magn 2001; 37: 2678
4. Y. Furuya. Mat. Res. Soc. Symp. Proc. 2000; 604: 109-116.
5. C.Saito, Y.Furuya,T.Okazaki,T.Matuzaki and T.Watanabe MATERIALS TRANSACTION, JIM, Vol.45,No.2, (2004) 193-198.
6. T .Okazaki, Y. Furuya, C. Saito, T. Matsuzaki, T. Tadao and M. Wuttig, Materials Research Society 785 (2004) pp427-432.
5. C.Saito,Y.Furuya and T.Okazaki,Mater.Trans.JIM, 46.8(2005)pp.1933-1937
6. C.Saito, Y.Furuya, T.Okazaki and M.Omori, On-line Proc. SPIE .vol.6170, 61700L(2006) L1-L8

Mater. Res. Soc. Symp. Proc. Vol. 1129 © 2009 Materials Research Society   1129-V01-05

Development of Fe-Ga-Al Based Alloys with Large Magnetostriction and High Strength by Precipitation Hardening of the Dispersed Carbides

Toshiya Takahashi, Teiko Okazaki, Yasubumi Furuya
Faculty of Science and Technology, Hirosaki University, Hirosaki 036-8561, Japan

ABSTRACT

$(Fe_{0.80}Ga_{0.15}Al_{0.05})_{99}X_{0.5}C_{0.5}$ [at.%] (X=Zr, Nb and Mo) alloys were developed in order to enhance the magnetostriction as well as strength by carbide reaction X-C. Carbide precipitation phases of XC were observed by XRD and laser-microscope in the arc-melted alloys. As a result, $(Fe_{0.80}Ga_{0.15}Al_{0.05})_{99}Zr_{0.5}C_{0.5}$ alloys showed maximum magnetostriction $\lambda_{max}=134$ppm and its tensile strength $\sigma_B=832$MPa. These alloys showed a transgranular fracture. It will be considered to cause the increase of strength due to the precipitation of carbide.

Before now, the polycrystalline Galfenol with such a high tensile strength more over 800MPa as well as large magnetostriction over 100 ppm has not yet been reported.

INTRODUCTION

Ferromagnetic materials that can exhibit large Joule magnetostriction in low magnetic fields are interest for use as sensors and actuators [1-3]. The earliest crystalline magnetostrictive alloys used in transducer were nickel based alloy for example, Ni-Co alloys and large magnetostriction values was obtained in RFe_2 (R=Rare-earth elements) intermetallic compounds with C15 structure such as Terfenol-D $(Dy_{0.7}Tb_{0.3})$ Fe_2 alloy [4]. These alloys show magnetostriction more over 1,000ppm. However, these alloys are expensive, very brittle and high fields required for magnetic saturation. Galfenol alloy (Fe-Ga) by Clark et al. [5,6] has A2 structure with (i) high mechanical strength $(\sigma_B=$over 500MPa)[7,8], (ii) good ductility[9,10], (iii) large magnetostriction value $(\lambda=100$ppm order), (iv) low magnetic saturation fields, (v) low material prices. However, Galfenol alloy has not had so sufficient strength as a robust actuator / sensor materials under severe environment for the machinery engineering uses.

Therefore, we have been studied the material design for enhancement of strength of Galfenol with carbon, boron element addition as well as the novel material process techniques such as rapid solidification method or spark plasma sintering (SPS) to Galfenol alloy[11,13]. For example, the $(Fe_{0.80}Ga_{0.15}Al_{0.05})_{98.5}C_{1.5}$ alloy which we developed has high strength (more over 600MPa) [11], but magnetostriction decreased to about 40-50 ppm because an A2 phase changes to $L1_2$ phase $(Fe_3GaC_{0.5})$ by carbon addition. $L1_2$ $(Fe_3GaC_{0.5}$ or $Fe_3Ga)$ is known for decreasing magnetostriction [11-15].

The purpose of this study is to enhance the magnetostriction of Galfenol alloy as well as strength over 700MPa in this research. Carbon and an element Zr, Nb, Mo with low formation energy (Zr-C, Nb-C and Mo-C) as shown in Fig. 1 [16] were added in small amount (<1at.%) to our former ternary $Fe_{80}Ga_{15}Al_5$ alloy. We intended to increase the strength by precipitation hardening with inhibiting a considerable reduction of A2 phase. Then, the alloys were annealed at 823K (Furnace cooling) in order to investigate the effect of annealing.

We investigated the crystal structures by X-ray diffraction (XRD), the magnetostriction measurements using by the strain gauge method. The strength property was studied by a tensile test and the fractography using a scanning electron microscope (SEM). Those developed materials have a potential of industrial applications such as high sensitive magnetic robust actuator and sensor materials.

$(Fe_{0.80}Ga_{0.15}Al_{0.05})_{99}Zr_{0.5}C_{0.5}$, $(Fe_{0.80}Ga_{0.15}Al_{0.05})_{99}Nb_{0.5}C_{0.5}$ and $(Fe_{0.80}Ga_{0.15}Al_{0.05})_{99}$ $Mo_{0.5}C_{0.5}$ alloys were made by the arc melting method using $Fe_{82}C_{18}$ alloy and the elements such as Ga, Al, Zr, Nb and Mo (purity >99.9%). They were annealed at 1,373K for 3.6ks to equalize into the A2 phase, and then quenched into iced water. They were again annealed at 823K for 9.0ks to stabilize into the equilibrium structure and remove the stress. They were cooled into room temperature in furnace.

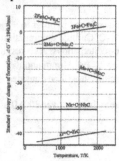

Fig.1. Relationship between standard formation energy and temperature of each carbide [16].

EXPERIMENT

X-ray diffraction (XRD) of these alloys was performed using CuKα radiation to determine the crystal structures. Cross sections of these alloys were observed by the laser microscope, after buffing with 0.1μm alumina abrasive coating. Surface was etched by 10% Nital solution. The crystal grain size, d, and the volume fraction of precipitation, f, were computed with image analysis software. An electron probe micro analyzer (EPMA) was used to observe the distribution of elements. Compositions of these alloys were performed by inductively coupled plasma (ICP) and infrared absorption method (IR).

Magnetostriction was measured by the strain gauge method. The measurement direction of magnetostriction was longitudinal one. Applied magnetic field was 10kOe (796kA・ m⁻¹) at room temperature. The size of 3.0mm in width, 5.0mm in length and 2.0mm in thickness.

Mechanical strength at room temperature was evaluated by the tensile test of a plate type strain rate of about 0.2sec⁻¹. Strain was measured using the strain gauge on the samples. Breaking point, σ_B, was obtained from the strain-stress curve. Fractography on the fracture surface was observed by a scanning electron microscope (SEM). The average values of magnetostriction, λ_{ave}, and breaking point, σ_{ave}, were estimated from the experimental results of three samples under same condition.

DISCUSSION

Table I shows the chemical compositions of alloys. Fig. 2 shows XRD patterns and the photographs of cross section of alloys. $(Fe_{0.80}Ga_{0.15}Al_{0.05})_{99}Zr_{0.5}C_{0.5}$ and $(Fe_{0.80}Ga_{0.15}Al_{0.05})_{99}Nb_{0.5}C_{0.5}$ alloys exhibited A2 (αFe(Ga,Al)) phase, B1 phase (ZrC and NbC phase) and the diffraction peaks of $L1_2$ phase. However, $((Fe_{0.80}Ga_{0.15}Al_{0.05})_{99}Mo_{0.5}C_{0.5}$ alloys have A2 and $L1_2$ phase. Table II shows crystal grain size, d, and the volume fraction of precipitation, f, of alloys. The B1 ZrC and NbC phase were verified in the cross section image (Fig.2) and area analysis by EPMA of $(Fe_{0.80}Ga_{0.15}Al_{0.05})_{99}Zr_{0.5}C_{0.5}$ and $(Fe_{0.80}Ga_{0.15}Al_{0.05})_{99}Nb_{0.5}C_{0.5}$ alloys. And grain size of these alloys is several times as small as grain of $(Fe_{0.80}Ga_{0.15}Al_{0.05})_{99}Mo_{0.5}C_{0.5}$ alloy. In addition, the volume fraction, f, of the alloy of this study is smaller than the reference sample of f=19%. The reason of this grain fining of $(Fe_{0.80}Ga_{0.15}Al_{0.05})_{99}Zr_{0.5}C_{0.5}$, $(Fe_{0.80}Ga_{0.15}Al_{0.05})_{99}Nb_{0.5}C_{0.5}$ alloys will be considered to be caused by a carbide precipitation with high melting point such as ZrC and NbC which have inhibitive effect on crystal grain growth [17-19].

Table I. Alloy composition.

Alloys	Desired Composition element							Analyzed Composition element						
	Fe	Ga	Al	C	Zr	Nb	Mo	Fe	Ga	Al	C	Zr	Nb	Mo
$(Fe_{0.80}Ga_{0.15}Al_{0.05})_{99}Zr_{0.5}C_{0.5}$	79.2	14.85	4.95	0.5	0.5	—	—	78.9	14.97	5.02	1.59	0.52	—	—
$(Fe_{0.80}Ga_{0.15}Al_{0.05})_{99}Nb_{0.5}C_{0.5}$	79.2	14.85	4.95	0.5	—	0.5	—	79.8	14.8	4.93	9.06	—	0.50	—
$(Fe_{0.80}Ga_{0.15}Al_{0.05})_{99}Mo_{0.5}C_{0.5}$	79.2	14.85	4.95	0.5	—	—	0.5	79.8	14.8	4.92	8.87	—	—	0.51

Table II. Grain size of matrix A2, d (mm), and volume fraction of precipitation, f (%), measurements of $(Fe_{0.80}Ga_{0.15}Al_{0.05})_{99}X_{0.5}C_{0.5}$ (X=Zr, Nb and Mo) alloys.

Alloys	d (μm)	f (%)
$(Fe_{0.80}Ga_{0.15}Al_{0.05})_{99}Zr_{0.5}C_{0.5}$	79	4.9
$(Fe_{0.80}Ga_{0.15}Al_{0.05})_{99}Nb_{0.5}C_{0.5}$	78	6.0
$(Fe_{0.80}Ga_{0.15}Al_{0.05})_{99}Mo_{0.5}C_{0.5}$	455	7.3

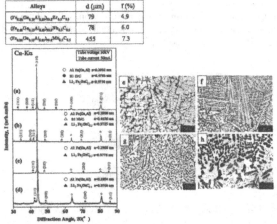

Fig.2. X-ray diffraction patterns of $(Fe_{0.80}Ga_{0.15}Al_{0.05})_{99}Zr_{0.5}C_{0.5}$ (a), $(Fe_{0.80}Ga_{0.15}Al_{0.05})_{99}Nb_{0.5}C_{0.5}$ (b), $(Fe_{0.80}Ga_{0.15}Al_{0.05})_{99}Mo_{0.5}C_{0.5}$ alloys (c), and $(Fe_{0.80}Ga_{0.15}Al_{0.05})_{98.4}C_{1.6}$ arc-melted alloys

(d). Cross section images of $(Fe_{0.80}Ga_{0.15}Al_{0.05})_{99}Zr_{0.5}C_{0.5}$ (e), $(Fe_{0.80}Ga_{0.15}Al_{0.05})_{99}Nb_{0.5}C_{0.5}$ (f), $(Fe_{0.80}Ga_{0.15}Al_{0.05})_{99}Mo_{0.5}C_{0.5}$ alloys (g), and $(Fe_{0.80}Ga_{0.15}Al_{0.05})_{98.4}C_{1.6}$ arc-melted alloys by laser microscope (h). Etching: 10% Nital solution (Nitric acid 10%+Ethanol 90%).

Fig.3 shows the magnetostrictive curves and the fractographs on the fracture surface of the four kinds of alloys. Fig.4 shows the result of average value, λ_{ave}, and, σ_{ave}, of $(Fe_{0.80}Ga_{0.15}Al_{0.05})_{99}$ $X_{0.5}C_{0.5}$ (X=Zr, Nb and Mo) alloys. From the result of magnetostriction in Fig.3 (a) and Fig.4, it is clear that the magnetostriction values of developed alloys are improved in comparison with the reference samples of arc-melted $(Fe_{0.80}Ga_{0.15}Al_{0.05})_{98.4}C_{1.6}$. As the reason of the improvement of magnetostriction from the reference samples, it is possibly thought that (i) reduction of large magnetostriction phase (A2) was controlled from the result in Fig.2 and Table II. (ii) The amount of L1$_2$ phase is lower than the reference samples $(Fe_{0.80}Ga_{0.15}Al_{0.05})_{98.4}C_{1.6}$.

From the strength property in Fig.3 (b)-(e) and Fig.4, $(Fe_{0.80}Ga_{0.15}Al_{0.05})_{99}Zr_{0.5}C_{0.5}$ alloy showed $\sigma_B=840MPa$, $\sigma_{Y0.1\%}=730MPa$, $(Fe_{0.80}Ga_{0.15}Al_{0.05})_{99}Nb_{0.5}C_{0.5}$ alloy showed $\sigma_B=840MPa$, $\sigma_{Y0.1\%}=740MPa$ and $(Fe_{0.80}Ga_{0.15}Al_{0.05})_{99}Mo_{0.5}C_{0.5}$ alloy showed $\sigma_B=880MPa$, $\sigma_{Y0.1\%}=870MPa$. Mechanical property of this study is improved in comparison with the conventional alloys $(Fe_{0.80}Ga_{0.15}Al_{0.05})_{98.4}C_{1.6}$, where, σ_B, was about 670MPa [11] and σ_B of $Fe_{83}Ga_{17}$ single crystal bulk [7] was about 580MPa.

Polycrystalline $Fe_{81.3}Ga_{18.7}$[10] have average grain size of about 400μm and exhibits an intergranular fracture, indicating a severe deterioration in ductility, but the developed alloys here indicates a clear improvement of ductility, because these alloys showed a transgranular fracture mode. It will be considered to cause increase of strength due to the segregation of carbide.

From the experimental results, the factors of improvement in strength will be as follows.
(1) Grain fining by formation carbide phase [17-19].
(2) Increase of the precipitation strengthening ability by carbide formation (ZrC, NbC and a slight amount of L1$_2$ phase) in the matrix.
(3) In $(Fe_{0.80}Ga_{0.15}Al_{0.05})_{99}Mo_{0.5}C_{0.5}$ alloys with maximum magnetostriction, solid solution hardening ability of matrix increased, because Mo can be easily saluted[16] to the matrix and the distributed state of carbide precipitates is very fine[17].

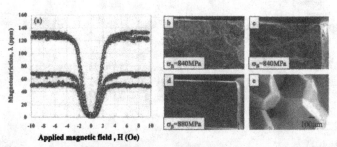

Fig.3. The magnetostrictive curves of $\circ(Fe_{0.80}Ga_{0.15}Al_{0.05})_{99}Zr_{0.5}C_{0.5}$, $\triangle(Fe_{0.80}Ga_{0.15}Al_{0.05})_{99}$ $Nb_{0.5}C_{0.5}$, $\square(Fe_{0.80}Ga_{0.15}Al_{0.05})_{99}Mo_{0.5}C_{0.5}$ alloys, and $\diamond$the reference samples $(Fe_{0.80}Ga_{0.15}Al_{0.05})_{98.4}C_{1.6}$ after annealed at 823K for 9.0ks (a). Observation of fracture side of test

piece of $(Fe_{0.80}Ga_{0.15}Al_{0.05})_{99}X_{0.5}C_{0.5}$ (X=Zr, Nb and Mo) alloys (b)-(d), and $Fe_{81.3}Ga_{18.7}$ arc-melted alloys [10] (e).

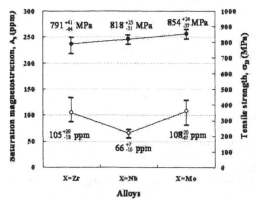

Fig.4. The result of average value, λ_{ave}, and, σ_{ave}, of $(Fe_{0.80}Ga_{0.15}Al_{0.05})_{99}X_{0.5}C_{0.5}$ (X=Zr, Nb and Mo) alloys. Solid and open circles show the resulted in the case of, λ_{ave}, and the case of, σ_{ave}, respectively.

CONCLUSIONS

From the metallurgical characterization, magnetostriction measurements and strength property of the developed $(Fe_{0.80}Ga_{0.15}Al_{0.05})_{99}X_{0.5}C_{0.5}$ [at.%] (X=Zr, Nb and Mo) alloys , the following experimental results are obtained.

(1) $(Fe_{0.80}Ga_{0.15}Al_{0.05})_{99}Zr_{0.5}C_{0.5}$ and $(Fe_{0.80}Ga_{0.15}Al_{0.05})_{99}Mo_{0.5}C_{0.5}$ alloys have very high tensile stress over 800MPa and large magnetostriction more over 100 ppm.
(2) $(Fe_{0.80}Ga_{0.15}Al_{0.05})_{99}Nb_{0.5}C_{0.5}$ alloy has about σ_B=840MPa and about λ_{max}=70ppm.
(3)   The excellent magnetostrictive and high strength alloys in this study have not been reported until now. The alloys in this study satisfied simultaneously to have the magnetostriction come up to polycrystalline Galfenol alloys [20-23] and high level strength over 800MPa.

ACKNOWLEDGMENTS

We would like to thank Prof. M. Matsumoto of Institute for Materials Research, Tohoku University for his help in discussion. And we are grateful to Prof. M. Shiba of Faculty of Science and Technology of Hirosaki University for his support of chemical composition analysis of samples.

REFERENCES

1.   P. R. Downey, A. B. Flatau, J. Appl. Phys., **97**, 10R505 (2005).
2.   S. Dong, J. Zhai, N. Wang, F. Bai, J. Li, and D. Viehland, T. A. Lograsso, Appl. Phys. Lett.,

87, 222504 (2005).

3. J.Atulasima, A. B. Flatau, J. R. Cullen, J. Appl. Phys., **103**, 014901 (2008).
4. K.B Hathaway and A. E. Clark, MRS Bull **18**, 34 (1993).
5. A. E. Clark, M. Wun-Fogle, J. B. Restorff, T. A. Lograsso, A. R. Ross, and D. L. Schlagel, IEEE Trans Magn., **36**, 3238 (2000).
6. A. E. Clark, M. Wun-Fogle, J. B. Restorff, T. A. Lograsso, A. R. Ross, and D. L. Schlagel, Proc. 7th Int. Conf. on New Actuators, Bremen, Germany, 111 (2000).
7. A. E. Clark, M. Wun-Fogle, J. B. Restorff, T. A. Lograsso, and J. R. Cullen. IEEE Trans Magn., **37**, 2678 (2001).
8. R.A. Kellogg, A.M. Russell, T.A. Lograsso, A.B. Flatau, A.E. Clark and M. Wun-Fogle, Acta Mater., **52**, 5043 (2004).
9. S. Guruswamy, N. Srisukhumbowornchai, A.E. Clark, J.B. Restorff, and M. Wun-Fogle1 Scripta mater., **43**, 239 (2000).
10. Suok-Min Na and A. B. Flatau, J.Appl.Phys., **103**, 07D304 (2008).
11. M. Hasegawa, K. Hashimoto, W. Yoshida, T. Okazaki, Y. Furuya and M. Shimada, Jour. Jpn. Soc. Mech. Eng., A,**73**, 1309 (2007).
12. T. Okazaki, Y. Furuya, C. Saito, T. Matsuzaki, T. Watanabe and M. Wuttig, Mat. Res. Soc. Symp. Proc., **785**, 427 (2004).
13. C. Saito, Y. Furuya, T. Okazaki, T. Matsuzaki and T. Watanabe, Mater. Trans., **45**, 193 (2004).
14. N. Srisukhumbowornchai and S. Guruswamy, J. Appl. Phys., **92**, 5371 (2002).
15. R. Wu, J. Appl. Phys., **91**, 7358 (2002).
16. Metal Data book, 4th edition, Japan Institute of Metal, Maruzen, (2004).
17. Metal Handbook, 6th edition, Japan Institute of Metal, Maruzen, (2000).
18. T.Mori, M. Tokizane and Y. Adachi, Tetsu-to-Hagane, **9**, 1061 (1962).
19. The Ninth Edition of Metals Handbook, ASM International Handbook Committee, (1989).
20. S.F. Xu, H.P. Zhang, W.Q. Wang, S.H. Guo, W. Zhu, Y.H. Zhang, X.L. Wang, D.L. Zhao, J.L. Chen and G.H. Wu, J. Phys. D, Appl. Phys., **41**, 015002 (2008).
21. R.S. Turtelli, C. Bormio-Numes, J.P. Sinnecker R. Grössinger, Physica B, **384**, 265 (2006).
22. C. Bormio-Nunes, M.A. Tirelli, R.S. Turtelli, R. Grössinger, H. Müller, G. Wiesinger, H. Sassik, and M. Reissner, J. Appl. Phys., **97**, 033901 (2005).
23. N. Srisukhumbowornchai and S. Guruswamy, J. Appl. Phys., **90**, 5680 (2001).

Active Polymers and Soft Matters

Mater. Res. Soc. Symp. Proc. Vol. 1129 © 2009 Materials Research Society   1129-V08-01

Durable Dielectric Elastomer Actuators via Self-Clearable Compliant Electrode Materials

Wei Yuan, Tuling Lam, Qibing Pei*
Department of Materials Science and Engineering, The Henry Samueli School of Engineering and Applied Science and California Nanosystems Institute, University of California, Los Angeles, CA 90095-1595, USA

* qpei@seas.ucla.edu

ABSTRACT

Dielectric elastomers are shown to exhibit greater than 200% strain with actuation energy densities and power densities greatly exceeding those of most existing actuators and muscles. Limited lifetime and manufacturability have hindered the commercialization of these otherwise remarkable polymer actuators. We have introduced fault-tolerance as a new feature to enable high production yield and actuation stability. The use of single walled carbon nanotube (SWNT) electrodes allows the dielectric elastomers to self-heal. During localized dielectric breakdown at a defect that is either intentionally introduced, inherent in the films, or newly formed due to fatigue or stress relaxation, the SWNT electrodes self-clear, isolating the defect from the rest of the active area which can then actuate normally. A coating of polyaniline nanofibers (Pani-NF) sprayed from aqueous dispersion was also found to be effective in achieving self-clearing during localized dielectric breakdown. The fault-tolerance significantly enhances the actuation lifetime during constant voltage and cyclic actuations. It also eliminates a major safety concern for high voltage actuators.

INTRODUCTION

Dielectric elastomers are a category of electroactive polymers exhibiting large actuation strain, high output stress and specific energy density [1-4]. A dielectric elastomer actuator comprises of a thin soft elastomer film sandwiched between two compliant electrodes. Upon application of a high voltage, the charges will flow into and store in the compliant electrodes across the dielectric film. The electromechanical force resulted from the attraction of opposite charges across the elastomer film and repulsion of like charges along the film surface will cause the film to shrink in thickness and expand in area while the overall volume remain unchanged. Once the voltage is turned off, the stored charges will recombine through the external circuit. Without electrostatic force, the strained elastomer film returns to its original shape. If opposite electrodes are shorted through the dielectric film either intentionally by the experimenter, or by a defect inherent to the film, the charges can no longer be stored on the electrodes and the actuator ceases function. Therefore, a dielectric fault generally translates into a terminal failure of the polymer actuator.

The dielectric strength of the elastomer film plays a key role in the performance of the actuators. High actuation performance has been obtained with 3M VHB acrylic elastomer films, silicone elastomers, and thermoplastic elastomers, among others [5-9]. Prestraining the soft elastomer films by a factor of 4-36 times in area expansion was found to be an effective method to increase the dielectric strength of an existing film. [10-12].

The electrodes which sandwich the elastomer film also play a key role in electromechanical strain performance. They must be highly compliant from both electrical and mechanical viewpoints. The electrical conductance on the surface of the electrodes is high. This conductivity remains high at large area strains such that charges flow into the electrodes and continuously deform the dielectric film. At the same time, the mechanical stiffness of the electrodes is low and does not impede the area expansion of the soft elastomer film. So far, carbon grease [2, 5], carbon graphite [13], carbon nanotube [14-16], salt-water solution [17], conducting polymers [18], cell culture medium [19], metal coatings on corrugated surfaces [20] have been used as electrodes for dielectric elastomer actuators. The actuation performance and stability of a dielectric elastomer actuator is also dependent on the quality and uniformity of the elastomer film. Defects, such as imbedded particles, pores, pockets of incompletely cured polymers or gel particles, un-even film thickness are more or less inevitable in a film that is soft and sticky. As the elastomer film fatigues over time during repeated expansion and contraction, new defects will form. All these defects may cause dielectric breakdown at a driving voltage lower than predicted by the polymer's intrinsic dielectric strength. A technique is therefore required to eliminate this problem caused by defects. This is critically important in the fabrication of actuators involving large-area elastomer films We have been developing self-clearable complaint electrodes to enhance the production yield and actuation stability of dielectric elastomers [14-16, 19]. Ultrathin coatings of single-walled carbon nanotube (SWNT) and polyaniline nanofiber (Pani-NF) have been found to degrade and loose electrical conductivity around the dielectric breakdown site. This degradation spreads concentrically from the breakdown site until the dielectric short is electrically isolated and the applied voltage on the remaining area is restored. This electrode self-clearing can significantly increase the actuation stability of the dielectric elastomers at large strains.

EXPERIMENTAL

Actuation and Self-clearing Measurement Set-up

The self-clearing measurement set-up is shown in Figure 1. The biaxially prestrained 3M VHB acrylic films coated with compliant electrodes (SWNT or Pani-NF) were connected in series with current-limiting resistors (5 MΩ or larger), Keithley 2000 multimeter, and a high-voltage power supply. The current-limiting resistors were used to protect the multimeter from damage due to high current spike during a dielectric breakdown. The chronoamperic curve was obtained by a Keithley 2000 multimeter through LABView software on a computer.

Area expansions of the films were recorded by a digital camera. The area strain was calculated by the equation $(a_1 - a_0)/a_0$, where a_1 is the actuated area under high voltage, and a_0, the area at 0 V. For films where self-clearing took place, the cleared circular areas around the defects were subtracted from both a_1 and a_0.

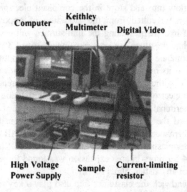

Figure 1: Photograph of the actuation measurement set-up

SWNT Electrodes:

Purified P3-SWNT (Carbon Solution, Inc.) was dispersed in an aqueous solution (mixture of water and isopropanol by 1:1 in weight) with aid of probe sonication for 30 min. The concentration

184

of the P3-SWNT solution was 2mg/mL. Actuators with SWNT/acrylic/SWNT sandwich structures were prepared by spraying the SWNT solution onto both surfaces of each acrylic film (3M VHB tape with 300% biaxial prestrain). Contact masks were used to form overlapping circular SWNT areas on the opposite surfaces to define the active area of the actuators. .

Pani-NF Electrodes:

Conducting polyaniline nanofibers (Pani-NFs) were synthesized according to the literature [21] and dispersed in isopropanol to a concentration of ~2 mg/mL. The Pani-NFs were sprayed with an airbrush by consistently sweeping horizontally across the entire electrode area. The electrode geometry (actual diameter 1.9 cm) was controlled with a contact mask. Each electrode was sprayed in thin layers at a time, allowing 10-15 seconds for the solvent to evaporate in between sprayed layers. Approximate thicknesses of the sprayed Pani-NF electrodes were determined by measuring the volume of the dispersed fibers.

RESULTS and DISCUSSIONS

SWNT Electrode Self-clearing

An actuator using 300% biaxally prestrained VHB 4905 film sandwiched by two sprayed P3-SWNT electrodes are shown in Figure 2. The surface conductance of the electrodes is 600 $\Omega/\square$. The circular active area before actuation is shown Figure 2 (a), during actuation at 190% strain in area expansion at 3.5kV in (b). This strain value is comparable to that of a similar actuator using highly compliant carbon grease electrodes.

Figure 2 (c) shows the active area intentionally punctured with a cactus pin. As the actuation voltage is turned on, the SWNT around the pinhole displayed localized degradation and became non-conductive. The induced strain however, was not affected significantly by the self-clearing event. The actuation voltage applied to the remaining active area produced large electromechanical strains, similar to those of an unpuctured elastomer actuator. The electrical short at the pin puncture was thus electrically isolated, the actuator's performance preserved by the self-clearable SWNT electrodes. Upon close examination of the deactivated area, it can be observed that this area appears darker than the rest of the electrode. This is because the elastomer film in this area was thicker as it was not strained (deactivated).

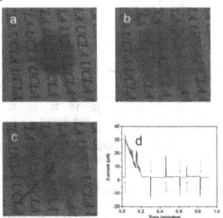

Figure 2: Self-clearing demonstration of a circular expansion actuator with SWNT electrodes, (a) at 0kV, (b) 190% area strain obtained at 3.5kV, (c) high strain actuation in the presence of cactus pin puncturing, (d) the chronoamperic curve during self-clearing and subsequent switching cycles of voltage on and off (upward arrows indicate voltage on, and downward arrows indicate voltage off).

185

The chronoamperic curve in Figure 2 (d) shows that upon re-application of the voltage after puncturing, the current jumps to a high value. This is due to a shorting of current through the punctured fault. As the SWNT self-clears, the resistance across the film increases and the current drops. When the cleared circle becomes large enough to sustain the applied voltage, the self-clearing is at completion. The current is thus decreased to the nominally low actuation current. Simultaneously, the actuator resumes its high actuation strain. From the subsequent charging and discharging curves as the actuation voltage is switched on and off, the actuator still functions normally even when punctured with a cactus pin.

Actuator Lifetime Test with SWNT Electrode

Circular expansion actuators similar to the one shown in Figure 2 (a) were actuated at a constant high voltage of 3kV to test the prolonged lifetime. The transient current and actuation strains are shown in Figure 3. The actuation current initially increases to a stable value of about 3.5 µA with a corresponding area strain of approximately 130%. Near the 2 minute mark of the actuation, the current suddenly jumps to a high value, indicating a dielectric breakdown. No pin puncture was applied. The breakdown was most likely caused by a defect within the film. After the self-clearing, the actuator quickly regains normal actuation, and the current falls back to a normal, steady value. There are five successful breakdown-clearing events, and the area strain remains a relatively steady value of 160% for about 30 minutes. After that, the current gradually decreases from 3.5 µA to about 0.2 µA at the 35 minute mark, where the strain decreases from 160% to about 20%. Raising the actuation voltage did not increase the strain. This means that this actuator has permanently lost part of its actuation function. A similar problem was also observed with an actuator driven under half-sine voltage signals [16]. It was thus shown that the self-clearing of SWNT electrodes can prevent sudden terminal failure due to dielectric breakdown; however, its effectiveness to significantly prolong the operation lifetime is still limited.

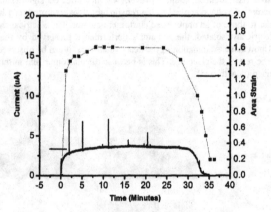

Figure 3: The current and strain of actuator with SWNT electrode during the lifetime test at constant 3kV

Actuator Lifetime Test with Modified SWNT Electrode

The decreasing actuation strain shown in Figure 3 was due to gradual electrode degradation and loss in electrode conductivity in an area which spread concentrically from the self-clearing site. To overcome this problem of unabated SWNT degradation on the electrode, dielectric oil was applied onto each of the SWNT electrodes. As a result, the actuation stability was dramatically improved. Figure 4 shows the actuation strain at a 3 kV constant voltage for 1100 minutes. No gradual degradation or failure was observed: both the actuation current and strain remained at a high, stable value during the extended actuation. In the course of this actuation, a number of breakdown-clearing events occurred. However, these events pose little impact on the current and strain except for the short intervals during the self-clearing process. This improvement may be attributed to the quenching of corona discharging of the SWNT nanotubes by the dielectric oil coating.

The actuators were also tested with a half-sine wave voltage, at 0 to 2.9 kV, a 50% duty cycle, and a 200 mHz frequency. The strains in each cycle were measured from the smallest area when the voltage was turned off to the largest area during the peak voltage. The results are plotted in Figure 5 as a function of operation duration. The actuator maintained an average strain of approximately 70% for over 2700 minutes.

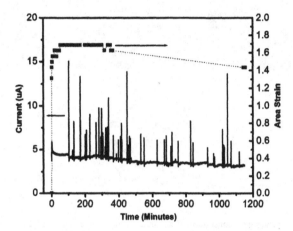

Figure 4: Actuation current and strain of an actuator with modified SWNT electrode during the lifetime test at a constant 3kV.

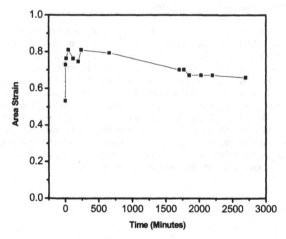

Figure 5: Actuation strain of an actuator with modified SWNT electrode during the lifetime test at a pulsed signal drive (half duty of sine wave, 2.9 kV peak voltage, 200 mHz).

Pani-NF Electrodes

Figure 6 shows an actuator based on Pani-NF electrodes, spray-coated with a 1.1 um thickness. The 300% biaxially prestrained VHB 4905 film is actuated to approximately 80% area strain at 3kV (a and b). After a self-clearing event due to an internal dielectric shorting, the area strain increased slightly to 90% at the same voltage (c).The self-clearing process is recorded as shown in (d). The clearing is marked with a current spike followed by a rapid decline until it returned to the nominal actuation current level.

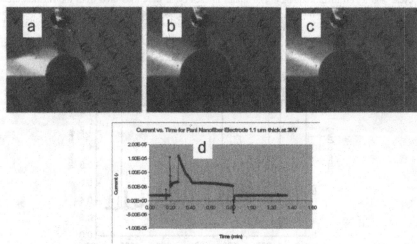

Figure 6: Self-clearing demonstration of a circular expansion actuator with 1.1 um thick Pani-NF electrodes, (a) at 0kV, (b) around 80% area strain obtained at 3 kV, (c) 90% area strain obtained after the self-clearing, (d) the chronoamperic curve of the self-clearing.

Actuator Lifetime Test with Pani-NF Electrodes

The best constant voltage test result was obtained on electrodes with coating thicknesses of 1.1 μm Pani-NF, which displayed a preserved strain of 93% for a duration of 10 minutes at 3kV, after which a hole was formed in the film and subsequent tearing of the dielectric film.

A cycling test with pulsed voltage was performed on the actuator with a 1.1 μm thick electrode, due to its long lifetime and high strain. The actuator was connected to an oscillating sine wave of 3 kV with a frequency of oscillation of 0.156 sec[-1] in order to determine the duration of actuation cycles. Figure 7 shows the number of switching cycles (increase and decrease in current) obtained with time. The actuator displayed a maximum strain of 91% and sustained over 700 actuation cycles for a total duration of approximately 75 minutes. The percent area strain however, was observed to gradually decrease as time progressed, as displayed by a decrease in current. This indicates a gradual dedoping of the nanofibers with actuation cycles. The final failure of the actuator was due to the Pani-NF electrodes, which eventually became nonconductive.

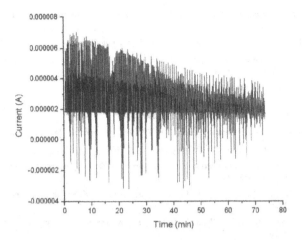

Figure 7: Current vs. time profile of actuator with Pani-NF electrodes recorded during pulse voltage cycles, under an oscillating sine wave of 3 kV with a frequency of 0.156 sec[-1]

CONCLUSION

It has been shown that ultrathin layers of SWNT and Pani-NF serving as compliant electrodes for dielectric eleastomer actuators can drive a 300% biaxially prestrained VHB 4905 film to over 100% area strain. Both electrodes also exhibited self-clearing, preventing premature failure while significantly improving actuation lifetime.. The self-clearing mechanisms around dielectric breakdown sites are different for SWNT and Pani-NF electrodes. For the SWNT, the electrode material around the defect was decomposed under plasma caused by high voltage. The Pani-NF material becames non-conductive due to a dedoping process around the defect.

During prolonged lifetime tests in constant or pulse driven voltage schemes, both SWNT and Pani-NF electrodes exhibited unabated degradation, resulting in limited improvement in actuation lifetime. SWNT electrodes with dielectric oil coatings resulted in prolonged lifetime by at least two orders of magnitude. This improves the fault tolerance of the actuator and will significantly improve its manufacturing yield and operation reliability at large strains.

ACKNOWLEDGMENTS

This work was prepared with partial financial support by the National Science Foundation grant # 0507294, General Motor Corporation, and the University of California Discovery Program.

REFERENCES

1. Electroactive Polymer (EAP) Actuators as Artificial Muscles: Reality, Potential, and Challenge, 2nd ed. (Ed: Y. Bar-Cohen), SPIE, Bellingham, WA 2004.

2. R. Perine, R. Kornbluh, Q. Pei, J. Joseph, *Science* **287**, 836, (*2000*).

3. Q. Pei, R. Pelrine, M. Rosenthal, S. Stanford, H. Prahlad, and R. Kornbluh, *Proc. of SPIE* **5385**, 41, (2004).

4. Q. Pei, M. Rosenthal, S. Stanford, H. Prahlad, R. Pelrine, *Smart Materials & Structures* **13**, N86, (2004)..

5. S.M. Ha, W. Yuan, Q.B. Pei, R.P. Pelrine, S. Stanford, *Adv. Mater.* **18**, 887, (2006).

189

6. F. Carpi, A. Khanicheh, C. Mavroidis, D. De Rossi, *IEEE-ASME Transactions on Mechatronics* **13**, 3, 370, (2008).

7. G. Mathew, J.M. Rhee, C. Nah, D.J. Leo, C. Nah, *Polymer Engineering and Science* **46**, 10, 1455, (2006).

8. R. Pelrine, R. Kornbluh, J. Joseph, R. Heydt, Q. Pei, S. Chiba, *Materials Science & Engineering C-Biomimetic and Supramolecular Systems* **11**, 2, 89, (2000).

9. C.G. Cameron, J.P. Szabo, S. Johnstone, J. Massey, J. Leidner, *Sensors and Actuators A-Physical* **147**, 1, 286, (2008).

10. X.H. Zhao, Z.G. Suo, *Applied Physics Letters* **91**, 061921, (2007).

11. G. Kofod, R. Kornbluh, R. Pelrine, P. Sommer-Larsen, *Proc. SPIE Int. Soc. Opt. Eng.* **141**, 4329. (2001).,

12. J.S. Plante, S. Dubowsky, *Smart Materials & Structures* **16**, S227, (2007).

13. P. Khodaparast, S.R. Chaffarian, M.R. Khosrosham, N. Yousefimehr, D. Zamani, *Journal of Optoelectronics and Advanced Materials* **9**, 11, 3585, (2007).

14. W. Yuan, L.B. Hu, Z.B. Yu, T. Lam, J. Biggs, S.M. Ha, D.J. Xi, B. Chen, M.K. Senesky, G. Gruner, Q. Pei, *Adv. Mater* **20**, 621, (2008).

15. W. Yuan, T. Lam, J. Biggs, L.B. Hu, Z.B. Yu, S.M. Ha, D.J. Xi, M.K. Senesky, G. Gruner, Q. Pei, *Proc. of SPIE* **6524**, 65240N-1, (2007),.

16. W. Yuan, L.B. Hu, S.M. Ha, T. Lam, G. Gruner, Q. Pei, *Proc. of SPIE* **6927**, 69270P-1, (2008).

17. F. Carpi, P. Chiarelli, A. Mazzoldi, D.D. Rossi, *Sensors and Actuators A* **107**, 85. (2003).

18. T. Lam, H. Tran, W. Yuan, Z.B. Yu, S.M. Ha, R. Kaner, Q. Pei, *Proc. of SPIE* **6727**, 69270O-1, (2008).

19. Q. Pei, Presentation in MRS Conferences, Fall 2008, Symposium BB.

20. M. Benslimane, P. Gravesen, P. Sommer-Larsen, *Proc. SPIE Int. Soc. Opt. Eng.* **4695**, 150. (2002).

21. J. Huang, R. Kaner, *Chem. Commum.* **4**, 367, (2006).

Mater. Res. Soc. Symp. Proc. Vol. 1129 © 2009 Materials Research Society   1129-V04-05

Bending Behavior of Polymer Films in Strongly Interacting Solvents

Jianxia Zhang and John B. Wiley*
Department of Chemistry and Advanced Materials Research Institute,
University of New Orleans, New Orleans, LA 70148 U.S.A.

ABSTRACT

Several polymers show bending behavior on exposure to selected solvent systems. Either on emersion or evaporation of particular solvents, films can bend or curl within seconds. The response depends on the interaction between polymer chains and solvent molecules as well as the film geometry and surrounding temperature.

INTRODUCTION

Shape-memory polymers (SMP) have attracted much attention in recent years [1-5]. The driving forces for actuators based on these materials are varied. Examples include structural changes driven via electroactive stimuli [3], optical processes (photomechanics) [6] and variations in pH [7]. Here we investigate a simple mechanical response based on solvent uptake (swelling) and release. Polymer swelling is a well-known process and is dependent on the strong interactions between the polymer chains and a particular solvent [8]. Here we investigate a series of polymer-solvent combinations and study the rapidity of response as well as cycling in these processes.

EXPERIMENTAL DETAILS

Synthesis.

The reagents 4-aminophenol, phenol, 6-chloro-1-hexanol, acryloyl chloride and polyethylene glycol 600 were purchased from Alfa Aesar and tetrabutylammoniumbromide from Sigma-Aldrich. All were used as received. 2,2'-azobisisobutyronitrile (AIBN) was purchased from either Sigma-Aldrich or Fisher Scientific and recrystallized from methanol before use.

Synthesis of 4,4'-Dihydroxyazobenzene. 4,4'-Dihydroxyazobenzene was synthesized following the method of Hu et al. [9] 16.232 g (0.149 mol) of 4-aminophenol was dissolved in a mixture of 41 mL water and 41 mL concentrated HCl to form a suspension. And then the mixture was cooled with an ice-water bath with vigorous stirring. The solution of 11.308 g (0.164 mol) of sodium nitrite/35 mL water was added with a dropping funnel. In the mean time, 14.023 g (0.149mol) of phenol was dissolved in 15 g sodium hydroxide/50 mL and cooled to 0°C. Diazotized solution (the first reacted solution) was added to phenol solution dropwise. Then, HCl (12 M) was used to adjust pH to around 1.0. The reaction continued for another 2 hours with stirring. The precipitate was filtered and washed with excess distilled water until a pH 7 was achieved. The crude product was recrystallized three times with acetone/water. Yield: 6.413 g (20.09%)

Synthesis of 4,4'-bis (6-hydroxyhexyloxy)azobenzene. 4,4'-bis (6-hydroxyhexyloxy) azobenzene was synthesized following method of Kumaresan et al. [10]. This reaction was taken under dry nitrogen. 3.032 g (14.15mmol) of 4, 4'-dihydroxyazobenzene, 10.891 g (70.75 mmol) of potassium carbonate and 0.178 g (5% amount to 4,4'-dihydroxyazobenzene) of tetrabutylammoniumbromide were dissolved in 30mL anhydrous Dimethylformamide (DMF). The mixture was heated to 90 $^{\circ}$C. 4.253 g (31.13 mmol, 4.15 mL) of 6-chloro-1-hexanol was added dropwise. The reaction kept 48h at 90°C. The reaction mixture was cooled to room temperature and slowly poured into 100ml 6M HCl. The precipitate was filtered and washed with excess distilled water until pH 7. The product was vacuum dried and then recrystallized with acetone. Yield: 3.623 g (61.8%)

Synthesis of 4,4'–bis (6-acryloyloxyhexyloxy)azobenzene. This reaction followed the published work of Angeloni et al. [11]. This reaction was also carried out under dry nitrogen. 2.039 g (4.919 mmol) of 4, 4' –bis (6-hydroxyhexyloxy)azobenzene was dissolved in 30 mL tetrahydrofuran (THF). 1.194 g (11.8mmol, 1.64mL) of triethylamine was added. 1.068 g (11.8 mmol, 0.96 mL) of acryloyl chloride was added dropwise. Reaction kept for 2h with violent stirring. The mixture was filtered and precipitate was washed with THF several times and combined to filtrate. THF was distilled off. The product was recrystallized twice with acetone. Yield: 0.682 g (26.5%)

Synthesis of diacyloyloxy polyethylene glycol. This reaction is similar with the preparation of 4,4'–bis (6-acryloyloxyhexyloxy)azobenzene. 18.538 g (0.03090 mol) polyethylene glycol 600 was dissolved in 100mL THF. 7.504 g (0.07416 mol, 10.31 mL) triethylamine was added. The mixture was cooled to 0 $^{\circ}$C. 6.712 g (0.07416 mol, 6.03 mL) was added dropwise. Reaction kept for another 4 h with vigorous stirring. Precipitate was separated by centrifuging. The residue was washed with THF twice and all solution combined. The combined solution was centrifuged again if necessary. THF was distilled off. Yield: 19.985 g (91.35%).

Polymer Films

Preparation of diazobenzene films. Films were cast following published procedures [11]. 0.0289 g (0.00553 mmol) of 4, 4' –bis (6-acryloyloxyhexyloxy) azobenzene and 0.011 g (0.0067 mmol) AIBN was dissolved in 1mL toluene in a flat bottom flask. After three freeze-thaw pump cycles, the solution was put into a 80 $^{\circ}$C oil bath. Polymerization was carried out over 48 h, the toluene solution was decanted off, and the polymer film was washed several times with methanol. The film was carefully separated from the bottom of the flask. The film was then cut into specific shapes with scissors.

Preparation of PEG films. 0.9935 g (1.391 mmol) diacyloyloxy polyethylene glycol, 0.0205 g (0.00392 mmol)4, 4' –bis (6-acryloyloxyhexyloxy)azobenzene and 0.026 g (0.016

mmol) AIBN was dissolved in 1mL toluene in a flat bottom flask. After three freeze-thaw pump cycles, the solution was put into a 90 °C oil bath. Polymerization was carried out 2.5 h, the toluene solution was decanted off, and the polymer film was washed several times with acetone. The film was carefully separated from the bottom of the flask. The film was then cut into specific shapes with scissors.

Other polymer films were taken directly from the materials (latex glove, tape, bag, water bottle) around us and were cut with scissors.

Characterization

Bending Test. Polymer films were cut to specific shapes, rectangle (ca. 4mm×2mm) or round (diameter 7 mm) and put flat on watch glass. Dichloromethane (DCM) or water was poured until the film was just immerged. The displacements of the films were monitored during solvent immersion or evaporation.

RESULTS

Figure 1 shows two examples of the polymer film bending behavior in dichloromethane. Films will start to expand after immersion into solvent and then bend. The bending behavior will happen immediately after the film partially immerses into the solvent and will continue until all the solvent evaporates away. If a polymer (e.g. latex) is exposed to a solvent under reflux conditions the polymer piece will show continuous active bending; we carried out these experiments for at least a half hour.

Table 1 lists the bending behavior of two different diazobenzene polymer films in dichloromethane. Both pieces of film were fixed on one end with glass slides. The thinner film will bend to a larger degree, though it may take a longer time to reach the largest point of deformation. Also worth noting is that the two sides of the film show different bending behavior due to a density gradient that occurred on polymerization; the bottom face of the film is of greater density. Figure 2 shows the two pieces of film's bending process. The bending behavior for this polymer is not limited to dichloromethane, but will also happen in some other strongly interacting solvents such as toluene and chloroform.

Table 2 lists bending behaviors of various common polymers in dichloromethane. With proper solvents, dichloromethane here, polymer elastomers will show bending behavior but different components and different thicknesses will result in different behaviors. Figure 3 shows the bending process of crosslinked polyethylene glycol (PEG) film when it was immersed into water. Besides water, PEG film will respond with acetone.

Latex Glove Piece (mainly Cis-1,4-polyisoprene) (in CH$_2$Cl$_2$, 0-136s)

Diazobenzene Polymer (in CH$_2$Cl$_2$, 0-56s)

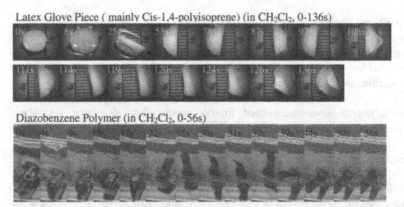

Figure 1. Two examples of film bending behavior in dichloromethane during solvent evaporation. This bending behavior will occur continuously in a reflux system.

Table 1. Diazobenzene polymer films bending behavior

Thickness (mm)	Length before(mm)	Maximum Length(mm)	Larger deformation (side A)*		Smaller deformation (side B)*	
			Bending degree	Time (to reach largest deformation status) (s)	Bending degree	Time (to reach largest deformation status) (s)
0.071	3.6	4.3	90	50	25	200
0.143	3.8	4.2	75	20	45	230

*Side A refers to top face of film as grown and side B is side grown in contact with the flask.

i) Side A of thick (left) and thin (right) films

| 0s | 12s | 23s | 31s | 37s | 49s | 126s | 172s |

ii) Side B of thick (left) and thin (right)

| 0s | 7s | 32s | 122s | 205s | 236s | 305s | 347s |

iii) Thin film (left) - side A; thick film (right) - side B

| 0s | 2s | 4s | 7s | 10s | 14s | 18s | 29s |

Figure 2. Dynamic response for bending process of diazobenzene films – variation versus film orientation and film thickness. (i) larger deformation side for both films (side A); (ii) smaller deformation side for both films (side B); (iii) larger deformation side for thinner film and smaller deformation side for thicker film. Side A refers to top face of film as grown and side B is side grown in contact with the flask. (*Thin film* is 0.071 mm thick and *thick film* is 0.143 mm thick.)

Table 2. Bending behavior of various polymers in dichloromethane

Polymer films	Thickness (mm)	Bending degree	Length before(mm)	Maximum Length	Response Time (s)	Time to reach largest deformation (s)
water bottle (PETE)	0.189	~30	2.9	(does not recover original shape)	immediate	180
Latex glove piece	0.105	180 (turn around)	4	6	immediate	11
Scotch tape (acrylic polymer)	0.059	90	1.9	2.5	immediate*	5
Clear plastic bag (polyethylene)	0.050	~30-40	2.2	2.3	immediate	35
*partially dissolved						

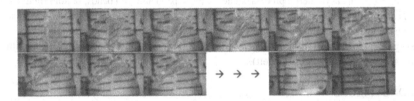

Figure 3. Bending behavior of crosslinked polyethylene glycol film (going left to right starting in upper row). Last 2 frames (lower right) are after several bending-unbending cycle, PEG film reaches an expanded (swollen) status and one hour after water removes away the film will go back to its starting status.

DISCUSSION

Polymer chains can strongly interact with effective solvents. When a polymer film is partially immersed in such a solvent, the polymer chains are rapidly solvated so that the polymer quickly expands. If solvation occurs unevenly in the polymer, polymer films will bend or curl due to an asymmetric volume expansion. The bending behavior will continue until the absorption and expansion is complete for all polymer chains. At the end of the solvent uptake, the film clearly shows increased volume (swelling). During solvent evaporation, there will be a reverse process. Solvent molecules leave the polymer chains and the chains contract to their normal status. This too can occur asymmetrically so the film bends again until all solvent molecules have gone. Commonly the film will return to its starting volume after all the solvent evaporates. Figure 4 shows a schematic process of the bending behavior.

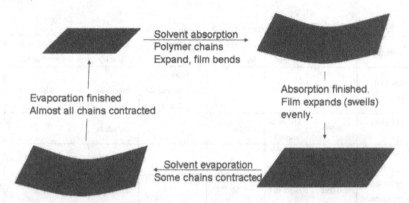

Figure 4. Schematic process of film bending in effective solvents. Top left: the starting point. Polymer chains coil before solvent absorption; top right: during solvent absorption, because of the asymmetric process, some chains expand and film bends; bottom right: after all chains finish interaction with solvent molecules, film expands (swells) evenly; bottom left: partial solvent molecules leave during evaporation and so partial chains coil while others keep expanded, and so film bend again. After all solvent molecules evaporate away, the film goes back to starting status (top left).

CONCLUSIONS

Many polymer materials, including water-responsive ones, show significant bending behavior in strongly interacting solvents. The mechanical response is based on asymmetric swelling of the polymer film and can be quite rapid (seconds). Since the driving force is based on a simple volume expansion, fabrication of polymer films with variable density, or layered or co-polymer films with asymmetric structures, could allow for the formation of new effective actuator systems.

REFERENCES

1. I. S. Gunes and S. C. Jana, *J. Nanosci. Nanotechnol.* **8**, 1616 (2008).
2. C. Liu, H. Qin and P. T. Mather, *J. Mater. Chem.* **17**, 1543 (2007).
3. J. Leng, H. Lv, Y. Liu and S. Du, *Appl. Phys. Lett.* **91**, 144105 (2007).
4. W. M. Huang, B. Yang, L. An, C. Li and Y. S. Chan, *Appl. Phys. Lett.* **86**, 114105 (2005)
5. J. Leng, H. Lv, Y. Liu and S. Du, *Appl. Phys. Lett.* **92**, 206105 (2008).
6. Y. Yu and T. Ikeda, *Angew. Chem. Int. Ed.*, **45**, 5416 (2006).
7. C. Baker, B. Shedd, P. Innis, P. Whitten, G. Spinks, G. Wallace and R. Kaner, *Adv. Mater.* **20**, 155 (2008).
8. B. LaVine, N. Kaval, D. Westover and L. Oxenford, *Analytical Letters*, **39**, 1773 (2006).
9. X. Hu, X. Zhao, L. Gan and X. Xia, *J. of App. Poly. Sci.* **83**, 1061 (2002).
10. S. Kumaresan and P. Kannan *J. of Polym. Sci.: Part A: Polym. Chem.* **41**, 3188 (2003).
11. A. Angeloni, D. Caretti and C. Carlini, *Liq. Cryst.*,**4**, 513 (1989).

Mater. Res. Soc. Symp. Proc. Vol. 1129 © 2009 Materials Research Society   1129-V04-07

Formation of Wrinkle Patterns on Porous Elastomeric Membrane and Their fabrication of Hierarchical Architectures

Yue Cui and Shu Yang
Department of Materials Science and Engineering, University of Pennsylvania, Philadelphia, PA 19104

ABSTRACT

We report the formation of wrinkle patterns on porous elastomeric membrane and their fabrication of hierarchical architectures through mechanical stretching and replica molding. The technique builds upon a buckling instability of a stiff layer supported by a porous elastomeric membrane which was induced by surface plasma oxidation of the pre-stretched porous elastomer followed by removal of the applied mechanical strain to form wrinkle patterns, and replica molding of the deformed features on the porous membrane into epoxy to form hierarchical architectures through casting the UV-curable epoxy prepolymer and UV curing. We find that due to the existence of micropores on the membrane, the formation of wrinkle patterns is different from that formed on a continuous elastomeric film, and by varying the applied mechanical stretching strain condition and plasma oxidation condition, the wrinkle patterns could be either confined by the micropores on the membrane to exhibit a wavelength equal to its pitch or form wrinkles with much large wavelength compared with that formed on a continuous elastomeric film. Therefore, the micropillar arrays fabricated by replica molding could stand on different types of wrinkle patterns to form different hierarchical architectures. The method we illustrate here offers a simple and cost-effective approach to fabricate various hierarchical structures, and provides possibilities for potential applications in various fields, such as microfluidics, micro- and nanofabrication of complex structures, crystal formation, cell attachment, superhydrophobicity and dry adhesion.

INTRODUCTION

It is well known that surface properties, such as hydrophobicity and adhesion, can be manipulated by both physical roughness (topography) and chemical heterogeneity. Nature offers excellent examples of hierarchical architecture with multi-faceted functions. Superhydrophocity found in aquatic plants, notably Lotus leaves, [2,3] and insect wings [4], and adhesion found in gecko's toe pads [5, 6] have been attributed to their micro- and nano-rough hierarchical architectures. Inspired by the bioorganisms, the control of interfacial properties have been affecting to human's daily life and scientific research. By mimic Nature, many researchers have studied wettability, adhesion, friction, biocompatibility, and etc and applied them for surface coating to make the surface to be protective, self-cleaning, fouling-releasing, antifogging, anti-static, antifriction, etc.

Patterning of soft materials provides an exceptional route for the generation of nano- to micro- scale hierarchical structures. Non-photolithographic methods have attracted great attentions not only due to their simplicity, low cost, versatility, and surface three-dimensionality. [7] Fabrication of nano-/micro- structures with high aspect ratios (ratio of height to lateral feature size) with defined size and shape, so-called nano-/micro- pillars, have attract increasing due to the advantages of large surface area, high linearity in mechanical transduction response, and etc. [8] Poly(dimethylsiloxane) (PDMS) which provides high fidelity replication, optical transparency, and widely suitable surfaces and thus having become ubiquitous in soft. Besides, PDMS could be easily tuned to form complex structures including the wrinkle patterns, resulting from a buckling instability of a stiff layer supported by an elastomeric membrane, induced by surface oxidation of a pre-stretched elastomer followed by removal of the applied mechanical strain. [9, 11]

Here, we report the formation of hierarchical architectures with confined wrinkle patterns through mechanical stretching and replica molding. The method offers a simple and cost-effective approach to fabricate various hierarchical structures, and provides possibilities for potential applications in various fields, such as microfluidics, micro- and nanofabrication of complex structures, crystal formation, cell attachment, superhydrophobicity and dry adhesion.

EXPERIMENT

Scanning electron microscopy (SEM) images were obtained on FEI Strata DB235 Focused Ion Beam under voltage of 5 keV. A spot size of 3 was chosen for high resolution and detailed topographical information. All samples were coated with a thin layer of platinum. The dimensions of micropillars were measured using the preinstalled software on SEM. The oxygen plasma process was performed with technics etcher. The polymerizations of epoxy replica were achieved with a UV lamp (UVP Black-Ray). The stretching stage was home made.

Poly(dimethylsiloxane) (PDMS, Sylgard 184) and its curing agent, epoxy resin DER 354, and cycracure UVI 6976 were purchased from Dow Corning (USA). Master of micropillar arrays was kindly provided.

The PDMS prepolymer and curing agent were thoroughly mixed with a weight ratio of 10:1, followed by degassing for 1 h to remove the air bubbles. The liquid PDMS prepolymer mixture was poured over the micropillar arrays of the epoxy master and cured at 65 °C for overnight, followed by carefully peeling off from the master to obtain the porous PDMS membrane.

The continuous PDMS film or porous PDMS membrane was clamped and stretched by 20% with a uniaxial stretching strain, followed by plasma oxidation for different periods at a power of 200 W and an oxygen pressure of 0.1 Torr. After the oxidation process, the strain was slowly fully released, and the continuous PDMS film or porous PDMS membrane contracted and placed the silicate layer under compressive stress to form various wrinkle patterns.

The epoxy resin prepolymer solution was prepared by mixing epoxy resin DER 354 and cycracure UVI 6976 as cationic photoinitiators (w/w 97:3), followed by the incubation in dark overnight for removing the trapped air bubbles during mixing. The air bubble-free prepolymer solution was poured over the deformed PDMS continuous film or porous membrane and allowed to sit around 2 min to ensure the complete filling of the microporous PDMS mold, which was then exposed to UV exposure for 1 h. After curing, the polymeric micropillar arrays were carefully peeled off from the PDMS membrane.

RESULTS AND DISCUSSION

As shown in figure 1, the soft polymer, poly(dimethylsiloxane) (PDMS) porous membrane (figure 1a) is clamped and stretched by uniaxial stretching strain (figure 1b), followed by plasma oxidation to form a thin and stiff silicate layer on its surface (figure 1c). When the strain is released, the PDMS mold contracts and places the silicate layer under compressive stress to form wrinkles (figure 1d), and by varying the applied mechanical stretching strain condition and plasma oxidation condition, we can form wrinkles with various shapes on the porous PDMS membrane (figure1e). Through casting UV-curable prepolymer on the deformed PDMS mold and UV curing (figure 1f), micropillar arrays could be fabricated (figure 1g). Depending on the angle between the stretching strain and the micropore lattice axis of the porous membrane (figure 1h), various structures of micropillar arrays with wrinkle patterns could be fabricated, including individual pillar standing on the peak of the wrinkle in parallel lattice structure (figure 1i), or several pillar arrays standing on one wrinkle with the direction of individual pillars varying in a sinusoidal wave (figure 1j).

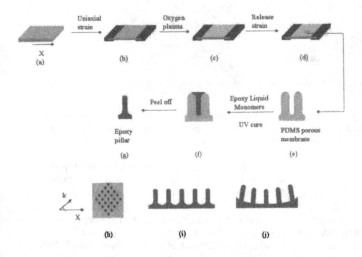

Figure 1. Schematic illustration for the fabrication of high-aspect-ratio polymeric micropillar arrays with wrinkle structures (a) porous PDMS membrane (b) stretched porous PDMS membrane (c) plasma oxidation on the stretched porous PDMS membrane(d) porous PDMS membrane with released strains (e) a typical pore on the PDMS membrane (f) PDMS pores with casted UV curable epoxy resin (g) epoxy micropillar (h) angle between the direction of

mechanical stretching strain and the lattice axis of the porous PDMS membrane (i) micropillar arrays with wrinkle patterns formed between the nearby micropillar arrays (j) micropillar arrays with wrinkle patterns in a large wavelength.

The conditions before the processes of mechanical stretching and oxygen plasma were studied. The dimensions of the pillars were: square pillars, $d = 0.8$ μm, $h = 2$ μm, $p = 2$ μm, and area density of 2.5×10^7 pillars per cm^2. In a large area, uniform micropillar arrays were formed by using soft lithography for fabrication. The pillars vertically stand on the substrate, and parallel to each other, and four nearby pillars forms the square lattice. The plasma oxidation process formed a hard and thin layer on the surface of stretched PDMS film, and after releasing the strain, wrinkle patterns could be formed with a direction the same as that of the stretching strain. Since both plasma power and time contributed to the formation of wrinkle patters, herein, we fixed one factor and varied the other: the plasma power was fixed at 200 W, and the oxidation time was varied which resulted in different surface energies by increasing either Young's modulus or thickness that makes the oxide layer harder to bend, and all of them were than the pitch of the micropillar arrays (p = 2 μm). With a plasma oxidation time of 5 min, the wrinkle wavelength was 4.05 μm, and with a plasma oxidation time of 20 min, the wrinkle wavelength was 4.34 μm, and the wavelengths of wrinkle patterns almost became saturated due to the saturation of Young's modulus and hard layer thickness.

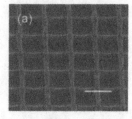 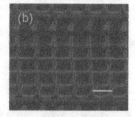

Figure 2. SEM images of micropillar arrays with confined wrinkle patterns by using porous PDMS membrane with 0 ° angled strain and plasma oxidation time of (a) 5 min (b) 20 min. Plasma power: 200 W. Oxygen pressure: 0.1 Torr. Stretching strain: 20%. Scale bar: 2 μm.

Figure 2 shows the two typical SEM images of micropillar arrays with wrinkle patterns by using a uniaxial stretching strain with a 0 ° angled direction to the lattice axis of PDMS micropore arrays. A shown in the figure, the direction of the wrinkles was the same as that of the stretching strains with a direction angle of 0 ° to the lattice axis of micropillar arrays. When the time plasma oxidation was 5 min on the porous PDMS membrane, the elastic energy was not high enough to form wrinkles with large wavelengths, and all the wrinkles were confined between the nearby micropillar arrays owing to the existence of the micropores on the PDMS membrane. Compared with the epoxy wrinkle patterns formed by using a continuous PDMS film, which showed quite larger wavelengths by using the same periods of plasma oxidation, the wavelengths of the epoxy wrinkle patterns formed by using the porous PDMS membrane was equal to the pitch of micropillar arrays (p = 2 μm), as shown in figure 2a. While as the plasma oxidation time was 20 min which is long enough, there were types of wrinkles patterns formed, as shown in figure 2b. One type of the wrinkles was formed between the nearby micropillar arrays with a wavelength equal to the pitch of the micropillar arrays. The other type of wrinkles was formed with a larger wavelength of around 13.3 μm, which was much larger than that formed by using a continuous PDMS film, to make several arrays of micropillars standing on it, and the structures of the wrinkles were a sinusoidal wave. With plasma oxidation time of 5 min (figure 2a), the fabricated micropillars stood vertically on their substrates and parallel to each other due to their standing on the same positions of the wrinkles. With a plasma oxidation time of 20 min (figure 2b), the micropillar arrays stood vertical on the substrate and due to the structure of substrate being in a sinusoidal wave, the direction of individual pillars varied in a sinusoidal wave. Besides, due to the deformation of the micropores of the PDMS membrane, the dimensions of micropillars were shrinked from a diameter of 0.8 μm to 0.4 μm, which were much thinner than the original micropillar arrays, and it was often observed that the heads of the micropillars were larger than the foots of the micropillars, which meant the deformation on cylinder pores of PDMS membrane were not uniform.

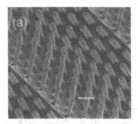 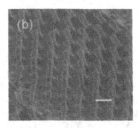

Figure 3. SEM images of micropillar arrays with confined wrinkle patterns by using porous PDMS membrane with 45 ° angled strain and plasma oxidation time of (a) 5 min (b) 20 min. Plasma power: 200 W. Oxygen pressure: 0.1 Torr. Stretching strain: 20%. Scale bar: 2 μm.

Figure 3 shows two typical SEM images of micropillar arrays with wrinkle patterns by using a uniaxial stretching strain with a 45 ° angled direction to the lattice axis of PDMS micropore arrays. As shown in the figure, the wrinkles were formed in the same directions as that of the stretching strains which also had a direction angle of 45 ° to that of the PDMS mold lattice axis. Due to the existence of the degree between the stretching strain and lattice axis of PDMS hole arrays, the micropores on the PDMS membrane could not confine the wrinkle patterns, which were not the same as the tendency for wrinkle formation on the porous PDMS membrane with 0 ° angled stretching strain. Therefore, there were wrinkle patterns formed with wavelengths larger than the pitch of micropillar arrays. With the plasma oxidation time increasing from 5 min to 20 min, the wavelengths of the wrinkles increased slightly. Besides, due to the deformation of the micropores of the PDMS membrane, the dimensions of micropillars showed the same tendency as that of a porous PDMS membrane with a 0 ° angled stretching strain, and the diameters were shrinked from 0.8 μm to 0.4 μm, which were much thinner than the original micropillar arrays, and it was also observed that the heads of the micropillars were larger than the foots of the micropillars, which meant the deformation on cylinder pores of PDMS membrane were not uniform.

CONCLUSIONS

In this work, we have demonstrated the formation of wrinkle patterns on porous elastomeric membrane and their fabrication of hierarchical architectures through replica molding and mechanical stretching. The structural details can be affected by the plasma oxidation condition, types of stretching strains, angles between stretching strain and hole lattice axis. The method we illustrate here could be used for the study of surface properties, such as wetting, adhesion, frication, biocompatibility, etc. Besides, it could also be used for fabrication of hierarchical architectures with various dimensions and topographies by using different stretching or releasing strains. Moreover, it provides new possibilities for the fabrication of periodic structures in other materials which may have potential applications in photonics, semiconductor, etc.

REFERENCES

1. Baumgärtner, A.; Muthukumar M. *J. Chem. Phys.* 1991, *94*, 4062.
2. Barthlott, W.; Neinhuis, C. *Planta* 1997, *202*, 1.
3. Neinhuis, C.; Barthlott. W. *Ann. Bot.* 1997, *79*, 667.
4. Wagner, T.; Neinhuis, C.; Barthlott, W. *Acta Zoologica* 1996, *77*, 213.
5. Hansen, W. R.; Autumn, K. *Proc. Nat. Acad.Sci. USA* 2005, *102*, 385.
6. Autumn, K.; Sitti, M.; Liang, Y. C. A.; Peattie, A. M.; Hansen, W. R.; Sponberg, S.; Kenny, T. W.; Fearing, R.; Israelachvili J. N.; Full, R. J. Proc. Nat. Acad. Sci. USA 2002, 99, 12252.
7. Y. Xia and G. M. Whitesides, Angew. Chem., Int. Ed., 1998, 37, 550.
8. Zhang, Y.; Lo, C.W.; Taylor, J.A.; Yang, S. Langmuir 2006, 22, 8595.
9. P. Lin and S. Yang, *Appl. Phys. Lett.* 2007, 90, 241903.
10. P. Lin, etc., *Soft. Matt.* 2008, In press.
11. J.Y. Chung, etc. *Soft. Matt.* 2007, 3, 1163.

Mater. Res. Soc. Symp. Proc. Vol. 1129 © 2009 Materials Research Society   1129-V11-12

Cross Linkable Phosphorescent Polymer-Leds Based on a Soluble New Iridium Complex ((Btp)2Ir(III)(acac)-Ox)

Young-Jun Yu[1, *], Olga Solomeshch[1], Vladislav Medvedev[1,] Alexey Razin[1], Nir Tessler[1], Kyu-Nam Kim[2], Dong Hoon Choi[2], Jung-Il Jin[2, †]

[1] Microelectronic & Nanoelectronic Centers, Electrical Engineering Department, Technion, Haifa 32000, Israel
[2] Department of Chemistry and Center for Advanced Materials Chemistry, Korea University, Seoul 136-701, Korea

ABSTRACT

PVK is a prototypical host for polymer based phosphorescent OLEDs. We report the photo- (PL) and electroluminescence (EL) properties of phosphorescent organic light emitting diodes (OLEDs) based on poly(vinylcarbazole) (PVK) derivatives doped with cyclometalated Ir(III) complexes. To allow for stable morphology as well as photopatternability, both the host and guest materials were rendered cross-linkable by attaching a cross linking oxetane moiety. Regarding the PVK matrix we also used a new derivative (PVK-Cin) of poly(vinylcarbazole) (PVK) that was prepared by tethering cinnamate pendants to the carbazole group through an octylene spacer.
While the use of the oxcetane moieties attached to the ligands of the Ir complex slightly improves the luminescence efficiency, the use of oxcetane derivatized PVK and the associated catalyst result in poor performance. In contrast, the use of the (catalyst free) cinnamate derivatized PVK results in high efficiency cross-linked PLEDs with efficiency higher than 4.5 cd/A.
Photopatternability of the new PVK based-polymer was investigated through photocrosslinking reaction under UV light illumination (= 254 nm). The blends of PVK-Cin and a soluble (Btp)2Ir(III)(acac)Ox were employed for studying a photocrosslinking behavior. We observed that PVK-Cin readily undergoes $2\pi + 2\pi$ cycloaddition when exposed to the UV light ($\lambda_{max} = 254$ nm) at 110 °C without photoinitiator.

INTRODUCTION

Recently, there have been a number of studies reported on organic light-emitting devices (OLEDs) using phosphorescent emitting materials.[1-5] The efficiency of organic light-emitting diodes (OLEDs) have been dramatically improved by the use of host materials mixed with Ir(III) phosphorescent emitters.[6-9] Polymeric OLEDs offer significant advantages as they can be directly processed from solution by spin casting, ink-jet printing and other techniques.
Until now, the investigations of various phosphorescent dyes in polymer based LEDs have been performed, since it can be easily fabricated to results in efficient electro-phosphorescence at low concentration of phosphorescent dye. Carbazole derivatives and polymers are the most popular host for the phosphorescent organometallic complexes.[10, 11] Their favorable electronic levels for efficient energy transfer to the complexes and good miscibility preventing the formation of aggregates of light emitting compounds appear to be the most important factors for their

selection. In order to achieve molecular dispersity of the complex in the host matrix, one can attach the phosphorescent organometallic moieties onto polymer chains as pendants.[12]
In this work we study two crosslinkable host polymers, PVK-Cin (Poly[(1-(9-carbazole)ethylene)-co-(3-cinnamoyloxyoctyl-9-carbazolyl)ethylene) and PVK-Ox (Poly[9-ethyl-3-(3-methyl-oxetan-3-ylmethoxy)-decyl]-9-carbazole): (PVK-Ox).[13, 15]

EXPERIMENT

Reagents and Synthesis

All reagents and solvents were commercially purchased from Aldrich and were used as supplied. PEDOT:PSS was Baytron P Purchased from H. C. Starck. Poly-TPD and BCP were purchased from American Dye Sources and Aldrich Chem. Co., and the PVK matrix we also used a new derivative (PVK-Cin) of poly(vinylcarbazole) (PVK) that was prepared by tethering cinnamate pendants to the carbazole group through an octylene spacer.[16]
The synthesis of the Btp$_2$Ir(acac)Ox (14-(3-Methyl-oxetan-3-ylmethoxy)-tetradecane-2,4-diol) is outlined in Figure 1. The chloride-bridged dimmer (0.50 g, 0. 42 mmol) of, and silver trifloromethansulfonate (2.15 g, 0.84 mmol) were mixed in 2-ethoxyethanol and added by transfer addition of a solution of 14-(3-Methyl-oxetan-3-ylmethoxy)-tetradecane-2,4-dione (6.56 g, 1.68 mmol) in 2-ethoxyethanol and the solution refluxed under Argon atmosphere for 6 h. The reaction mixture was cooled to room temperature, and the solution was filtered with water and ethylacetate then organic layer was separated. The organic layer was dried with MgSO$_4$. The solvent was removed in vacuum and the residue was purified by silica-gel chromatography (ethyl acetate: hexane 1:10). The product was isolated as red compound. The impure product was recrystallized from hexane and methanol. This procedure was repeated 3 times (0.18 g, yield 45 %). The chemical structures of the hosts and guest were shown in Figure 1.

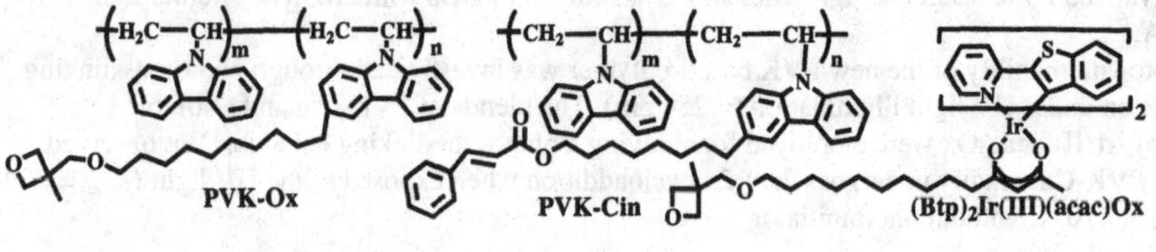

Figure 1. The chemical structures of the hosts and guest.

Thin Films Preparation

All the following steps were performed inside an inert glove box. To make electron blocking layer as a followed: C60F36 was dissolved in mono-chlorobenzene and then mixed with PVK-Cin:Poly-TPD. The solution was filtered through a 5 μm pore size PTFE filter. All the following steps were carried out inside inert glove box (~1ppm oxygen & water). Electron blocking layers (EBL) were prepared with the following composition: PVK-Cin: Poly-TPD: C60F36, 45: 45: 10 W%. The mixed solution was stirred for 3 hr and the solution was spin coated at 5000 rpm on UV/ozone treated ITO substrate. The PVK-Cin/Poly-TPD/C60F36 coated substrate was crosslinked at 110 °C for 10 min using 254 nm UV. After that, the EBL was washed with THF for 1min. The emitting layer (EML) materials were spin coated on top of EBL layer and washed with THF for 1 min. The typical thickness of the emitting layer was around 40 nm, as measured with an Alfa step 350 surface profiler. The UV-Vis absorption spectra and PL of the blend films were dissolved in monochlorobenzene, which solution of the blending polymer was spin cast onto quartz substrate at 4000 rpm under nitrogen conditions, resulting in a 70 nm thick film. The LEDs were prepared on patterned glass/ITO substrates purchased from Psiotec Ltd.. Top contacts were 50 nm of BCP and 50 nm of the Ca followed by 150 nm of Al both evaporated at 0.1 nm/s and at a pressure of ~5 × 10^{-7} Torr.

DISCUSSION

Photoluminescence Spectra of Blends

Figs.2 shows the UV-vis absorption and photoluminescence (PL) spectra of Btp$_2$Ir(acac)Ox and Figs.4 shows it for the mixtures of Btp$_2$Ir(acac)Ox with the two different hosts, PVK-Ox, and PVK-Cin. As the figures show, Btp$_2$Ir(acac)Ox is a red emitter with absorption that overlaps the PL emission of the PVK hosts.

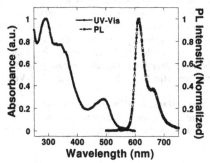

Figure 2. UV-Vis absorption and PL of the pure Btp$_2$Ir(acac)Ox; The circle symbol is UV-Vis absorption of the Btp$_2$Ir(acac)Ox and the square symbol is PL spectrum (~50 nm film thickness).

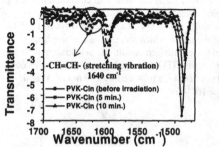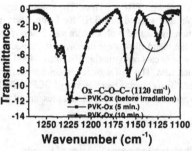

Figure 3. The a) and b) were IR spectra of PVK-Cin and PVK-Ox after crossilinking reaction ether by irradiated with UV-light (λ_{max} = 254 nm wavelength) for 10 min and by washed to use THF.

Figure 3 compares the IR-spectra of the PVK-Cin (a)) and PVK-Ox (b)) after irradiation at 110 °C from 5 to 10 minutes. The most noticeable change after crosslinking (10 min.) of the PVK-Cin and PVK-Ox were the appearance of a week absorption peak at about 1640 cm^{-1} and 1120 cm^{-1} arisen from the –CH=CH– stretching vibrational mode and Ox–C–O–C– stretching vibrational mode.[17] We also studied that is not the only optical micrograph and scanning electron micrograph (SEM) of the patterned images but also the general properties used to crosslinkable polymers formed by UV irradiation which is already published.[13,15] Moreover, the Fig.4 shows that already for low concentration of dopant there is a strong red emission indicating the Forester transfer mediated through the spectral overlap between the PL of the polymers and UV-Vis absorption of the Btp$_2$Ir(acac)Ox.

Fig. 4 also suggests that PVK-Cin is a better cross-linkable host[18] which we attribute to two effects: a) the PI used for PVK-Ox cross-linking reaction quenches the luminescence b) the energy transfer appears to be less efficient (this could also be due to the PI but would required further study).

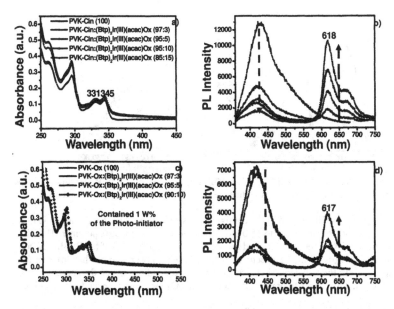

Figure 4. The a) and B) were UV-Vis and PL spectra of PVK-Cin + Btp$_2$Ir(acac)Ox blend films after crossilinking reaction ether by irradiated with UV-light (λ_{max} = 254 nm wavelength) for 10 min and by washed to use THF. The c) and d) were UV-Vis and PL spectra of PVK-Ox + Btp$_2$Ir(acac)Ox with 1 wt% photo initiator blend films after crossilinking reaction ether by irradiated with UV-light (λ_{max} = 254 nm wavelength) for 10 min and by washed to use THF.

EL Performance of Blend Films

We constructed electroluminescence(EL) devices having the following four configurations: A; ITO/PEDOT-PSS (100 nm)/EBL (50 nm)/PVK-Cin + Btp$_2$Ir(acac)Ox (40-50 nm)/ BCP (50 nm)/ Ca (50 nm)/ Al (150 nm), B; ITO/PEDOT-PSS (100 nm)/EBL (50 nm)/PVK-Ox + Btp$_2$Ir(acac)Ox (40-50 nm)/ BCP (50 nm)/ Ca (50 nm)/ Al (150 nm), C; ITO/PEDOT-PSS (100 nm)/EBL (50 nm)/ PVK-Cin (10 nm) (HTL)/PVK-Cin + Btp$_2$Ir(acac)Ox (40-50 nm)/ BCP (50 nm)/ Ca (50 nm)/ Al (150 nm), and D; ITO/PEDOT-PSS (100 nm)/EBL (50 nm)/ PVK-Cin (10 nm) (HTL)/ PVK-Ox + Btp$_2$Ir(acac)Ox (40-50 nm)/ BCP (50 nm)/ Ca (50 nm)/ Al (150 nm).

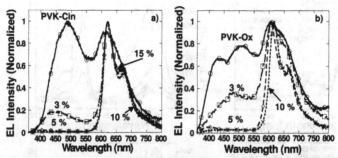

Figure 5. EL Spectra of the PVK-Cin/ Btp$_2$Ir(acac)Ox in structure A (a);, and PVK-Ox/ Btp$_2$Ir(acac)Ox blend films contained 1 wt% of the photo-initiator (PI) in structure B (b). All were measured at 12 V.

Figure 5 a) shows the EL spectra of the configuration structure A where the doping ratio of the red iridium complex was from 3.0 to 15.0 mol% relative to the host matrix polymer. The EL spectra of the PVK-Ox blend films doped from 3.0 mol% to 10.0 mol% of the Btp$_2$Ir(acac)Ox are shown in Fig. 5 b). All the EL spectra of hosts and blend films are somewhat different from their respective PL. First, the emission at 10 % is purely red indicating that the energy transfer (Fig. 4) is assisted by charge trapping at the Btp$_2$Ir(acac)Ox (see energy level diagram in Fig. 6).

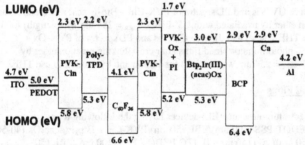

Figure 6. Energy Levels.

Second, the pristine film based LEDs exhibit strong additional emission bands which are typically attributed to exciplex emission occurring at the interface between hole transporting molecules (as PVK, PVK-Cin. PVK-Ox, or poly-TPD) and electron transporting molecules (as BCP). [19-21] Adding the Btp$_2$Ir(acac)Ox seems to drain the energy not only from the host-bulk but also from the exciplexes formed at the interface. Since the energy of the exciplex emission is relatively low we attribute this apparent energy transfer to a shift of the recombination zone from the BCP hole blocking layer interface where the exciplexes are localized. This shift seems to be more effective in the case of PVK-Cin and is probably due to a better spread of the Btp$_2$Ir(acac)Ox in it.

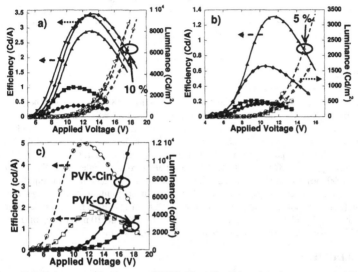

Figure 7. Efficiency and luminance of PVK-Cin + $Btp_2Ir(acac)Ox$ in structure A (a)), the PVK-Ox + $Btp_2Ir(acac)Ox$ is type of the structure B shown in (b)), and c) the dashed line and empty symbols denote the efficiency of the multi layer devices and the dashed line and full symbols are for luminance of the multi layer device, at 10 mol% and 5 mol% doped into hosts, respectively.

We have recently found[22] that the fluorinated derivative of C60 (C60F36) acts as p-type dopant when added to the polymer blend of PVK-Cin with Poly-TPD. The use of such doped electron blocking layer (EBL) enabled the fabrication of hybrid LEDs having improved turn-on voltage and brightness. In this paper, we have used the same three layer LED structure (structures A&B) and studied also the effect of capping the doped EBL with thin (~10 nm) undoped PVK-Cin (structures C&D). All solution processed layers were crosslinked and tested for their efficiency and luminance (see Figure 7). Fig. 7 a) shows the efficiency and luminance as a function of applied voltage for three layer LEDs at a various concentration from 3.0 to 15 mol% of $Btp_2Ir(acac)Ox$ in PVK-Cin matrix, and Fig. 7 b) shows the same for PVK-Ox + $Btp_2Ir(acac)Ox$. The diamond symbols denote the efficiency and luminance for the EML containing of 10 % $Btp_2Ir(acac)Ox$ doped into PVK-Cin, the maximum efficiency and luminance to reach around 3.5 cd/A and 9100 cd/m², respectively. However, the luminance begins to decrease for $Btp_2Ir(acac)Ox$ concentrations larger than 10 mol% in PVK-Cin and 5 mol% in PVK-Ox, indicating that a large concentration of red iridium would result in phase separation and aggregation dots in a matrixes.
Figure 7c) shows the results for the LED where the EBL was capped by ~10 nm PVK-Cin. Here the dash and solid lines are the efficiency and luminance, respectively. The circle symbols present the efficiency and luminance for the multi layer hybrid LED of 10 % $Btp_2Ir(acac)Ox$ doped into PVK-Cin and square symbols are the 5 % doped into PVK-Ox, respectively.

Examining the LEDs performance we note that adding the thin capping layer has improved both the turn-on voltage and the efficiency of the devices. In other cases (not shown here) where different dopant molecules were used we found the capping layer to have no effect at all. Hence, we attribute this effect to a specific interaction between the $Btp_2Ir(acac)Ox$ and the EBL causing degradation of the $Btp_2Ir(acac)Ox$ near the EBL interface.

CONCLUSIONS

In summary, we report a study of the photo crosslinking PVK-Cin blend with new crosslinkable $Btp_2Ir(acac)Ox$ phosphorescent molecule. We have investigated the synthesis, characterization and the device performance of the $Btp_2Ir(acac)Ox$ and have shown it can be successfully used in crosslinkable host polymers. We found, in the case of LEDs, that the high red emission from the red dopant is due to both energy and charge transfer. One of the important parameters enabling good performance LEDs was to develop cross-linkable HTM (Poly-TPD, PVK-Cin, and C60F36) and EML that could be crosslinked without photo initiator. Finally, we found that a specific interaction between the In addition, we make multiple layer devices to used thin film (hole transporting layer) of the $Btp_2Ir(acac)Ox$ and the EBL may degrade device performance and that by adding a thin intermediate layer of PVK-Cin (10 nm) on top of the electron blocking layer can recover the improved device performance.

ACKNOWLEDGMENTS

We acknowledge support by the Israel Ministry of Science & Technology, The Israel Science Foundation. Young-Jun Yu was supported by the Korea Research Foundation Grant funded by the Korean Government (MOEHRD).

REFERENCES

[1]. I. D. Parker and Q. Pei, *Appl. Phys. Lett.*, **65**, 1272 (1994).
[2]. Q. Huang, G. A. Evmenenko, P. Dutta, N. R. Armstrong, and T. J. Marks, *J. Am. Chem. Soc.*, **127**, 10227 (2005).
[3]. Y. Xin, G. A. Wen, Y. Cao, and W. Huang, *Macromolecules*, **38**, 6755 (2005).
[4]. S. J. Su, H. Sasabe, T. Takeda, and J. Kido, *Chem. Mater.*, **20**, 1691 (2008).
[5]. J. Kido, Y. Okamoto, *Chem. Rev.*, **102**, 2357 (2002).
[6]. A. B. Tamayo, B. D. Alleyne, M. E. Thompson, *J. Am. Chem. Soc.*, **125**, 7377 (2003).
[7]. W. Zhu,Y. Mo, and Y. Cao, *Appl. Phys. Lett.*, **80**, 2045 (2002).
[8]. M. C. DeRosa, D. J. Hodgson, and G. D. Enfight, *J. Am. Chem. Soc.*, **126**, 7619 (2004).
[9]. A. Hameed, A. Attar, Martin R. Bryce, *Appl. Phys. Lett.*, **86**, 121101 (2005).
[10]. X. Gong, J. C. Ostrowski, D. Moses, G. C. Bazan, A. J. Heeger, *Adv. Funct. Mater.*, **13**, 439 (2003).
[11]. W.-Y. Lai, Q.-Y. He, Q.-Q. Chen, and W. Huang, *Adv. Funct. Mater.*, **18**, 265 (2008).
[12]. X. Chen, J.-L. Liao, Y. Liang, S.-A. Chen, *J. Am. Chem. Soc.*, **125**, 636 (2003).
[13]. Y.-J. Yu, N. Tessler, and J.-I. Jin, *Macromol. Res.*, **15**, 142 (2007).

[14]. O. Solomeshch, V. Medvedev, P. R. Mackie, D. Cupertino, A. Razin, and N. Tessler, *Adv. Funct. Mater.*, **16**, 2095 (2006).

[15]. O. Solomeshech, Y.-J. Yu, and N. Tessler, *Synth. Met.*, **157**, 841 (2007).

[16]. F. Huang, Y.-J. Cheng, Y. Zhang, M. S. Liu, and K.-Y. Jen, *J. Mater. Chem.*, **18**, 4495 (2008).

[17]. A. Harrane, N. Naar, *Materials Letters*, **61**, 3555 (2007).

[18]. J. Chang, Y.-J. Yu, J. H. Na, J. An, D. H. Choi, J.-I. Jin, and C. Im, *J. Poly. Sci. Part B: Poly. Phys.*, **46**, 2395 (2008).

[19]. G. Mauthner, M. Collon, E. J. W. List, F. P. Wenzl, M. Bouguettaya, and J. R. Reynolds, *J. Appl. Phys.*, **97**, 063508 (2005).

[20]. X. H. Yang, and D. Neher, *Appl. Phys. Lett.*, **14**, 2476 (2004).

[21]. C.-L. Ho, W.-Y. W, Z.-Q. Gao, C.-H. Chen, and Z. Lin, *Adv. Funct. Mater.*, **18**, 319 (2008).

[22]. Y.-J. Yu, O. Solomesch, H. Chechnik, A. A. Goryunkov, R. F. Tuktarov, D. H. Choi, J.-I. Jin, and N. Tessler, *J. Appl. Phys.*, **104**, 124505 (2008).

Mater. Res. Soc. Symp. Proc. Vol. 1129 © 2009 Materials Research Society　　　　1129-V06-05

Variable Thickness IPMC: Capacitance Effect on Energy Harvesting

Rashi Tiwari, Sang-Mun Kim and Kwang J. Kim
Active Material Processing Laboratory, Mechanical Engineering Department, University of
Nevada, Reno, NV 89511, U.S.A.

ABSTRACT

Ionic Polymer Metal Composites (IPMCs) are manufactured by electroless deposition of metal on Nafion. This deposition method results in the IPMCs with thickness between 0.17mm to 0.20mm with the electrode thickness of around a few μm each. It is now generally accepted that on mechanical deformation IPMC produces charge thus making these materials potentially suitable for energy harvesting applications. Due to thin metal plating and inherited flexibility of the Nafion film the IPMCs suffer in stiffness that may be required for some energy harvesting applications. Also earlier works have shown that 0.20mm thick IPMC produce better battery charging than 0.17mm thick one. Hot pressing, using metal mold, Nafion films was employed to produce thicker and comparatively stiffer IPMCs electroded with Palladium metal. Palladium was used because of shorter manufacturing time. This IPMC shows improved energy harvesting. Due to the increased thickness these IPMCs also function as better capacitors than their conventional counterparts. On application of voltage, these IPMCs show charging and discharging effects of a capacitor. This property of IPMC may be useful for storing charge.

INTRODUCTION

Power source for integrated electronics has been a growing concern with the ever-increasing use of portable devices. Recent advancements in the smart materials have led scientists to explore the application of these materials for energy harvesting applications.

Ambient energy harvesting has many emerging applications in structural health monitoring, self-assembling devices, pacemakers, wireless devices and self-sustaining senor and actuators. IPMCs have previously not been given much attention for the same [1]. This provides a unique opportunity to characterize electromechanical property of IPMCs. Ionic polymers offer several advantages as energy harvester. Since the technology is comparatively new, there is potential for advancement as electromechanical transducers. Ionic polymers can sustain large strains making them ideal for the high stress environment [2]. High stress capabilities in turn allow the polymers to have a large stroke. Inherited flexibility, chemical stability and long life makes IPMC suitable for energy harvesting even in harsh environment [3]. Despite of these advantages, due to low output charge and time variant characteristics these are restricted in their application. The focus of this study is to increase the charge output from the IPMC.

IPMC samples are typically manufactured from Nafion® 117 [4] as shown in Figure 1. The membrane is hydrated in D.I water for 24hrs and cleaned in water bath. Cleaned sample is then immersed in a metal salt solution for couple of hours. The reduction process using ammonia and sodium borohydride follows, which are put every half hour with temperature increasing from 40^0C to 70^0C. Reduction process forms electrode layer to form on the surface of the membrane. Adsorption and reduction process is repeated till appropriate surface resistance is achieved.

It should be noted that the thickness of the samples hence produced is limited due to the available Nafion® membrane. This deposition method results in the IPMCs with thickness between 0.17mm to 0.20mm with the electrode thickness of around a few μm each.

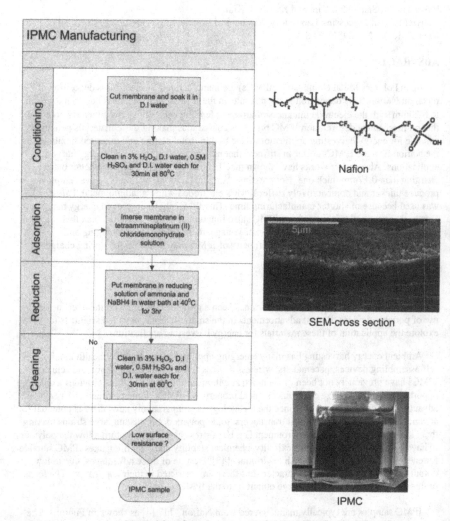

Fig. 1: Manufacturing process of typical IPMC.

Vibration based energy harvesters have gained momentum due to the advancements in wireless technology. Most of the wireless networks require power in micro to mille watts range [5] while most of the ambient energy sources generate vibration below 200 Hz [6]. Figure 2 compares power generated from vibration with solar and batteries. Vibration based energy harvesters show long life and good power generation capability in the required range. Vibration based energy harvesting devices mainly consists of three parts as shown in Figure 3: transducer that converts vibration into electrical energy, harvesting circuit to convert the generated charge into signal suitable for storage and storage device so that the generated energy may be stored for future use.

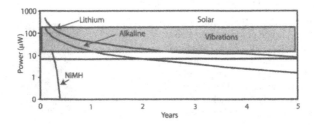

Fig. 2: Power generated from vibration, solar and batteries [5].

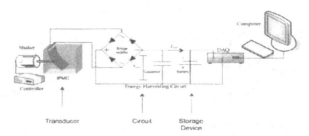

Fig. 3: Vibration based energy harvesting device.

Past research has focused on studying the impact of free-falling ball or vibration of small plate [7-8]. Energy available from leg motion of human and other motion has also been studied using both piezo-ceramics and piezo-film [9-10]. It has also been shown that the maximum efficiency occurs in a low frequency region (lower than the resonance) for PZT stacks [11]. Due to hysteresis of PZT the efficiency is also dependent on the amplitude of the input force. A $1cm^2$ piezo-ceramic wafer can power microwatt range MEMS device [12]. The feasibility of piezoelectric based energy harvesters for charging and re-charging battery has been studied [13]. Charging time and the amount of power generated impose a limitation on the application of these devices. In addition, low efficiency from the circuit and the material itself is a critical issue. One of the drawbacks of using piezo-ceramics is that it is brittle. The power source must be flexible, small, low maintenance and easy to integrate in the existing set-up.

SCALABILITY OF IPMC

Manufacturing

This research deals with the development of thicker IPMC from the polymer film readily available in the market. Nafion® 117 was cut into required dimensions and stacked into the mold with the Teflon film. For the purpose of the experiment three films were used. The mold was heated to the temperature of 180 °C for 20min with the Nafion® film. After 20min the press was pressurized to 50MPa at the same temperature as above for 10min followed by gradual cooling at room temperature. The film hence produced was cleaned with hydrogen peroxide (1 hour), sulfuric acid (1hour) and D.I. water (0.5 hour) in that order.

The cleaned thicker Nafion® film than conventional ones hence produced is coated with palladium metal as electrode. Palladium was used because of shorter manufacturing time and good sensing output. Length and width of the manufactured IPMC was 2.5cm and 1.25cm, respectively. The thickness achieved was 0.5mm. The sample is shown in Figure 4.

Fig. 4: An IPMC with 0.5 mm thickness.

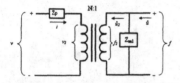

Fig. 5: Transformer circuit to represent the electro-mechanical model of IPMC.

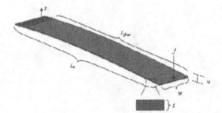

Fig. 6: Mechanical specification of IPMC.

Role of Thickness

Mechano-electric modeling helps in better understanding of the system and its response. IPMC was first represented as transformer circuit, such as one shown in Figure 5, by Newbury and Leo [14]. For the purpose of simplicity it is assumed that the electrical and the mechanical component of the system are coupled linearly. Applied mechanical force to the IPMC is represented by f and $\dot{u}$ is the velocity of application of force. The electrical domain consists of electrical voltage, v, and current, i, produced by IPMC. The mechanical specification of the IPMC sample is provided in Figure 6. The derived mechano-electric model [15] using the transformer circuit is:

$$\left\{ \begin{array}{c} v \\ f \end{array} \right\} = \left[\begin{array}{cc} (Z_p + Zm_1N^2) & Zm_1N \\ Zm_1N & Zm_1 \end{array} \right] \left\{ \begin{array}{c} i \\ \dot{u} \end{array} \right\} \tag{1}$$

The model helps in understanding macroscopic behavior. The mechano-electric model highlights thickness as one of the important characteristics. The charging of the battery increases with increase in thickness. This is also evident from the following equation for open circuit voltage:

$$\frac{v}{u} = \left(\frac{Yt^3}{2\varepsilon L_{free}^4} \right)^{0.5} \tag{2}$$

where Y and ε are Young's modulus and permittivity of the IPMC.

Capacitance

IPMC consists of electrode and polymer with interfacial layer in between. All three layers play an important role in energy harvesting. Single layer IPMC can be represented as an RC circuit as shown in Figure 7.

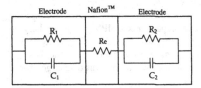

Fig. 7: Equivalent RC circuit model of IPMC.

Electrical component, Z_p (impedance), for a conventional IPMC can be given as:

$$Z_p = 1 / \sum_{i=1}^{n} \frac{R_i}{R_i C_i s + 1} + R_e \tag{3}$$

where,

$$C_{1,2} = \frac{\varepsilon_i A}{t} \tag{4}$$

where, ε_i is the dielectric constant of IPMC, A is the area of IPMC and t is the thickness of IPMC. The capacitance of IPMC is affected by the dielectric property of the material hence impedance of thick IPMC with respect to frequency of applied voltage was studies.

$$C(j\omega) = \frac{Q(j\omega)}{V(j\omega)} \tag{5}$$

From EIS (Electrochemical Impedance Spectroscopy) data we can get the real and imaginary component of capacitance. Hence,

$$C(j\omega) = C(j\omega) - jC(j\omega) \tag{6}$$

Imaginary component of capacitance represents the loss. Similar to capacitance, dielectric of the material may also be represented as:

$$\varepsilon(j\omega) = \varepsilon(j\omega) - j\varepsilon(j\omega) \qquad (7)$$

Thus

$$C(j\omega) = \frac{\varepsilon(j\omega)A}{d} \qquad (8)$$

Capacitance is related to the frequency and impedance as:

$$C = \frac{1}{j\omega Z(j\omega)} \qquad (9)$$

The capacitance of IPMC can be utilized in two ways. The higher the capacitance of IPMC better is the battery charging. Same capacitance may also be used as the storage device.

Experiments

EIS: Impedance was measured using potentiostat with three-electrode system in 1M H_2SO_4. The frequency range was from 100mHz to 100KHz.

Self-discharge: When voltage is applied thick IPMCs show charging and discharging effects of a capacitor. This property of IPMC may be used for storing energy in sensing applications. In order to determine electrical capacitance of IPMC, self-discharging experiment of IPMC was performed. Sampling frequency was 9.1743 Hz using PDaq 55/56. Thick IPMC sample dried in oven was charged with different DC voltages (1V, 2V and 3V) for 1 min. After charging the supply voltage was turned off and the sample was discharged for further minute.

Energy Harvesting: Set-up illustrated in Figure 8 is employed for energy harvesting using IPMC. Shaker assembly (TIRA Vibration System, TV52110) was excited using a sine wave (open-loop) signal with the amplitude of 1V and frequency of 10 Hz. The displacement of the shaker was adjusted at 0.5 inch for all the experiments. The clamp is used to capture the sensor signal of the IPMC and is transmitted to the energy harvesting circuit consisting of full bridge rectifier and storage device. The voltage stored on the 30 mAh rechargeable NiMH battery is connected to an IOTech Personal Daq 56 personal data acquisition system.

Fig. 8: Experimental set-up for energy harvesting.

218

RESULT & DISCUSSION

Capacitance

The result of the electrochemical test is shown in Figure 9. The impedance curve is related to the RC circuit. Capacitance measurement at fixed voltage of 1V and frequency changing from 100KHz to 100mHz with sample in 1M H_2SO_4 solution is shown in Figure 10. For the frequency below 2Hz, there is steep change in sample capacitance. Based on the EIS data it is acceptable to say that the capacitance of a 0.5 mm thick IPMC for DC voltage is around 25mF.

Charging and dis-charging of IPMC is simulated using SimEletronic in Matlab. The model is shown in Figure 11. The values of R and C in the circuit are derived from the impedance curve above. The simulation was performed for 3V charging. Initial voltage on the capacitor in the discharge model is set to 3V. Simulation result of the model is shown in Figure 12(a).

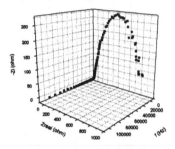

Fig. 9: Result of electrochemical analysis: EIS.

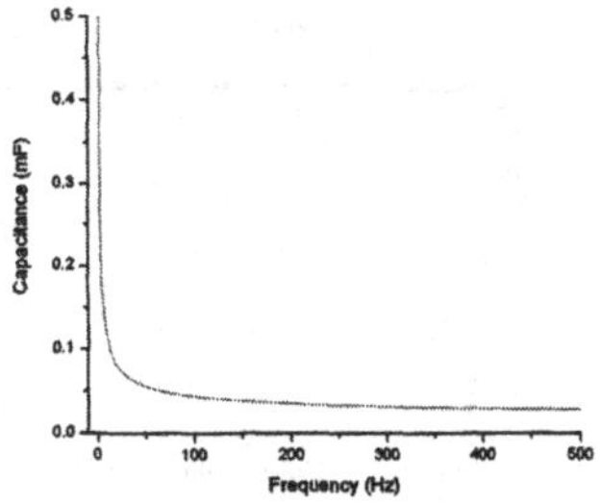

Fig. 10: Capacitance measurement in 1M H_2SO_4 solution.

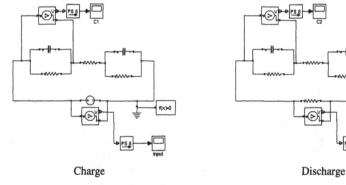

Charge   Discharge

Fig. 11: Model for simulating IPMC capacitive behavior.

219

Experiments were performed using a 0.5 mm thick IPMC sample. The result of experiments is shown in Figure 12(b). Further analysis is required from the discharge result to calculate capacitance of IPMC. Time constant (for the discharge cycle) is defines as the time it takes the capacitor to discharge to 37% of its initial value. Both simulation and experimental results show a time constant of 12 sec.

Method 1: Let the voltage on the capacitor, V_c, be represented using two exponential terms as:

$$V_c = V_0(A_1 e^{-B_1 t} + (1 - A_1)e^{-B_2 t}) \tag{10}$$

where, A_1, B_1 and B_2 are the coefficients that need to be calculated and V_0 is the initial voltage on the capacitor. The coefficient of the equation can be found by fitting the data using least square method in MATLAB as shown in Figure 13.

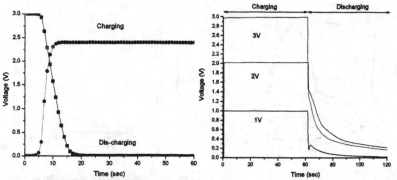

Fig. 12: (a) Simulation and (b) experimental results demonstrating IPMC capacitive behavior.

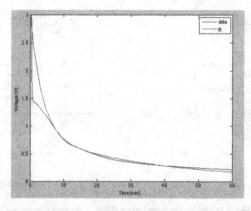

Fig. 13: Data fitting using least square method for coefficient determination.

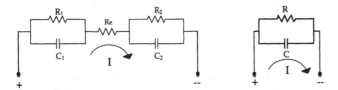

Fig. 14: Capacitor voltage during discharge.

The value of coefficient hence determined with initial voltage on the capacitor being 3V, are $A_1=0.25$, $B_1= 0.026$ and $B_2 = 0.272$. Voltage is applied across RC circuit shown in Figure 7. Using Kirchhoff's law in the mesh: The circuit may be modified as shown in Figure 14. Hence:

$$V_C = IR$$

$$q = CV \tag{11}$$

$$\frac{dq}{dt} = \frac{q}{RC}$$

The time constant of the circuit shown in Figure 14 is:

$$\tau = RC$$

$$R_1 = R_2 \tag{12}$$

$$C_1 = C_2$$

Time constant is:

$$\tau = \left(\frac{A_1}{B_1} + \frac{1-A_1}{B_2} \right) \tag{13}$$

Then:

$$RC = \left(\frac{A_1}{B_1} + \frac{1-A_1}{B_2} \right) \tag{14}$$

Substituting the values of the coefficients into Equation (14), the value of RC was found to be 0.0925F. From the EIS this value was found to be 0.025F.

Method 2: Another method to calculate capacitor is using Equation (4). The value of relative permittivity used is of the order 10^5. This value can be calculated for IPMC with capacitance of 0.02F having area 1cm by 5cm and thickness of 0.2mm. Capacitor value for thick IPMC was calculated to be 0.005F.

Method 3: Another method to calculate capacitor is using Equation (12). The value hence calculated was found to be 0.0821F.

The value calculated using second method is lower as compared to other methods. Due to the particulate nature of the electrode, the actual surface area is much higher. However, area term in Equation (4) does not take into account the actual surface area.

The value of capacitance is higher as compared to typical IPMC. This is due to the electrode property. IPMC electrodes are particulate in nature and hence the actual surface area is much higher. Due to the increased thickness the role of dielectric property of the material is much more as compared to typical IPMC.

Energy Harvesting

The model in Equation (2) was simulated using SIMULINK as shown in Figure 15. Simulation was performed for typical IPMC as well as thicker IPMC (0.5 mm thickness). The result of the simulation is shown in Figure 16. As expected, charging from thicker IPMC is much higher than typical IPMC. Simulation results were then compared with the experimental results as shown in Figure 16. In addition, to substantiate the results, energy harvesting using commercially available PVDF was performed at the similar conditions.

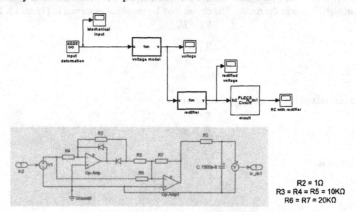

R2 = 1Ω
R3 = R4 = R5 = 10KΩ
R6 = R7 = 20KΩ

Fig. 15: SIMULINK model of IPMC

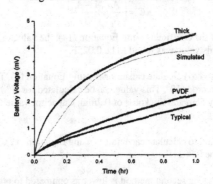

Fig. 16: Measured battery charging

222

CONCLUSIONS

Reported research demonstrated the capacitive behavior or IPMC that can be controlled by controlling IPMC thickness. Thick IPMC was successfully manufactured using hot pressing and electrode deposition was achieved using electroless method. The thickness hence achieved was 0.5mm. Increasing the thickness increases overall capacitance of IPMC. This has two-fold implications. Firstly, thicker IPMC shows improved battery charging as compared to conventional IPMC. Secondly, due to higher capacitance the sample may be used as an energy storage device. Further study is required to probe into the energy storage characteristics of IPMC.

ACKNOWLEDGEMENTS

The project was partially supported by U.S. National Science Foundation and Nevada System of Higher Education.

REFERENCES

1. M. Addington, and D. L. Schodek, "Smart Materials and Technologies in Architecture," *Architectural Press*, MA (2004).

2. M. Shahinpoor, K. J. Kim and M. Mojarrad, "Artificial Muscles: Applications of Advanced Polymeric Nano-composites," *Taylor and Francis*, New York (2007).

3. M. Shahinpoor, Y. Bar-Cohen, J. O. Simpson, and J. Smith, "Ionic Polymer Metal Composites as Biomimmetic Sensors and Actuators," presented at SPIE 5th Annual Symposium on Smart Structures and Materials (1998).

4. K. Oguro, "Recipe-IPMC," (2001).

5. M. J. Guan and W.H. Liao, "Characteristics of Energy Storage Devices in Piezoelectric Energy Harvesting Systems", *J. Of Intelligent Material Systems and Structures*, Vol. 9, pp. 671-680 (2008).

6. S. Roundy, P. K. Wright and J. Rabaey, "A Study of Low Level Vibrations as the Power Source for Wireless Sensor Node", *Computer Communications*, Vol. 26, pp. 1131-1144 (2003).

7. M. Umeda, K. Nakamura and S. Ueha, "Analysis of Transformation of Mechanical Impact Energy to Electrical Energy Using a Piezoelectric Vibrator", *Japanese Journal of Applied Physics*, Vol. 35, pp. 3267-3273 (1996).

8. M. Kimaru, "Piezoelectric Generation Device", US Patent# 5,801,475 (1998).

9. T. Sterner, "Human-Powered Wearable Computing", *IBM Systems Journal*, Vol. 35, pp. 618 (1996).

10. J. Kymissis, C. Kendall, J. Paradiso, N. Gershenfeld, "Parasitic Power Harvesting in Shoes", *2nd IEEE International Conference on Wearable Computing*, pp. 132-139 (1998).

11. M. Goldfarb and L. D. Jones, "On the Efficiency of Electric Power Generation With Piezoelectric Ceramics", *Journal of Dynamic Systems, Measurement and Control,* Vol. 121, pp. 566-571 (1999).

12. W. Clark and M. J. Ramsay, "Smart Material Transducers as Power Sources for MEMS Devices", *International Symposium on Smart Structures and Microsystems,* Hong Kong (2000)

13. H. A. Sodano, E. A. Magliula, G. Park and D. J. Inman, "Electric Power Generation Using Piezoelectric Devices", *13th International Conference on Adaptive Structures and Technologies,* Germany (2000).

14. K. Newbury and D. J. Leo, "Electromechanical Modeling and Characterization of Ionic Polymer Benders," *J. Of Intelligent Material Systems and Structures,* Vol. 13, pp. 51–60 (2002).

15. R. Tiwari, K. J. Kim, and S.-M. Kim, "Ionic Polymer Metal Composites as Energy Harvesters," *Smart Structures and Systems,* Vol. 4, No. 5, pp. 549-563 (2008).

Mater. Res. Soc. Symp. Proc. Vol. 1129 © 2009 Materials Research Society   1129-V04-06

Preparation of asymmetric thermosensitive double-layer gel

Takashi Iizawa[1], and Akihiro Terao

[1] Department of Chemical Engineering, Graduate School of Engineering, Hiroshima University, 1-4-1 Kagamiyama Higashi-Hiroshima 739-8527, Japan.

ABSTRACT

Heterogeneous amidation of poly(acrylic acid) gel-1,8-diazabicyclo-[5,4,0]-7-undecene salt (DAA) in N-methyl-2-pyrrolidone containing an excess of alkylamine and triphenylphosphite occurred from the surface to give the corresponding DAA-poly(N-alkylacrylamide) (PNAA) core-shell type gel, consisting of an unreacted DAA core and a quantitatively amidated shell layer. Further amidation of the DAA-PNAA core-shell type gel with a second alkylamine afforded a novel core-shell type gel consisting of two PNAA layers: PNAA(2) and PNAA(1). The resulting cylindrical PNAA(2)-PNAA(1) core-shell type gels were resistant to marked deformation caused by swelling/de-swelling because of their axial symmetry. This paper proposes a new approach to the preparation of asymmetric thermosensitive PNAA(2)-PNAA(1) double-layer gels by several procedures using the synthetic method of the core-shell type gels containing of poly(N-isopropylacrylamide) and poly(N-n-propylacrylamide) layers. Among the obtained asymmetric double-layer gels the model I type gel (cylindrical grooved PNNPA-PNIPA core-shell type gel) was markedly bent in water at temperatures between the lower critical solution temperatures of both layers.

INTRODUCTION

Poly(N-alkylacrylamide) (PNAA) gels containing C2-C3 alkyl groups such as ethyl, n-propyl, isopropyl, allyl, and cyclopropyl groups are thermosensitive with lower critical solution temperatures (LCST).[1] They swell in water at temperatures below the LCST and de-swell at temperatures above the LCST. The LCST varies, depending on the alkyl groups. They can be used in a variety of applications such as polymer actuator,[2] extraction,[3] absorption,[4] drug-delivery,[5] and enzyme immobilization.[6] Since the swelling/de-swelling of these simple gels occur isotropically in response to the water temperature changes across the LCST, their volume changes; however, their shapes are not markedly deformed. On the other hand, thermosensitive composite gels consisting of two different PNAA layers, which have similar structure to bimetal, would be suitable for actuator because they can be bent effectively. However, few studies have been reported on thermosensitive double-layer actuator, because it is difficult to bond two kinds of hydrogel plates together inseparably.

Recently, we reported a new synthetic method of PNAA(2)-PNAA(1) core-shell type gels containing two different PNAA: poly(N-n-propylacrylamide) (PNNPA) and poly(N-isopropylacrylamide) (PNIPA) layers by the heterogeneous two-step amidation of the poly(acrylic acid) gel-1,8-diazabicyclo-[5,4,0]-7-undecene (DBU) salt (DAA) with alkylamines using triphenylphosphite (TPP) as an activating reagent.[7] Since the resulting core-shell type gels have essentially the same main chain structure and shape as the original **DAA** samples, 1) a desired shape and size of core-shell type gel can be prepared as well as fine particles, and 2) it has a defined structure—a continuous network and a boundary between both layers. The structure may have strong resistance against destruction on the boundary. However, the resulting cylindrical **PNAA(2)-PNAA(1)** core-shell type gel cannot be markedly deformed by swelling/de-swelling because of the axial symmetry. We reported the preparation of semi-cylindrical, thermosensitive, double-layer gels; the cylindrical core-shell type gels are cut into two halves. The gel was markedly bent in response to the water temperature changes. However, the shell layer peeled off gradually when the stepwise temperature changes were repeated.

This study reports a new approach to the preparation of asymmetric thermosensitive double-layer gels as shown in **Scheme 1** and their bending behavior in response to the water temperature changes.

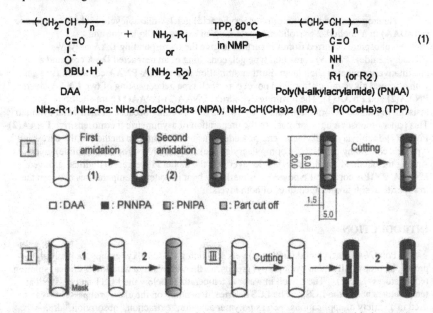

Scheme 1 Typical synthesis procedures of the asymmetric **PNIPA-PNNP** double-layer gels

EXPERIMENT

Materials

Acrylic acid and solvents were distilled prior to use. Commercial alkylamines such as *n*-propylamine (NNA) and isopropylamine (IPA) were used as purchased. *N,N'*-methylene*bis*-acrylamide (MBAA), DBU, and TPP were used without further purification. Cylindrical DAA (diameter: 5.3 mm, length: about 21 mm) was prepared via a two-step procedure, involving the copolymerization of acrylic acid with 0.5 mol% of MBAA in Teflon tubes (internal diameter: 6 mm), and neutralization of the resulting gel with excess DBU in methanol, as described previously.[8]

Synthesis of PNAA(2)-PNAA(1) core-shell type gels (asymmetric thermosensitive double-layer gels)

A typical synthesis of **PNAA(2)-PNAA(1)** was as follows. 200 mL of a mixed solution of TPP (1.0 mol L^{-1}) and NPA (2.0 moll^{-1}) in *N*-methyl-2-pyrrolidone (NMP) was charged into a 300 mL Erlenmeyer flask in an 80°C water bath. Dozens of **DAA** samples ($2R_0 = 5.3$ mm) were soaked in this solution for 200 minutes until the thickness of the shell layer ($1 - r_t/R_0$) reached 0.30, where $2R_0$ and $2r_t$ are the external diameters of the original DAA sample and core after $t = t$ min, respectively. The gels were taken out of the solution and put in a large quantity of methanol to terminate the reaction. The resulting **DAA-PNNPA** core-shell type gels were washed with methanol in a Soxhlet extractor. They were dried *in vacuo* at 60°C to a constant weight. The second amidation was carried out under the conditions similar to the first amidation described above. Free carboxylic acid in the gel was determined by neutralization titration of the filtrate with 0.01 mol L^{-1} HCl solution. However, it was not detected.

Measurement of IR spectra of PNNPA-PNIPA core-shell type gels

A **PNNPA-PNIPA** core-shell type gel sample was swollen in water at 25°C, and was cut into round slices. The swelled PNIPA shell layer was peeled off to divide into two groups, the swelled shell layer and unswelled core. After drying, every group was grinded down with an agate mortar. IR spectra (KBr) of the grinded powders were obtained using a Perkin Elmer model IR-700 spectrophotometer.

DISCUSSION

Synthesis of asymmetric PNAA(2)-PNAA(1) core-shell type gels

The heterogeneous reaction of **DAA** with IPA, using TPP, proceeded quantitatively via a mechanism that was similar to the unreacted-model;[9] the amidation occurred on the surface of the unreacted **DAA** core, since **DAA** did not swell in the solvent and the amidation reaction was very fast. The reaction from the surface gave **PNIPA** in the form of a core-shell type gel that consisted of an unreacted **DAA** core and a swelled shell (**Scheme 1**). Based on these previous papers,[8] the amidation of cylindrical **DAA** (diameter: 5.3 mm, length: about 21 mm) with alkylamines (2.0 mol L^{-1}), such as NNA and IPA, was carried out in NMP, using TPP, at 80°C. Typical disappearance rates for the core are shown in **Fig. 1**. The rate was not affected by the type of alkylamine used, because the reaction rate was independent of the concentration.[15b] The core disappeared after about 1200 min. The reaction of DAA with IPA and NPA gave **PNIPA** and **PNNPA** gels, respectively, when the core disappeared. The thickness of the shell layer was controlled by the reaction time.

DAA-PNNPA core-shell type gels and **DAA-PNIPA** core-shell type gels [$r_t/R_0 = 0.70$] were synthesized by the first amidation of **DAA** with the corresponding alkylamine in NMP, using TPP, at 80°C for 200 min. The resulting **DAA-PNAA** core-shell type gels were washed with methanol in a Soxhlet extractor. They were dried *in vacuo* at 60°C to constant weight. The second amidation of the dried core-shell type gels with another alkylamine was carried out for 24 h under similar conditions to the first amidation. After rapid initial swelling of the amidated shell layer, the second amidation, which showed similar characteristics to the first amidation, occurred. The IR absorption spectra of **PNIPA** core and **PNNPA** shell were clearly consistent with those of **PNIPA** and **PNNPA** gels prepared by the radical polymerization of the corresponding monomers, respectively. These IR spectral data indicated that the second amidation of **DAA-PNAA** core-shell type gels with an alkylamine was nearly quantitative, giving the corresponding **PNAA(2)-PNAA(1)** core-shell type gels. Since unreacted **DAA** cores in the core-shell type gels can react with various reagents, they are useful for their chemical modification and the synthesis of a wide variety of core-shell type gels. We propose three modified synthesis procedures of asymmetric thermosensitive double-layer gels as shown in **Scheme 1**. The first procedure is cutting a groove in the cylindrical core-shell type gel. The second procedure is using grooved **DAA**. The third procedure is using masking tape in the first amidation. These procedures give asymmetric thermosensitive double-layer gels (**Model I, II, and III types**).

Thermal swelling/de-swelling behavior of the gels

In the synthesis of the **PNAA(2)-PNAA(1)** type gels, attention was paid to the thermal swelling/de-swelling behavior of **PNNPA** and **PNIPA** gels. The equilibrium swelling ratio for **PNNPA** and **PNIPA** gels was measured in water over a wide temperature range (**Fig. 2**). **PNNPA** and **PNIPA** had LCSTs of approximately 21 and 31°C, respectively. Previous paper[7] reported that **PNNPA** and **PNIPA** gels prepared by this method showed higher thermosensitivity than the corresponding gels obtained by polymerization of the monomers. When the temperature was lower than the LCST, the gels swelled as they absorbed water, and the equilibrium swelling ratio increased drastically. The $(Ww + Wp) / Wp$ values for **PNNPA** and **PNIPA** gels, where Ww and Wp were the weights of the absorbed water and the dried gel, respectively, were 21 at temperatures below the LCST. In contrast, the gels barely swelled at temperatures above the LCST. The difference in the equilibrium swelling ratios of **PNNPA** and **PNIPA**, at temperatures below and above the transition, was negligible. However, **PNNPA** de-swelled, and **PNIPA**

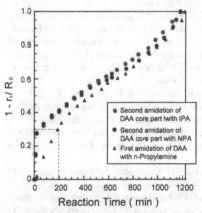

Fig. 1 Reaction of **DAA** with some
alkylamines in NMP at 80°C gels

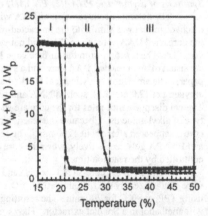

Fig. 2 Equilibrium swelling ratio of (▲)
PNIPA and (■) PNNPA in water

228

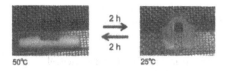

Fig. 3 Bending motions of the **PNIPA**-**PNNPA** double-layer gel (model I type)

swelled at temperatures between their LCSTs. These results suggest that thermal swelling/de-swelling behavior of **PNNPA** and **PNIPA** gels was divided into three regions: 1) at a lower temperature region (< 21°C, region I), **PNNPA** and **PNIPA** swelled, 2) at a middle temperature region (between 21-31°C, region II), **PNNPA** and **PNIPA** de-swelled and swelled, respectively, and, 3) at a higher temperature region (> 31°C, region III), both **PNNPA** and **PNIPA** de-swelled. The differences among the three regions were so large that the properties of **PNNPA**-**PNIPA** and **PNIPA**-**PNNPA** type gels would be expected to change dramatically in response to changing temperatures. The swelling/de-swelling behavior of **PNAA(2)**-**PNAA(1)** type gels was measured in water at temperatures (15, 25, and 35°C) in the three regions.

Bending behavior of asymmetric, thermosensitive, double-layer gel
The resulting asymmetric double-layer gels would be bending in water at 25°C, because the **PNIPA** layer swelled and the **PNNPA** layer shrunk. Typical bending behavior of the asymmetric thermosensitive double-layer gels was shown in **Fig. 3**. The model I type gel (cylindrical grooved **PNIPA**-**PNNPA** core-shell type gel) containing a **PNNPA** shell layer and **PNIPA** core slowly swelled in water at 15°C, but maintained its cylindrical shape. The swollen gel de-swelled slowly in water at a temperature above 35°C. When the de-swollen gel was immersed in water at 25°C, the milky **PNNPA** core part immediately turned transparent and expanded independently of the shell, curling markedly towards the outside. The reversible bending behavior was observed in response to stepwise temperature changes cross the LCSTs. It bent more rapidly and effectively than the other double-layer gels, the model II and III type gels. Although the shape of the bent former gel was extremely distorted, it was not damaged when the stepwise temperature changes were repeated several times. However, it took 2 hours;

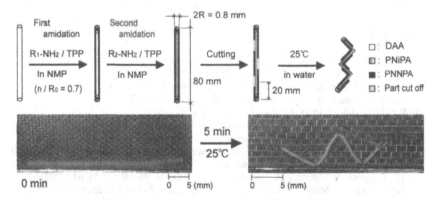

Scheme 2 Synthesis of a thin, long, asymmetric, thermosensitive, double-layer gel and its bending behavior

229

an extremely long time to reach the equilibrium. The swelling/de-swelling and bending rates should depend on their diameters because diffusion is the rate-determining factor governing their process. Therefore, we tried the synthesis of a thin, long, asymmetric, thermosensitive, double-layer gel to prepare a gel bent in zigzag with high-speed response. As shown in **Scheme 2**, the bending of 0.8 mm diameter gel is about 25 times faster than that of 4.8 mm diameter gel.

CONCLUSIONS

The synthesis of asymmetric double-layer gels has been investigated by the two-step amidation of DAA with alkylamines and the following conclusions have been obtained:

(1) The asymmetric double-layer gels containing **PNIPA** and **PNNPA** layers were prepared by some synthetic methods
(2) The swelling / de-swelling of shell layers and cores in the double-layer gels occurred almost independently in response to the temperature changes.
(3) The asymmetric double-layer gels were bent in response to the temperature changes
(4) Among the obtained asymmetric double-layer gels the model I type gel (cylindrical grooved **PNNPA-PNIPA** core-shell type gel) was markedly bent in water at temperatures between the LCST of both layers.
(5) The bending rate of the asymmetric cylindrical double-layer gels increased with decreasing their diameters since diffusion is the rate-determining factor governing this process.

ACKNOWLEDGMENTS
This work was supported by a grant from the Ministry of Education, Science and Culture of Japan (No. 18550195 and No. 20550189), which is gratefully acknowledged.

REFERENCES

1. S. Ito, *Kobunshi Ronbunshu,* **46,** 437 (1989).
2. Y. Osada, H. Okuzaki, H. Hori, *Nature,* **355,** 242 (1992)
3. (a) R. F. S. Freitas and E. L. Cussler, *Chem. Eng. Sci.,* **42,** 97 (1987).
   (b) W. Cai, and R. B. Gupta, *Ind. Eng. Chem. Res.,* **40,** 3406 (2001).
4. Y. Seida and Y. Nakano, *J. Chem. Eng. Jpn.,* **29,** 767 (1996).
5. L. C. Dong and A. S. Hoffman, *J. Controlled Release,* **15,** 141 (1990).
6. T. G. Park and A. S. Hoffman, *J. Biomed. Res.,* **24,** 21 (1990).
7. T. Iizawa, A. Terao, M. Ohuchida, Y. Matsuura and Y. Onohara, *Polymer J.,* **39,** 63 (2007).
8. (a) T. Iizawa, N. Matsuno, M. Takeuchi, and F. Matsuda, *Polymer J.,* **31,** 1277 (1999).
   (b) T. Iizawa, N. Matsuno, M. Takeuchi, and F. Matsuda, *Polymer J.,* **34,** 63 (2002).
   (c) T. Iizawa, Y. Matsuura, K. Hashida, and Y. Onohara, *Polymer J.,* **35,** 815 (2003).
   (d) T. Iizawa, Y. Matsuura, and Y. Onohara, *Polymer,* **46,** 8098 (2005)
9. S.Yagi and T. Kunii, *Kogyo Kagaku Zashi,* **56,** 131 (1953)..

Mater. Res. Soc. Symp. Proc. Vol. 1129 © 2009 Materials Research Society 1129-V05-03

Rheology and Electrorheology of Nanorod-Loaded Liquid Crystalline Polymers

Ana R. Cameron-Soto, Sonia L. Aviles-Barreto, and Aldo Acevedo-Rullán
Department of Chemical Engineering, University of Puerto Rico, Mayagüez
P.O Box 9046, Mayagüez, PR, 00681

ABSTRACT

The effect of carbon nanotube concentration and dispersion on the rheology of liquid crystalline solutions of hydroxypropylcellulose (HPC) has been experimentally studied. The rheology of nanocomposites of HPC and multiwalled carbon nanotubes (MWCNT) in m-cresol was characterized in steady-state and transient dynamic tests. The rheology as particle loading increases shows a very distinct response in the magnitude and scaling of the steady-state viscosity, and the storage and loss modulus. The liquid crystalline phase was characterized by direct observations by reflected polarized light microscopy. Additionally, an electric-field effect was observed on the rheology of the HPC/MWCNT in m-cresol soft composites. The HPC in m-cresol matrix is non-responsive, thus the electrorheological effect is due to the presence of the carbon nanotubes. The mechanism for this effect is still uncertain, since it does not follow the scaling predicted by simple models for heterogeneous or homogeneous ER fluids.

INTRODUCTION

Inclusion of nanoparticles in a polymer matrix may provide value-added properties, such as increased electrical conductivity, increased thermal conductivity, low flammability, or reduced permeabilities, to mention a few, in addition to modifying the mechanical properties [1, 2]. Addition of anisotropic nanorods results in very different properties primarily due to the ability to tune the physical properties by controlling the size, aspect ratio, and assembly of the rods. A special case of field-assisted alignment includes the use of a liquid crystalline (LC) matrix and its ability to orient along an applied field, which allows for control of the orientation [3-5]. Lower applied fields and thus, lower energy requirements would be necessary to obtain orientation in a LC phase-mediated alignment.

Liquid crystalline phase behavior have been observed in rodlike nanoparticle dispersions, such as MWNT [6], SWNT [7], CdSe nanorods [8], goethite [9], and natural clays [10], upon increase of the concentration (lyotropic LC). Similar to the phase behavior of rodlike polymers, as rod length increases the onset of liquid crystallinity decreases and the biphasic gap increases with increasing length distribution. Duran and coworkers (2005) reported an enhancement of the isotropic-nematic phase transition temperature of low molecular weight liquid crystal E7 with inclusions of MWCNT [11]. No changes were observed when spherical nanoparticles were added to E7 under similar conditions. These experiments suggest a significant contribution of inclusions' anisotropy to the entropy of the nematic system. Therefore, inclusion of rodlike nanoparticles into an LC polymer matrix may not have a detrimental impact on the orientational order and phase behavior of the mixture.

In this work, the effect of nanorod inclusions on the LC structure is experimentally determined through dynamic oscillatory rheology of multi-walled carbon nanotubes (MWCNT) dispersions in liquid crystalline hydroxypropylcellulose (HPC) solutions. Additionally, the effect of an applied external DC electric field on the steady-state viscosity is elucidated. HPC was chosen since it is a non-responsive liquid crystalline polymer, and the electric field response is mainly attributed to the nanorod inclusions.

EXPERIMENTAL SECTION

Hydroxypropylcellulose (HPC) (M_w = 100 kDa) and multiwalled carbon nanotubes (>95%, O.D. = 20-30 nm, L = 0.5-2 mm) were purchased from Sigma Aldrich, while m-cresol (97% purity) was purchased from Fisher Scientific. Phase behavior was determined in a Micromaster II polarized optical microscope. A completely liquid crystalline phase was observed above 35 wt%. The HPC/MWCNT/m-cresol solutions were prepared by mixing of HPC in MWCNT/m-cresol stock solutions. Solutions were left to equilibrate for at least one week. Stock solutions and final mixtures were dispersed by ultrasonification in a Branson 450W sonicator.

Rheological characterization was performed on a Reologica StressTech HR (*ATS RheoSystems*) stress controlled rheometer using stainless steel 35 mm parallel plate fixture or the electrorheological (ER) cell equipped with a TReK high voltage power supply. After loading, the sample was allowed to relax for 10 to 15 min. Dynamic viscoelastic moduli were measured at a strain of 0.5% over a frequency range of 0.1 to 100 Hz. Steady-state viscosity values were obtained from transient experiments at constant temperature of 25 °C and shear rates (either 1.0 or 10 s^{-1}). Reported viscosities correspond to the average measurement over a 30-second interval after reaching steady-state, while the error corresponds to the standard deviation.

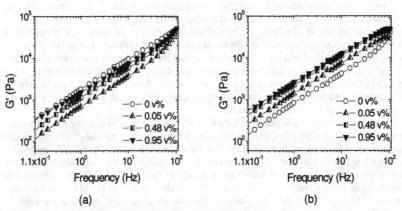

(a)   (b)

FIGURE 1. Effect of MWCNT loading on the storage (a) and loss (b) modulus of a liquid crystalline 45 wt% hydroxypropylcellulose (HPC)/m-cresol solution.

RESULTS AND DISCUSSION

Rheology

Figure 1 shows the effect of particle loading on the storage (G') and loss (G'') moduli of a 45 wt% HPC in m-cresol solution. The storage modulus of the MWCNT nanocomposites initially decreased with the addition of particles followed with an increase upon increase of the loading, while the loss modulus increases with the addition and increase of particle loading. Nevertheless, the same scaling was conserved in all solutions. G' and G'' scales as the 0.74 and 0.70 power of frequency, respectively. These suggest that the nanorods do not have an impact on the internal structure of the liquid crystalline matrix.

Electrorheology

Figure 2 shows the change of the steady-state viscosity, i.e. difference between the viscosity with field on and field off, as function of the DC electric field strength for an HPC solution without and with nanoparticles at a constant shear rate of 1 s^{-1}. The HPC solution without particles did not show a significant change in the viscosity, thus, it was considered non-responsive to electric fields. The particle loaded solutions showed a decrease in the viscosity, i.e. negative ER effect, with increasing field strength. At lower fields, the effect was more pronounced for the 0.48 vol% loading. Nevertheless, at higher fields the difference is not as pronounced. Additionally, for fields higher than 1 MV/m an almost linear dependence of the viscosity change with increasing field strength is observed. The viscosity at a field of 2 MV/m has decreased by a factor of 14 from that at zero-field, for both cases.

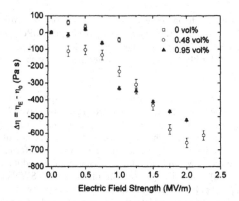

FIGURE 2. Effect of MWCNT loading on the steady-state viscosity of a liquid crystalline 45 wt% hydroxypropylcellulose (HPC) in m-cresol solution. All reported values correspond to the average ± the standard deviation of the steady-state measurements.

The shear rate effect was also studied at 1 and 10 s⁻¹, as shown in Figure 3, for the HPC solution with a 0.48 vol% MWCNT loading. For the 10 s⁻¹ shear rate, the ER effect was not as strong as that for the lower shear rate. Furthermore, it oscillated from positive to negative randomly as field strength increased. For the system presented in this work, the negative effect is not due to orientation along the flow direction. It has been previously reported that as shear rate increases, the shear forces dominate and overcome the orientation of the director or chain-formation perpendicular to flow, in homogeneous or heterogeneous ER fluids, respectively. If orientation along the flow direction was occurring as electric field is applied, increasing shear rate would facilitate it and a more pronounced effect would be observed as it increases.

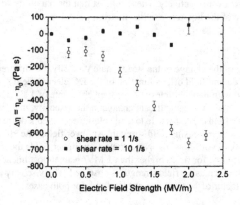

FIGURE 3. Effect of shear rate on the steady-state viscosity of a 0.48 vol% MWCNT dispersion in a liquid crystalline 45 wt% hydroxypropylcellulose (HPC) in m-cresol solution. All reported values correspond to the average ± the standard deviation of the steady-state measurements. Error bars for black squares are smaller than the symbol sizes.

The ER effect on the linear viscoelastic rheology of suspensions of carbon nanotubes and $Pb_3O_2Cl_2$ nanowires in silicone oil have been reported [12]. A negative ER effect on the storage modulus of the carbon nanotube suspension which dissapears as electric field increase was observed. The effect was attributed to the formation of layered nanostructures, which decrease the modulus, followed to migration of particles to the electrodes and formation of column-like structure along the electrodes, which in turn increases the storage modulus. In a related work, Park and coworkers (2006) oriented single wall carbon nanotubes (SWNT) through the use of an AC electric field in a photopolymerizable matrix [13]. It was proposed that the SWNT not only aligned with the field, but also migrated laterally to form thick columns between the electrodes. The proposed mechanism suggests an initial orientation towards the field due to the electric torque, followed by an electrophoretic effect. In both cases, at higher fields (but lower than those used in this work) the nanotubes formed chain-like structures between the electrodes which should affect positively the rheological response, reminiscence of ER fluids composed of semiconductor particles. Nevertheless, in this work a negative change in viscosity over the whole studied electric field range was measured and no direct observation of chain-like structures between the parallel plates was distinguished. The difference between the reports found in the literature and our results is attributed to the liquid crystalline matrix. The negative ER effect may be due to formation of small aggregates or bundles as suggested on the works discussed above. Since the effective size of these bundles is much higher than for single nanotubes, they may cause a distortion of the liquid crystal phase. Thus, it displaces the solution into a biphasic regime which may explain the decrease in viscosity. On the other hand, in a liquid crystal polymer the translational diffusion of molecules is negligible [14]. Hence, migration of the bundles to the electrodes and chain formation may be restricted by the LC phase. In the case of higher shear rates, shear forces do not allow for significant bundle formation, thus no significant effect would be expected.

CONCLUSIONS

Dynamic linear viscoelastic rheology suggests that inclusion of rodlike nanoparticles to a liquid crystalline matrix do not affect the internal structure, producing a homogeneous solution. In addition, a negative ER effect was observed for the first time in an LCP-based nanocomposite. The negative ER effect was observed to have a similar effect with increasing particle loading, and increases with decreasing shear rate. The effect cannot be explained by simple ER mechanisms for heterogeneous or homogeneous electrorheological fluids. Results suggest that distortion of the liquid crystalline phase due to aggregation of the nanoparticles causes the negative ER effect. Nevertheless, further studies of the reported phenomena have to be conducted.

ACKNOWLEDGMENTS

The authors gratefully acknowledge the support of the Institute for Functional Nanomaterials (IFN) Seed Money Grant (NSF-PREPSCoR). The authors (SLAB & ARCS) were supported through the Puerto Rico Louis Stokes Alliance for Minority Participation (PR-LSAMP) Bridge to the Doctorate Fellowships.

REFERENCES

1.   Koo, J.H., *Polymer Nanocomposites: Processing, Characterization and Applications*. 1st ed. 2005, New York: McGraw-Hill. 272 p.
2.   Moniruzzaman, M. and K.I. Winey, *Polymer nanocomposites containing carbon nanotubes*. Macromolecules, 2006. **39**(16): p. 5194-5205.
3.   Kimura, F., T. Kimura, M. Tamura, A. Hirai, M. Ikuno, and F. Horii, *Magnetic alignment of the chiral nematic phase of a cellulose microfibril suspension*. Langmuir, 2005. **21**(5): p. 2034-2037.
4.   Baik, I.S., S.Y. Jeon, S.H. Lee, K.A. Park, S.H. Jeong, K.H. An, and Y.H. Lee, *Electrical-field effect on carbon nanotubes in a twisted nematic liquid crystal cell*. Applied Physics Letters, 2005. **87**(26).

5.   Kawasumi, M., N. Hasegawa, A. Usuki, and A. Okada, *Nematic liquid crystal/clay mineral composites.* Materials Science & Engineering C-Biomimetic and Supramolecular Systems, 1998. **6**(2-3): p. 135-143.
6.   Song, W.H. and A.H. Windle, *Isotropic-nematic phase transition of dispersions of multiwall carbon nanotubes.* Macromolecules, 2005. **38**(14): p. 6181-6188.
7.   Davis, V.A., L.M. Ericson, A.N.G. Parra-Vasquez, H. Fan, Y.H. Wang, V. Prieto, J.A. Longoria, S. Ramesh, R.K. Saini, C. Kittrell, W.E. Billups, W.W. Adams, R.H. Hauge, R.E. Smalley, and M. Pasquali, *Phase Behavior and rheology of SWNTs in superacids.* Macromolecules, 2004. **37**(1): p. 154-160.
8.   Li, L.S., M. Marjanska, G.H.J. Park, A. Pines, and A.P. Alivisatos, *Isotropic-liquid crystalline phase diagram of a CdSe nanorod solution.* Journal of Chemical Physics, 2004. **120**(3): p. 1149-1152.
9.   Vroege, G.J., D.M.E. Thies-Weesie, A.V. Petukhov, B.J. Lemaire, and P. Davidson, *Smectic liquid-crystalline order in suspensions of highly polydisperse goethite nanorods.* Advanced Materials, 2006. **18**(19): p. 2565-+.
10.   Davidson, P. and J.C.P. Gabriel, *Mineral liquid crystals.* Current Opinion in Colloid & Interface Science, 2005. **9**(6): p. 377-383.
11.   Duran, H., B. Gazdecki, A. Yamashita, and T. Kyu, *Effect of carbon nanotubes on phase transitions of nematic liquid crystals.* Liquid Crystals, 2005. **32**(7): p. 815-821.
12.   Lozano, K., C. Hernandez, T.W. Petty, M.B. Sigman, and B. Korgel, *Electrorheological analysis of nano laden suspensions.* Journal of Colloid and Interface Science, 2006. **297**(2): p. 618-624.
13.   Park, C., J. Wilkinson, S. Banda, S. Ounaies, K.E. Wise, G. Sauti, P.T. Lillehei, and J.S. Harrison, *Aligned Single-Wall Carbon Nanotube Polymer Composites Using an Electric Field Effect.* Journal of Polymer Science Part B-Polymer Physics, 2006. **44**: p. 1751-1762.
14.   Doi, M. and S.F. Edwards, *The Theory of Polymer Dynamics,* in *The Theory of Polymer Dynamics.* 1986, Clarendon Press: Oxford, UK. p. 289-380.

Mater. Res. Soc. Symp. Proc. Vol. 1129 © 2009 Materials Research Society   1129-V05-02

Cellulose Electroactive Paper (EApap):
The potential for a novel electronic material

Joo-Hyung Kim*, Kwangsun Kang, Sungryul Yun, Sangyeul Yang,
Min-Hee Lee, Jung-Hwan Kim and Jaehwan Kim

Creative Research Center for Electroactive Paper (EApap) Actuator,
Dept. of Mechanical Engineering, INHA University
Young-Hyun Dong 253, Nam Gu, Incheon, South Korea

ABSTRACT

Cellulose electro-active paper (EApap) has attracted much attention as a new smart electronic material to be utilized as mechanical sensors, bio compatible applications and wireless communications. The thin EApap film has many advantages such as lightweight, flexible, dryness, biodegradable, easy to chemically modify, cheap and abundance. Also EApap film has a good reversibility for mechanical performance, such as bending movement, under electric field. The main actuation mechanism governed by piezoelectric property can be modulated by material direction and stretching ratio during process. In this paper we present the overview as well as fabrication process of cellulose EApap as a novel smart material. Also we propose the method to enhance the piezoelectricity, its mechanical and electromechanical properties. In addition, the fabrication of high quality metal patterns with Schottky diode on the cellulose surface is an initiating stage for the integration of the EApap actuator and electronic components. The integration of flexible actuator and electronic elements has huge potential application including flying magic carpets, microwave driven flying insets and micro-robots and smart wall papers.

INTRODUCTION

Recently, cellulose has a spotlight as a new smart material. It has been discovered that cellulose materials have many attractive properties such as large deformation, light weight, inexpensive fracture tolerance. A cellulose based Electroactive Paper (EApap) film has been investigated as an attractive smart material due to its characteristics of lightweight, biodegradable, low cost, large bending displacement, low power consumption and piezoelectricity.[1,2] It consists of β-D-glucopyraonosyl units with a (1–4)-β-D-linkage and forms a linear chain through many inter- and intramolecular hydrogen bonding. The linearity of cellulose makes it easy for the molecules to produce parallel arrays and cause a high degree of crystallinity. It was reported the maximum tip displacement of 4.3 mm out of 40 mm long beam was occurred under 0.25 MV/μm of electric field and the electric power consumption of EApap actuator was 10 mW/cm^2 when an electric field is applied across the thickness direction of the EApap films. The main actuating performance depends on the combined effects of piezoelectricity and ion migration associated with dipole.

Cellulose itself has a hydrophilic property due to its –OH to make a weak hydrogen bond with water in environment. Therefore, as a sensor, cellulose EAPap film is an interesting flexible electrical substrate due to high electrical sensitivity according to changes of relative humidity and/or temperature in living condition.[3]

For the potential applications, it can be suggested for many industrial, medical, military and telecommunications such as micro-insect robots, flying objects, smart wall papers, pressure sensor and surface acoustic wave device used in microwave communication and bio-sensors.[4]

In this paper, we present the overview of cellulose EAPap film as well as fabrication process for a novel smart material. Also we propose the method to enhance the piezoelectricity, its mechanical and electromechanical properties.

EXPERIMENTS

To make a thin EAPap film, natural cellulose, LiCl and N, N-Dimethylacetamide (DMAc) were prepared. Because the EAPap film is sensitive to ambient conditions such as humidity and temperature, it is acceptable to use DMAc/LiCl as a non-degrading solvent for cellulose.[1,3]

The cellulose was mixed with LiCl and dried at 150 °C, low pressure for 30 min. Using DMAc(N, N-dimethyl acetamide) solvent, the cotton cellulose was dissolved by heating process. A transparent cellulose solution was obtained at room temperature. Finally, re-generated thin cellulose films were prepared by spin-coating and drying process using ultra-violet light. Au electrodes were deposited on both sides of the EAPap film by physical vapor deposition method. For actuating performance test and sensor application, the size of the EAPap sample was 10 mm x 40 mm. The thickness of the gold electrodes was so thin (0.1 μm) that the gold electrodes do not significantly affect the bending stiffness of the cellulose paper. Also, as a potential sensor applications such as surface acoustic wave (SAW) device, RF identification (RF ID) and wireless communication, an inter digit transducer (IDT) structure was fabricated on the EAPap film by lift-off technique.

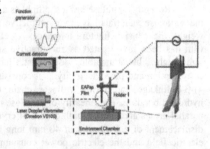

Fig. 1 cross-sectional image of EAPap film.   Fig. 2 Actuation performance measurement system of EAPap film.

Figure 1 shows the cross-sectional image after film process. The well ordered cellulose fibrils were observed from the image. Figure 2 present the actuation performance measurement system of EAPap film under the different humidity and temperature conditions.

Unconventional lithography process including micro-contact printing process was utilized to fabricate rectenna pattern on the cellulose film. The processes comprised of polydimethylsilane (PDMS) mold fabrication, gold evaporation on the mold, self-assembled monolayer (SAM) fabrication on the gold surface, and contact printing process. Mercappropyltrimethoxysilane (MPTMS) was used for SAM layer fabrication. Various materials and structures of Schottky diodes were fabricated on the rectenna to convert microwave signal to electrical power. Al and Au electrodes were used as metal electrodes, and poly(3,4-ethylenedioxythiophene)poly(styrenesulfonate) (PEDOT:PSS) doped with pentacene, TiO_2 nanoparticels, and ZnO was used as semiconducting layer.

RESULTS AND DISCUSSION

Figure 3 shows the bending performance of EAPap thin films. Fig. 3 (a) presents that the bending displacement of EAPap films depends on the humidity level due to hydrophilic property. As relative humidity increases, the bending displacement increases while the operating frequency to actuate the maximum displacement decreases. To realization of the working prototype using bending displacement, a dragon fly shaped act-

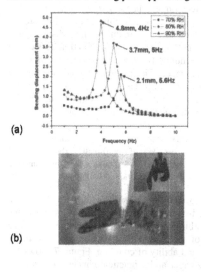

(a)

(b)

Fig. 3 (a) The measured bending displacement of EAPap film under different humidity levels (b) A prototype of the EAPap actuator. The inset presents the bending performance by the applied field.

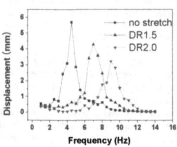

Fig. 4 Drawing ratio (DR) effects on bending displacement performance at 90 % relative humidity.

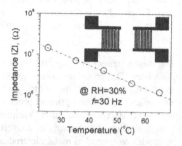
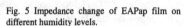
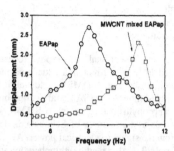

Fig. 5 Impedance change of EAPap film on different humidity levels.

Fig. 6 Displacement performance change of a MWCNT mixed EAPap film.

uator was fabricated. The images in Fig. 3 (b) are the snapshot of the actuating performance of the actuator. The inset presents bending performance of EAPap film when the electric field was applied. This bending performance also can be moderated by a mechanical method during EAPap film process.

Figure 4 shows the wet drawing effect on the bending displacement of thin EAPap films. If the extra pulling force (= drawing ratio) on the thin EAPap film during a drying process, the operating frequency, where the maximum bending displacement occurs, increases while the bending displacement decreases. Humidification/desiccation of cellulose is strongly related to the relative humidity which can affect the bonding strength between –OH of cellulose and water vapor. The potential application can be addressed as a humidity sensor. Figure 5 shows the impedance changes of EAPap film using IDT structure. The good linear relation between the impedance change and relative humidity level indicates that the EAPap has a potential as a humidity sensor.[1]

In order to enhance the actuation performance as a EAPap actuator, it was revealed that the mixing of small amount (0.1 wt. %) of multi-wall carbon nanotube (MWCNT) in EAPap increased 35% resonance frequency and 76% Young's modulus than pure cellulose.[5] Figure 6 showed the measurement data of the MWCNT mixed EAPap actuator.

Another approach using a thin EAPap film is a microwave communication in order to overcome the requirements of hard-wire circuits and power supply for the practical applications of smart materials. Flexible dipole rectenna devices appeared to be attractive for various applications because of the adaptability on complex structures; possibility for higher power density features, and ability of coupling. Figure 7 shows the micro fabricated rectenna devices on EAPap paper and a practical concept for wireless communication.[1,6] The concept of potential application of EAPap rectenna film was presented in Figure 8.

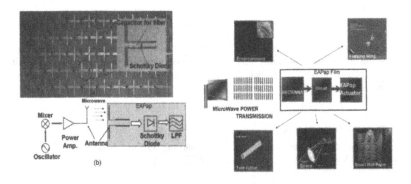

Fig. 7 (a) The image of EAPap rectenna and (b)
a concept of microwave communication.

Fig. 8 Applications of the EAPap rectenna film.

To realize the EAPap rectenna, the arrayed rectenna patterns must be fabricated on the cellulose paper in terms of three-layered Schottky diode structure. Fig. 9 shows the three-layered Schottky diode structure comprised of Al/PEDOT:PSS-pentacene. To make thick gold layer, six-times deposited the gold electrode on the top of the PEDOT:PSS-pentacene layer. The current-voltage (I-V) characteristics of PEDOT:PSS-pentacene with different doping levels were shown in Figure 10. As the pentacene concentration increased onto the PEDOT:PSS semiconducting polymer layer, the forward current drastically increased. This enhancement of the forward current might be due to the high mobility of the pentacene. [7]

CONCLUSIONS

The cellulose EAPap film is considered as a smart sensor material. The actuation principle is based upon the combination of piezoelectric effect and ion migration effect

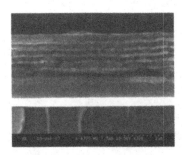

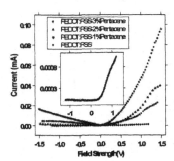

Fig. 9 The image of three layered Schottky diode structure for EAPap rectenna applications.

Fig. 10 Current-voltage (I-V) plots of Schottky diode as a function of different PEDOT:PSS doping levels.

associated with constituent of cellulose material. The material performance of EAPap is sensitive to the electric field and humidity levels. These material properties of EAPap film can be modulated by the film fabrication process as well as a mixing of functionalized material such as MWCNT. A linear relation of the impedance change as a function of different humidity level indicates the potential of EAPap as a humidity sensor. Also thin film rectenna EAPap structure can be utilized for wireless communication devices.

ACKNOWLEDGEMENTS

The work is performed under the financial support of the Creative Research Initiatives Program (EAPap Actuator) of KOSEF/MEST.

REFERENCES

[1] J. Kim, S. Yun and Z. Ounaies, Macromolecules 39 (2006) 4202.
[2] C-H Je, K. J. Kim, Sensors and actuators A 112 (2004) 107.
[3] J. Kim, N. Wang, Y. Chen, S. Lee, C. Yang, Key Eng. Mater. 326-328 (2006) 1375.
[4] J. Kim, S. Yun, S.-K. Lee, J. Intell. Mater. Sys. & Struct. 19 (2008) 417.
[5] S. Yun, J. Kim and Z. Ounaies, Smart Mater. Struct. 15 (2006) N61.
[6] J. Kim, S-Y Yang, K. D. Song, S. Jones, J. R. Elliott, S. H Choi, Smart Mater. Struct. 15 (2006) 1243.
[7] K. S. Kang, H. K. Lim, K. Y. Cho, K. J. Han, J. Kim, J. Phys. D: Appl. Phys. 41 (2008) 012003.

Mater. Res. Soc. Symp. Proc. Vol. 1129 © 2009 Materials Research Society 1129-V12-03

Shape-memory Polymer Composite and Its Application

Yanju Liu[1], Haibao Lu[2], Jinsong Leng[2] and Shanyi Du[2]

[1]P.O. Box 301, Department of Aerospace Science and Mechanics, No. 92 West DaZhi Street, Harbin Institute of Technology (HIT), Harbin 150001, P.R. China.

[2]P.O. Box 3011, Centre for Composite Materials, No. 2 YiKuang Street, Science Park of Harbin Institute of Technology (HIT), Harbin 150080, P.R. China

ABSTRACT

Smart materials can be defined as materials that sense and react to environmental conditions or stimuli. In recent years, a wide range of novel smart materials have been developed in biomaterials, sensors, actuators, etc. Their applications cover aerospace, automobile, telecommunications, etc. This paper presents some recent progresses of polymeric smart materials and their applications. Special emphasis is laid upon electroactive polymer (EAP), shape memory polymers (SMPs) and their composites as well as the applications in morphing structures. For the electroactive polymer, an analysis of stability of dielectric elastomer using strain energy function is derived, and two types of electroactive polymer actuator are presented. For the shape memory polymer, a novel method is developed to use infrared laser to actuate the SMP through the optical fiber embedded into the SMP. Furthermore, a series of fundamental investigations of electroactive SMP are conducted. Electrically conductive fillers are utilized as the fillers to improve the electrical conductivity of polymer, and then the shape recovery is performed. Moreover the application of SMP in morphing is discussed.

INTRODUCTION

Smart materials are materials that have one or more properties that can be significantly changed in a controlled fashion by external stimuli, such as, temperature, moisture, pH, electric or magnetic fields [1-5]. Varieties of smart materials already exist, and are being researched extensively. These include piezoelectric materials, shape memory materials, magneto-rheostatic materials, electro-rheostatic materials and so on. The property that can be altered influences what types of applications the smart material can be used for [1-2]. In the last decade, a wide range of novel polymeric smart materials have been produced in biomaterials, bioinspired, multiscale structured materials [6], sensors, actuators, smart biomaterials, etc [7,8]. The applications cover aerospace, automobile, telecommunications, such as actively moving polymers, neural memory devices, smart micro/nanocontainers for drug delivery, various biosensors, dual/multi-responsive materials, biomimetic fins [1,9,10].

This paper presents some recent progresses of shape memory polymers and their composites as well as the applications. For SMP, a novel method is developed to use infrared laser to actuate the SMP. Then, the electrically conductive fillers are used as the fillers to improve the electrical conductivity of polymer. Moreover, the application of SMP in future is presented.

RECENT DEVELOPMENT OF SMP ACTUATION

Infrared laser-activated SMP

The actuation of shape memory styrene copolymer is proposed through an infrared optical fiber carrying infrared laser [11]. Figure 1 is the schematic plan of the SMP embedded with optical fiber, it shows that the treated optical fiber and treated fiber was embedded in the shape memory polymer. The infrared laser was chosen to actuate the SMP through the optical fiber embedded into the SMP. The working frequency of infrared laser was installed in 3-4μm, the working band of optical fiber was 1-6μm. An optical fiber was embedded into the SMP for delivery of 3-4μm laser light for activation. The surface of the optical fiber was etched by the aqueous solution of sodium-hydroxide in order to increase transmission efficiency of the optical fiber. We synthesized a thermoset SMP based on styrene copolymer, and the glass transition temperature (T_g) of this SMP is about 53.7 °C tested by a dynamic mechanical analysis (DMA). The thermally activated SMP is possible to initiate an originally shape by a touchless and highly selective infrared laser stimulus. The increase in temperature of SMP indicated that the infrared laser irradiated into the SMP and heated the SMP. The infrared laser stimulation of SMP was a candidate for actuator systems as well as medical applications.

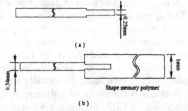

Figure 1. Illustration of infrared laser-activated shape memory polymer (a) Treated optical fiber; (b)The treated surface of optical fiber were embedded in the shape memory polymer

A typical shape recovery sequence of infrared laser-activated shape memory effect is shown in Figure 2(the red wire is the infrared optical fiber and the curving silvery white wire is the SMP).

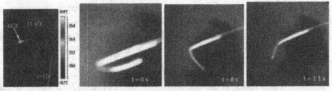

Figure 2. Shape recovery behavior of shape memory polymer

Electro-triggered SMP composites

In order to get rid of external heaters, thermoresponsive SMPs composites incorporated with various types of electrically conductive fillers (e.g., carbon nanotubes [3], carbon black [12], Ni powders [13,14], chopped or continuous carbon fiber [15-17]) have been developed, so they can be actuated by means of Joule heat (i.e., by passing an electrical current, just like that of NiTi shape memory alloys).

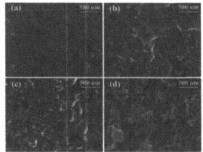

Figure 3. SEM images (SEI, 20KV, 20000.0X) of the distributions of nano-carbon powders in the SMP matrix (a), CB2; (b), CB4; (c), CB6; (d), CB10

This section focuses on the progress of electro-activate SMP composites. Special emphases are given on the filler types that affect the conductive properties of these composites. Then, the mechanisms of electric conduction are addressed. A series of fundamental investigations of electroactive SMP are conducted. Electrically conductive fillers are utilized as the fillers to improve the electrical conductivity of polymers. With incorporating conductive fillers into SMP matrix, not only it mechanics properties improve a lot, but also it can be actuated by passing an electrical current. The conductive fillers include: SMP/CB/short-carbon-fiber, CB, Ni, carbon nanotubes, carbon-fiber. We systematically compared the conductivity and performed the shape recovery properties for different conductive fillers. This is a good foundation to improve the electroactive performances of SMP, which maybe become a more smart material.

An electroactive thermoset styrene-based shape-memory polymer (SMP) nanocomposite filled with nano-sized (30 nm) carbon black (CB) is presented [3]. With an increase of the incorporated nano-carbon powders of the SMP composite, its T_g decreases and storage modulus increases. Due to the high micro-porosity and homogeneous distributions of nano-carbon powders aggregations in SMP matrix (Figure 3), the SMP composite shows a good electrical conductivity with a percolation of about 3.8 %. This percolation threshold is slightly lower than that of many other carbon based conductive polymer composites.

Figure 4 presents the experimental results of electrical resistivity of SMP composites filled with different volume fractions of nano-carbon powders. It shows that the electrical resistivity of composite with less than 3% volume fraction of nano-carbon powders is extraordinary high (10^{14}-10^{13} Ω.cm). By contrast, a sharp transition of electrical conductivity occurs between 3% to 5% , which is called as a percolation threshold range. As the fillers contents is larger than 5%, the resistivity reduce to a low and stable level (10^3-10^1 Ω.cm).

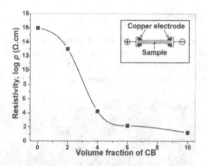

Figure 4. Resitivity vs. volume fraction of carbon black

The electric field-triggered shape recovery of the sample CB10 is shown in Figure 5. Due to the relatively high electrical conductivity, a sample filled with 10 vol% nano-carbon powders shows a good electroactive shape recovery performance heating by a voltage of 30 V above a transition temperature of 56-69°C.

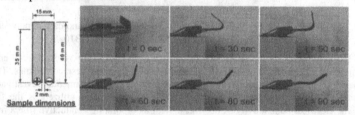

Figure 5. Sequences of the shape recovery of sample CB10 by passing an electrical current (voltage, 30V)

SMP in response to solution

On the basis of polymer science, it is obvious that as polymer is immersed in solution, the solution molecule has plasticizing effect on polymer firstly, followed with chemical or physical interaction, resulting in transition temperature reduced until to the ambient temperature. Thus, shape recovery of SMP can be induced by its solution which has an indirect effect on the transition temperature [18-20].

Figure 6. Shape recovery of a 2.88 mm diameter SMP wire in DMF. The wire was bent into "n" like shape

Originated from these, two methods were used to exemplify that the actuation of SMP can be triggered by chemical interaction and physical swelling effect with corresponding solvent. The styrene-based thermosetting SMP can recover from the pre-deformed 'n' shape sequentially after

246

being immersed in N,N-Dimethylformamide (DMF) for forming the chemical interaction shown in Figure 6. And with following tests, as FT-IR, differential scanning calorimeter (DSC) and dynamic mechanical analysis (DMA), it is validated that the experimental results agree well with the theoretical analysis.

Subsequently, the rubber elastic theory is used to testify the feasibility of SMP induced by physical swelling effect. Different from chemical interaction, the swelling effect can be observed that shape recovery actuation of styrene-based SMP induced by toluene solvent along with volume swelling as shown in Figure 7. However, the FT-IR experimental result reveals that there are no characteristic groups being involved into chemical interaction [21].

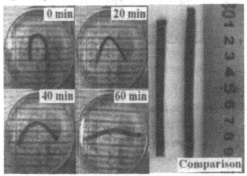

Figure 7. Series photos of shape recovery of rectangle SMP sample (bent into "n" like shape) in Toluene; comparison of length change between unswollen and swollen samples

In summary, styrene-based SMP in response to solvent, or to the extension, SMP in response to solution (namely solvent or mixture), the mechanism behind it is solution molecule has plasticizing effect on polymeric materials; and then increases the flexibility of macromolecule chains. These two effects make the transition temperature of materials reduced until shape recovery occurs. There are many extension results and achievements are based on their conclusion.

CONCLUSIONS

This paper concerns some recent progresses of SMP and its applications. For the shape memory polymer, a novel method is developed to use infrared laser to actuate the SMP through the optical fiber embedded into the SMP, which show good shape recovery performance. Furthermore, a series of fundamental investigations of electroactive SMP are performed. Electrically conductive fillers were incorporated into SMP matrix. Finally, the actuation of SMP driven by solution also has been realized by chemical interaction and physical swelling effect. Due to a good electrical conductivity of polymer, the shape recovery of SMP was good. These approaches should be applicable to other SMPs and their composites, and many extension applications and achievements can be based on these outcome.

REFERENCES

1. P. R. Buckley, G. H. McKinley, T. S. Wilson, W. IV. Small, W. J. Benett, J. P. Bearinger, M. W. McElfresh and D. J. Maitland, *IEEE Trsns. Biomed Eng.*, 53, 2075 (2006).

2. A. M. Schmidt, *Macromol. Rapid Commun.*, 27, 1168 (2006).

3. J. W. Cho, J. W. Kim, Y. C. Jung and N. S. Goo, *Macromol Rapid Commun.*, **26**, 412 (2005).

4. F. Xia and L. Jiang, *Adv. Mater.* **20**, 2842 (2008).

5. S. Mondal, *Appl. Therm. Eng.*, **28**, 1536 (2008).

6. C. A. Orme, A. Noy, A.Wierzbicki, M. T. McBride, M. Grantham, H. H. Teng and P. M. Dove, J. J. DeYoreo, *Nature (London)* **411**, 775 (2001).

7. D. Ratna and J. Karger-Kocsis, *J. Mater. Sci.* **43**, 254 (2008).

8. Y. Lu and J. Liu, *Acc. Chem. Res.* **40**, 315 (2007).

9. K. Kinbara and T. Aida, *Chem. Rev.* **105**, 1377 (2005).

10. J. Liu and Y. Lu, *Adv. Mater.* **18**, 1667 (2006).

11. N. Willet, J. F. Gohy, L. Lei, M. Heinrich, L. Auvray, S. Varshney, R. Jerome and B. Leyh, *Angew. Chem. Int. Ed.* **46**, 7988 (2007).

12. X. Lan, W. M. Huang, J. S. Leng, Y. J. Liu and S. Y. Du, *Adv. Mater. Res.* **47-50**, 714 (2008).

13. J. S. Leng, X. Lan, W. M. Huang, Y. J. Liu, N. Liu, S. J. Phee, Q. Yuan and S. Y. Du, *Appl. Phys. Lett.* **92**, 014104 (2008).

14. J. S. Leng, W. M. Huang, X. Lan, Y. J. Liu, and S. Y. Du, *Appl. Phys. Lett.* **92**, 204101 (2008).

15. J. S. Leng, H. B. Lv, Y. J. Liu, and S. Y. Du. *Appl. Phys. Lett.* **91**, 144105 (2007).

16. C. S. Zhang and Q. Q. Ni. *Compo. Struct.* **78**, 153 (2007).

17. H. B. Lv, J.S Leng and S. Y. Du, *15th SPIE international conference on Smart Structures/NDE, San Diego, March, 9-13th*, SPIE 6929, 6929L-1 (2008).

18. J. S. Leng, X. Lan, W. M. Huang, Y. J. Liu, N. Liu, S. J. Phee, Q. Yuan and S. Y. Du, *Appl. Phys. Lett.* **92**, 014104 (2008).

19. H. B. Lv, J. S. Leng, Y. J. Liu, and S. Y. Du, *Adv. Eng. Mater.* **10**, 592 (2008).

20. H. B. Lv, Y. J. Liu, D. X. Zhang, J.S. Leng and S. Y. Du, *Adv. Mater. Res.* **47-50**, 258 (2008).

21. J. S. Leng, H. B. Lv, Y. J. Liu, and S. Y. Du. *Appl. Phys. Lett.* (unpublished)

Other Materials, Devices and Characterization

Mater. Res. Soc. Symp. Proc. Vol. 1129 © 2009 Materials Research Society   1129-V03-06

Development of Multifunctional Structural Material Systems by Innovative Design and Processing

Hiroshi Asanuma, Jun Kunikata and Mitsuhiro Kibe
Mechanical Engineering Course, Graduate School of Engineering, Chiba University
1-33, Yayoicho, Inage-ku, Chiba-shi, Chiba, 263-8522, Japan.

ABSTRACT

Innovative designing concepts and fabrication processes to realize smart structural material systems which will lead to be reliable, energy saving, environmentally friendly and cost effective are proposed and several examples are introduced in this paper. Two types of developments are main contents, which are (1) Type I that uses highly functional ceramics in a metallic matrix as protective environment, and (2) Type II that uses a couple of competitive structural materials without using those sophisticated functional materials to generate functions. The Type I could be realized by embedding fragile functional fibers such as FBG (Fiber Bragg Grating) sensor and piezoelectric ceramic fiber in an aluminum matrix by the Interphase Forming/Bonding Method. The Type II can be explained that composite of competitive structural materials may have not only high mechanical properties, but also functional properties generated by their inconsistent secondary properties. As examples of this type, CFRP/Al active laminates, active FRMs and Ti fiber/Al multifunctional composites were successfully developed.

INTRODUCTION

Conventional composite materials are generally known as high performance structural materials. The typical one is fiber reinforced plastic (FRP) such as glass fiber reinforced plastic (GFRP) and carbon fiber reinforced plastic (CFRP). Some FRPs have been also laminated with metal or alloy sheets known as fiber-metal laminates (FMLs) [1]. Aiming at higher temperature applications, fiber reinforced metals (FRMs) have been developed. FRPs, FMLs and FRMs have chances to incorporate sophisticated phases such as optical fiber sensors, shape memory alloys, piezoelectric ceramics for sensing, health monitoring, actuation, and other capabilities.

An important purpose of developing the above mentioned multifunctional structural material systems is simplification of mechanical systems just by using them [2-4] as shown in figure 1. They may have sensing, health monitoring, actuation, healing, and/or other capabilities.

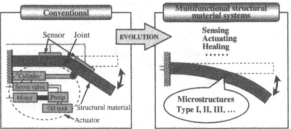

Figure 1. Simplification of conventional complicated mechanical systems by using multifunctional structural material systems.

Most of the conventional complicated mechanical systems need joints, lubrication, heavy actuators, delicate sensors, and so on. The multifunctional structural material systems may be able to eliminate them because of their intelligently designed microstructures.

The main method to develop multifunctional structural material systems is embedding functional materials in structural materials. This is shown in figure 2 as Type I, of which examples can be easily found. The sophisticated functional materials are usually fragile, heavy and/or expensive and have negative effects on mechanical properties. So it is not easy to develop highly reliable multifunctional structural material systems by this method. In order to overcome this problem, Asanuma invented a special method to enable embedding fragile functional materials in aluminum based materials as protective environment.

Asanuma also proposed another effective method to design multifunctional structural material systems without using functional materials. This idea is also shown in figure 2 as Type II and can be explained as follows: There are a couple of competitive structural materials which normally compete with each other because of their similar and high mechanical properties such as high strength and high modulus, or high specific strength and high specific modulus. They have secondary properties which may be different from each other or opposite among them. If they are combined together to make a composite, the similar properties, normally high mechanical property, can be maintained, and the other dissimilar properties conflict with each other, which will generate functional properties. In order to validate this concept, a couple of examples are introduced in this paper.

IF/B METHOD AS AN INNOVATIVE PROSESSING TO DESIGN MULTIFUNCTIONAL STRUCTURAL MATERIAL SYSTEMS

Development of IF/B Method [4-7]

The special technique named as "IF/B (Interphase Forming/Bonding) Method" to enable multifunctionality and fabrication of smart and robust aluminum based materials and devices has originated from an idea for forming interfacial layer to prevent fiber/matrix chemical reaction and galvanic corrosion between carbon fiber and aluminum matrix not by coating fibers with high cost and environmentally harmful materials and chemical agents but by using slightly

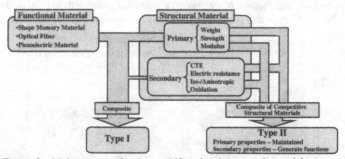

Figure 2. Main routes to develop multifunctional structural material systems.

oxidizing environment such as in the air instead of vacuum to enable matrix oxidation at the interface to form the barrier coating from the matrix side as shown in figure 3. [8-10]

The above mentioned technique led to an idea of forming an interphase to enable multifunctionality in FRMs. To enable fiber-matrix sliding for secondary forming as a new functionality of FRMs, fibers coated with lower melting temperature interphases and embedded in higher melting temperature matrix are necessary. If the coating is undertaken first and then followed by embedding, the coated interphases disappear during the following high temperature embedding process. In order to solve this problem, the IF/B Method as shown in figure 4 was proposed by Asanuma and can be explained as follows: U-grooves are formed on a piece of matrix (M) plate with an insert (I). The grooves are made a little larger than the diameter of the fiber. Next, the fiber filaments are placed in the grooves and covered with another piece of the matrix plate. Finally, the layered materials are hot-pressed at a low pressure in a low vacuum. The hot-pressing process can be separated into two steps. When the sample is heated to just above the eutectic temperature of the M-I alloy, the insert reacts with the matrix very quickly, and the molten alloy flows in the voids between the U-grooves and the fiber filaments (Step 1). If the material is cooled from this stage, the fiber is surrounded with M-I alloy with a composition close to that of its eutectic point. If the material is promptly heated, the molten alloy can be squeezed out smoothly by additional hot pressing (Step 2).

The tensile strength of the SUS304 stainless steel fiber/aluminum composite fabricated by the IF/B Method using zinc insert are shown in figure 5. The composite is sufficiently

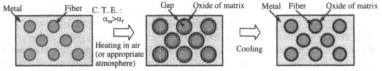

Figure 3. Formation of interphase to prevent fiber/matrix reaction and galvanic corrosion.

Figure 4. The Interphase Forming/Bonding Method.

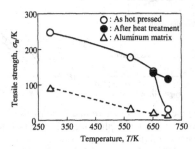

Figure 5. Temperature dependences of the composite and the matrix tensile strengths.

reinforced up to around 653 K, and then, it remarkably loses its strength above 653 K down close to that of the matrix at 703 K. On the other hand, the tensile strength of the heat treated composite at 703 K is as high as the reinforced level. Fiber pull-out is not observed at 653 K, but the fibers are completely pulled out at 703K After the heat treatment, fiber pull-out disappears again even at 703K. In this composite, the fiber-matrix interfacial shear strength remarkably reduces with increasing test temperature above its eutectic temperature. By taking advantage of this loose state of the interphase, secondary forming of the composite without fiber breakage becomes possible. The once remarkably decreased interfacial shear strength or the composite tensile strength can be recovered by mutual diffusion between the interphase, and the composite becomes sufficiently strengthened even at the temperatures higher than the eutectic temperature.

IF/B Method to enable embedding fragile functional materials in aluminum as a protective environment

As successful examples of the Type I multifunctional structural material systems, FBG sensor/aluminum composites and piezoelectric ceramic fiber/aluminum composites were developed as follows.

1) FBG sensor/aluminum composites [11]

As commercially available FBG (Fiber Bragg Grating) sensors are not suitable for high temperature and/or pressure processing, the above mentioned IF/B Method was applied for embedding them in aluminum matrix without losing their functions. Using copper insert and under the hot pressing condition of the temperature of 843 K, the pressure of 2.2 MPa and the period of 0.3 ks in a vacuum of 1×10^2 Pa, a FBG sensor was successfully embedded in pure aluminum by the IF/B Method. As shown in figure 6 (a), drastic peak shift of the reflected light from the embedded FBG sensor caused by the high residual stress generated during cooling after the hot pressing is clearly observed. The fabricated specimen was tensile tested under monitoring the peak wavelength shift and the level of optical transmission loss. The results given in figure 6 (b) indicate that the embedded FBG sensor is working as a strain sensor in the aluminum matrix.

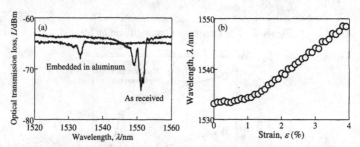

Figure 6. Wavelengths of the lights reflected from the FBG sensor before and after embedded in the aluminum matrix (a) and the wavelength of the reflected light from the embedded FBG sensor as a function of tensile strain (b).

2) Piezoelectric ceramic fiber/aluminum composites [12,13]

Piezoelectric ceramic fibers are very fragile and usually reactive with aluminum. In order to overcome these problems, the IF/B Method was applied. The piezoelectric ceramic (lead zirconate titanate) fiber of 0.2 mm in diameter and having a platinum core of 0.05 mm in diameter produced by National Institute of Advanced Industrial Science and Technology [14], aluminum plates as a matrix and copper foil as an insert were arranged as shown in figure 7 (a) to obtain the piezoelectric ceramic fiber/aluminum composite by the IF/B method. The applied hot pressing condition was the temperature of 873 K, the pressure of 2.2 MPa and the period of 2.4 ks in a vacuum of 1x10^2 Pa. A cross section of the fabricated composite was observed by SEM and given in figure 7 (b). According to this figure, the piezoelectric ceramic fiber is successfully embedded in the aluminum matrix without fracture.

A NEW CONCEPT TO DESIGN MULTIFUNCTIONAL STRUCTURAL MATERIAL SYSTEMS WITHOUT USING FUNCTIONAL MATERIALS

As already explained above and shown in figure 2, the Type II multifunctional structural material systems are intended to be realized without using functional materials. In order to validate this innovative concept, a couple of successful examples such as the CFRP/Al active laminate [15-19], the active FRMs [20-22] and the Ti fiber/Al multifunctional composites [23] were selected in this paper.

CFRP/Al active laminates [15-19]

In the case of the CFRP/Al active laminate as shown in figure 8, the main and common properties, that is, light weight and high strength are shown in the overlapped region of the ovals, and the other properties such as thermal expansion and electrical resistance are shown in the rest parts of them. Aluminum has high and isotropic CTE (coefficient of thermal expansion) and very low resistivity, but CFRP has anisotropic CTE, that is, very low in the fiber direction and very high in the transverse direction, and relatively high resistivity. When they are laminated as shown in figure 9 with an electrical insulation layer, a very unique and useful material system can be obtained. This simple laminate is one of the most useful types because of its unidirectional actuation. The laminate does not bend in the transverse direction because the CTEs of the CFRP and aluminum layers are close to each other, but remarkably bends in the fiber direction due to the large difference of them in this direction. As the CFRP layer works as a heater by application of voltage between its both ends, the curvature of the laminate can be controlled by the applied voltage. If this material system is regarded as a living body, the

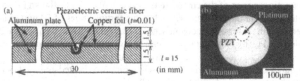

Figure 7. Cross section of the materials prepared for consolidation by the IF/B Method (a) and a cross section of the piezoelectric ceramic fiber/Al composite (b).

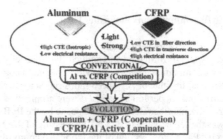

Figure 8. CFRP/Al active laminate as a typical Type II multifunctional structural material system.

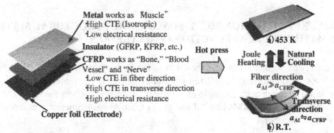

Figure 9. Structure of the CFRP/metal active laminate and its unidirectional actuation.

aluminum having a high CTE works as muscle, and the CFRP having a nearly zero CTE, especially the carbon fiber itself having a negative CTE, works as bone. In addition, the carbon fiber works as blood vessel to supply energy to the material system for actuation because it efficiently generates electric resistant heat, and it also works as nerve because it reflects temperature change as its electric resistivity change.

The materials used in this study are summarized in table I. The metal plate was used as the high CTE material. The CFRP prepreg was used as the low CTE/heating material. The KFRP prepreg was used as the low CTE material and the insulator between the metal or alloy plate and the CFRP layer. The copper foil was used to form the electrodes for electric resistance heating of the CFRP layer. The metal plate, the CFRP prepreg, the KFRP prepreg and the copper foil were cut into pieces. Bonding surfaces of the metal or alloy plate and the copper foil were roughened with a #320 and a #600 abrasive papers, respectively, and degreased with methyl ethyl ketone. They were prepared as shown in figure 10 and consolidated by hot pressing under the condition of 393 K, 0.1 MPa and 3.6 ks.

A schematic diagram of the curvature measurement set-up is shown in figure 11 (a). The active laminates were put on the block which has supporting edges of 50 mm span. Their shapes were measured with the laser displacement sensor attached on the x-y stage. The curvatures of them were calculated using those data. A schematic diagram of the output force measurement set-up is also shown in figure 11 (b). The active laminate was put on the same block used for curvature measurement, and fixed with the punch, which was heated and the force generated against the punch was measured with the load cell.

As the active laminates were fabricated by hot pressing in a flat die, they bent during

Table I. Materials used in this study.

Material	Function	Thickness, t [mm]	CTE, α [10⁻⁶K⁻¹]	Young's modulus, E [GPa]	Comments
Pure aluminum plate	High CTE	0.2	23.6	72	-
Pure titanium plate		0.2	8.9	116	-
CFRP prepreg	Low CTE/Heater	0.1	0.7	130	*1
KFRP prepreg	Insulator	0.15	-1.5	77	*2
Pure copper foil	Electrode	0.02	-	-	-

*1 : P3052S-10 (V_f=0.60) produced by Toray Co., Ltd. (Fiber : Torayca T700S. Resin : 393K cure type epoxy resin #2500.)
*2 : PC05-F-14 (V_f=0.56) produced by Toray Co.,Ltd. (Fiber : Kevlar 49 type 968. Resin : 393K cure type epoxy resin #2500.)

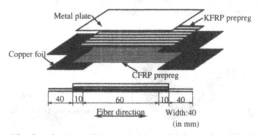

Figure 10. Lamination of the materials to fabricate the active laminates.

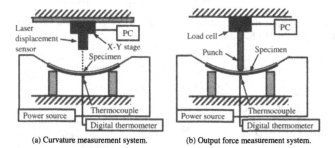

(a) Curvature measurement system. (b) Output force measurement system.

Figure 11. Schematic diagrams of the (a) curvature and (b) output force measurement systems.

cooling due to large contraction of the metal plate compared with that of the FRP (CFRP/KFRP) layer. To evaluate the performances of the active laminates, they were heated by electric resistance heating of the CFRP layers, and the effect of temperature on the curvatures of the active laminates using the pure aluminum, and the pure titanium plates in the fiber and transverse directions were obtained as shown in figure 12. Their shapes at 313 K are also shown in figure 13. According to these figures, the aluminum type laminate is better than the titanium one. The curvature of the aluminum type laminate linearly decreases with increasing temperature up to around the hot pressing temperature 393 K and the curvature in the transverse direction is kept at almost zero irrespective of the temperature change. In the case of the titanium type, it has a double curvature at 313 K due to the CTE mismatch existing not only in the fiber direction but also in the transverse direction. As for the generated forces shown in figure 14, the aluminum type laminate generates higher force than that of the titanium type. In the case of the aluminum type, from 313 K up to around 360 K, the force increases linearly with increasing temperature,

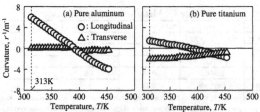

Figure 12. The curvature changes of the active laminates using (a) the pure aluminum plate and (b) the pure titanium plate in the longitudinal and transverse directions with increasing temperature.

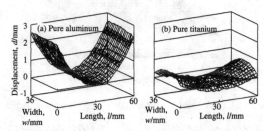

Figure 13. The shapes of the active laminates using (a) the pure aluminum plate and (b) the pure titanium plate at 313 K.

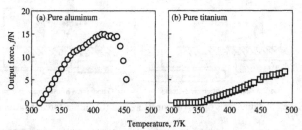

Figure 14. The output force changes of the active laminates under the constant metal thickness using (a) pure aluminum and (b) pure titanium with increasing temperature.

whereas the titanium type does not generate force due to its complicated shape change.

Taking advantage of the above mentioned actuation characteristics of the aluminum type active laminate, five types of demonstrators as shown in figure 15, that is, a hatch, stack, coil, lift type and one more additional type, were fabricated and demonstrated. In the case of the hatch type shown in figure 15 (a), a part of the body was cut out and this part was actuated by applying a voltage, where the initial current slightly increased with electric resistance decrease caused by temperature increase. This electrical resistance change can be used for monitoring its temperature and control of its curvature. Its actuation speed drastically increases with increasing input power. The voltage and current can be optimized by changing the resistance of the CFRP

258

layer. As an idea to obtain a larger displacement, the units of the active laminate were stacked as shown in figure 15 (b) and the height was changed by applying a voltage and measured as a function of temperature. The height can be changed almost linearly as a function of temperature, which is a very useful behavior for its application. A coil type demonstrator was also made and its diameter and pitch can be decreased with increasing temperature as shown in figure 15 (c). The active laminate is a high strength active material, so it does not need to generate force against something, but it can also do so as shown in figure 15 (d). One more additional example is given in figure 15 (e).

Active FRMs [20-22]

The concept was applied to SiC/Al composites and SiC/Ni composites to realize higher temperature active laminates, that is, a laminate of aluminum plate and SiC/Al composite plate, and a laminate of nickel plate and SiC/Ni composite. Thermal deformations of the laminates caused by non-uniform distribution of the SiC fibers were examined as a useful function. The former material system is introduced in detail in this paper. The materials were arranged as shown in figure 16 (a) to obtain a SiC/Al active composite having non-uniform distribution of the SiC fibers by the IF/B Method. The applied hot-pressing condition was the temperature of 873 K, the pressure of 2.7 MPa and the period of 1.2 ks in a vacuum of 1×10^2 Pa. The hot pressed sample was cooled in the furnace from hot-pressing temperature and its room temperature curvature was measured. The cooling was undertaken after removal of hot-pressing pressure. The cooling rate at 873 K was about 0.03 K/s. The pressure applied on the sample by the punch of the die-set during cooling was as low as 2.7 kPa. Temperature dependence of the curvature of the SiC/Al active composite was examined by thermal cycles of heating in an electric furnace and cooling in it. The heating rate was kept constant at 0.2 K/s and the cooling rate at the beginning (580K) was about 1 K/s.

In figure 16 (b), the appearances of the SiC/Al active composite at R.T. and 580 K are shown. According to this figure, the curvature decreases by heating and becomes around zero at 580K. It was thermally cycled between R.T. and 580 K and it was found that the curvature becomes reproducible after a couple of the thermal cycles as shown in figure 17.

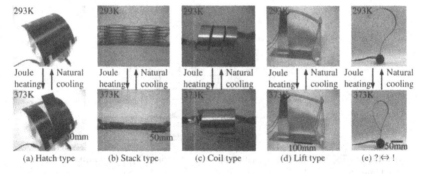

Figure 15. Five types of demonstrators using the CFRP/Al active laminate.

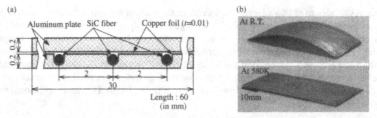

Figure 16. Arrangement of the materials for consolidation (a) and photographs of the SiC/Al active composite at room temperature and 580K (b).

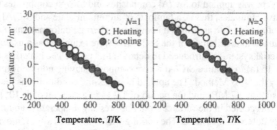

Temperature, T/K   Temperature, T/K

Figure 17. The curvature changes of the active composite during the Nth thermal cycles between room temperature and 813 K.

Ti fiber/Al multifunctional composites [23]

Ti fiber/Al multifunctional composite, is another interesting and useful example, and is summarized in figure 18. The basic concept is common to that of the CFRP/Al active laminate. The material system is very simple, but it can generate many useful functions such as heating, actuation, temperature sensing and strain sensing as shown in figure 19. This material system is under development.

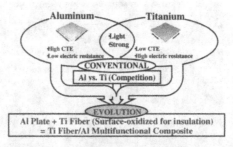

Figure 18. Ti fiber/Al multifunctional composite as a typical Type II multifunctional structural material system.

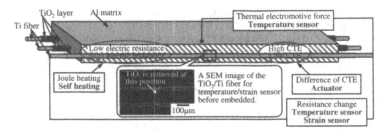

Figure 19. The functions generated by the Ti/Al multifunctional composite.

CONCLUSIONS

Multifunctional structural material systems have been successfully developed by innovative designing concept and processing as follows:

1) As the type of multifunctional structural material system of which functional properties are realized by embedding sophisticated functional materials in a metallic matrix as protective environment, FBG sensor/aluminum composite and piezoelectric ceramic fiber/aluminum composite were successfully developed by the IF/B Method.

2) As the type of multifunctional structural material system of which functional properties are realized by the innovative designing concept without using sophisticated functional materials, CFRP/Al active laminate was successfully developed.

REFERENCES

1. G. Lawcock, L. Ye and Y.W. Mai, SAMPE J. **31**, 175 (1995).
2. H. Asanuma, *Intelligent Materials*, edited by M. Shahinpoor and H.J. Schneider (RSC publishing, UK 2007), p. 478
3. H. Asanuma, *Metal and Ceramic Matrix Composites*, edited by B. Cantor, F. Dunne and I. Stone (Institute of Physics Publishing, UK 2003), p. 367
4. H. Asanuma, JOM, **52**, 21 (2000).
5. H. Asanuma et al., J. Jpn. Inst. of Light Metals, **40** (7), 527 (1990).
6. H. Asanuma et al., Proc. 6th Japan-U.S. Conference on Composite Materials, 434 (1992).
7. H. Asanuma et al., J. Intelligent Material Systems and Structures, **7** (3), 307 (1996).
8. H. Asanuma and A. Okura, Proc. 4th Intl. Conf. on Composite Materials Progress in Science and Engineering of Composites, **2**, 1435 (1982).
9. H. Asanuma, A. Okura, J. Jpn. Inst. Metals, **48** (11), 1119 (1984).
10. H. Asanuma, A. Okura, J. Jpn. Inst. Metals, **48** (12), 1198 (1984).
11. H. Asanuma et al., Proc. 11th Materials and Processing Conference (M&P2003, JSME), 63 (2003). (in Japanese)
12. H. Asanuma, N. Takeda, T. Chiba, H. Sato, Proc. 14th Materials and Processing Conference (M&P2006, JSME), 21 (2006). (in Japanese)
13. H. Asanuma, Proc. 10th Jpn. Intl. SAMPE Symposium & Exhibition, Smart Materials-1-1, 1 (2007).

14. H. Sato, Y. Shimojo, T. Sekiya, The 12th Intl. Solid State Sensors, Actuators and Microsystems, 512 (2003).
15. H. Asanuma, O. Haga, N. Naito and T. Tsuchiya, Proc. Annual Meeting of the Jpn. Society for Composite Materials, 19 (1996). (in Japanese)
16. H. Asanuma, O. Haga, J. Ohira, G. Hakoda and K. Kimura, Science and Technology of Advanced Materials , 3, 209 (2002).
17. H. Asanuma, O. Haga, J. Ohira, K. Takemoto and M. Imori, JSME Intl. J., Series A , 46, 478 (2003).
18. H. Asanuma, O. Haga and M. Imori, JSME Intl. J., Series A, 49, 32 (2006).
19. T. Nakata, H. Asanuma, T. Tanaka, M. Komori and O. Haga, J. Advanced Science, 18, 6 (2006).
20. H. Asanuma, G. Hakoda, and T. Mochizuki, JSME Intl. J., Series A, 46, 473 (2003).
21. H. Asanuma, G. Hakoda, H. Kurihara and Y. Lu, Materials Transactions, 46, 3, 691-696 (2005).
22. H. Asanuma, K. Kato and T. Mochizuki, J. Advanced Science, 18, 1&2, 16-19 (2006).
23. H. Asanuma, T. Ishii and G. Hakoda, Proc. 7th Jpn. Intl. SAMPE Symposium & Exhibition, 119 (2001).

Mater. Res. Soc. Symp. Proc. Vol. 1129 © 2009 Materials Research Society   1129-V11-01

Finite Element Analysis of Aluminum Nitride Bimorph Actuators – The Influence of Contact Geometry and Position

V. R. Pagán[1] and D. Korakakis[1,2]

[1] Lane Department of Computer Science and Electrical Engineering, West Virginia University, P.O. Box 6109, Morgantown, WV 26506-6109, U.S.A.
[2] National Energy Technology Laboratory, 3610 Collins Ferry Road, Morgantown, WV 26507-0880, U.S.A.

ABSTRACT

In this work, the results of 3-dimensional finite element analysis (FEA) of Aluminum Nitride (AlN) homogeneous bimorphs (d_{31} mode) are shown. The coupled-field FEA simulations were performed using the commercially available software tool ANSYS® Academic Research, v. 11.0. The effect of altering the contact geometry and position on the displacement, electric field, stress, and strain distributions for the static case is reported.

Piezoelectric beams are commonly used in microelectromechanical systems (MEMS) and have many possible applications in smart sensor and actuator systems. For example, they have been used as the active element in microfluidic and microactuator MEMS devices. In the actuator mode, they employ the converse piezoelectric effect to couple electrical energy into mechanical deformation. Aluminum Nitride (AlN) based devices have attracted much interest because AlN is a piezoelectric material with high thermal stability, high dielectric strength, a reasonable electromechanical coupling coefficient, and a perfect compatibility with standard silicon processing techniques.

INTRODUCTION

The geometry and position of the contacts on piezoelectric bimorphs are design parameters that have not been addressed much in the literature. Q. Q. Zhang et al. reported on the fabrication and simulation of d_{33} cantilever actuators that used interdigitated (IDT) contacts to produce electric fields in the plane of the piezoelectric layer [1] while O. J. Aldraihem et al. derived analytical solutions for cantilever beams with piezoelectric patches [2]. However, generally it is assumed that the top and bottom contacts span the entire length and width of the bimorph structure. This undoubtedly produces the highest tip displacement in d_{31} mode bimorphs and leads to the simplest analytical solutions. Nonetheless, the displacement profile produced is approximately parabolic, and as such, may not be optimal for certain applications.

This work builds on ongoing work being conducted in the area of surface acoustic wave (SAW) devices. Bulk micromaching techniques were used to create free standing silicon dioxide (SiO_2) diaphragms with dimensions $l = w \approx 290\mu m$ and $h \approx 400nm$ on a silicon (Si) substrate. Aluminum nitride (AlN) was deposited on the top surface of the diaphragm structure using DC magnetron reactive sputtering to form AlN/SiO_2 diaphragms (see figure 1a). AlN can be deposited on the bottom surface of the diaphragm structure using the same sputtering process. The top and bottom AlN film will then have opposing polarizations effectively creating a bimorph structure around an elastic layer. Simulations indicate that the presence of the elastic layer reduces the deflection of the bimorph (not shown). Thus, reducing the thickness of the elastic layer will improve the performance of the structure. This could be accomplished by etching the SiO$_2$ in hydrofluoric acid (HF) after depositing the top AlN layer and then depositing the bottom AlN layer. The cantilevers could be released by etching patterned diaphragms using inductively coupled plasma (ICP) etching. The resulting bimorph cantilever structure is shown in figure 1b.

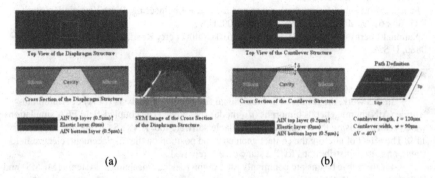

(a)   (b)

Figure 1. (a) Bulk micromachined diaphragm structure and (b) cantilever structure simulated using finite element analysis.

The deflection of a cantilever with $l >> w$ is given by equation 1 where f is a uniformly distributed load, E is the modulus of elasticity, $I = (1/12)wh^3$ is the moment of inertia, and l is the cantilever length.

$$\delta(y) = \frac{fy^2}{24EIl}\left(y^2 - 4ly + 6l^2\right)$$ (1)

For a given cantilever, the dimensions and material properties are fixed. Additionally, the resulting displacement profile is typically parabolic and is not a function of the cantilever width. Therefore, 3-d FEA was performed and the results are reported to show that the contact geometry and contact position can be optimized for a particular MEMS/smart sensor or actuator system.

EXPERIMENTAL DETAILS

The maximum deflection of a cantilever before breaking is given by equation 2 where S is the flexure strength, E is the modulus of elasticity, and l and h are the length and height of the cantilever, respectively.

$$\delta_{max} = \frac{2Sl^2}{3Eh} = 8.41\mu m$$ (2)

This value is an approximate maximum tip deflection (assuming a parabolic displacement profile) that an AlN cantilever with $l = 120\mu m$ and $h = 1\mu m$ can undergo before breaking.

264

Therefore, a dc voltage of $\Delta V = 40V$ was chosen because the simulation results indicate a maximum deflection of $\delta_{max(40V)} = 2.31\mu m < \delta_{max} = 8.41\mu m$ and because this voltage is readily available in the laboratory.

The finite element analysis was performed using the commercially available software tool ANSYS® Academic Research, v. 11.0. The model consisted of two SOLID226 coupled-field solid elements with their KEYOPTs set to 1001. This combination provided the model with displacement (UX, UY, UZ) and voltage (V) degrees of freedom (DOF). The elements were defined using two relative coordinate systems so that the polarization of the material would be opposing. The stiffness matrix coefficients, piezoelectric stress constants, dielectric permittivity, and density of AlN from literature were used with all values converted to μMKSV units [3,4]. The model was meshed with hexahedral-shaped elements using mapped meshing. This produced a mesh with 30,545 nodes and zero error/warnings.

The contact geometries simulated in this work are given in table I. The electric potential plot implies the shape of the top contact used. The bottom contact geometry was not varied and was defined to cover the entire length and width of the cantilever due to fabrication constraints.

Table I. ANSYS® Academic Research, v. 11.0 Results Viewer plots of the electric potential and total strain intensity of the contact geometries simulated in this work.

	Electric Potential	Total Strain Intensity
Full Contact (FC)		
Half Contact (HC)		
Diamond Shaped Contact (DC)		

T Shaped Contact (TC)		
Cantilever Shaped Contact (CC)		
Vertical Contacts (VC)		
Vertical Contacts $(4/3)^n$ (VC43)		
Vertical Contacts $(3/2)^n$ (VC32)		

DISCUSSION

The ANSYS® Academic Research, v. 11.0 Results Viewer was used to plot the stress intensity, total mechanical strain intensity, electric potential, and displacement of the bimorph in 3-d. The electric potential and total mechanical strain intensity 3-d plots are shown in table I. Three paths were defined and were named "Mid", "Edge", and "Tip" as shown in figure 1b. These paths correspond to $(x = w/2, y, z = 0)$, $(x = w, y, z = 0)$ and $(x, y = l,$ and $z = 0)$, respectively. The displacements in the z-direction of the cantilever along these paths for various contact geometries are shown in figure 2a-c.

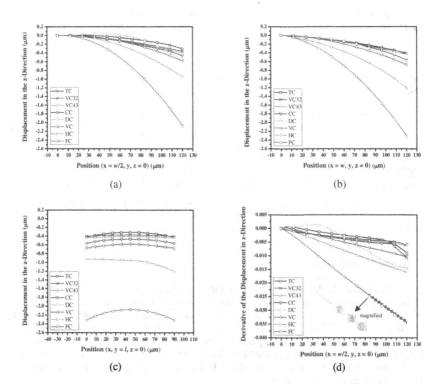

(a)

(b)

(c)

(d)

Figure 2. (a) Deflection along the "Mid" path (b) deflection along the "Edge" path (c) deflection along the "Tip" path and (d) derivative with respect to position of the deflection along the "Mid" path (paths are defined in Figure 1b).

It can be observed from figure 2a-c that the displacement varies as a function of contact geometry. The maximum deflection at $\Delta V = 40V$ was obtained from the structure with the "Full Contact" while the minimum deflection was obtained from the "T Contact". The displacement profiles of the tip were 2nd order for the "Full Contact" and "T Contact"; 3rd order for the "Half Contact"; 4th order for the "Cantilever Contact" and "Vertical Contact"; 6th order for the "Diamond Contact"; and, 8th order for the "Vertical Contacts (4/3)rd" and "Vertical Contacts (3/2)rd" as given by curve fitting with an adjusted R value ≥ 0.999. Additionally, the derivative of the "Mid" path was taken to show the slope of the displacement as a function of position. Linear curves indicate parabolic deflections.

CONCLUSIONS

This work has shown that the deflection of a piezoelectric bimorph in 3-d is a function of contact geometry and position. The displacement profile can be optimized for a specific application by engineering the geometry of the top contact. The maximum values of deflection, stress, strain, and contact area found in this study are provided in table II.

Table II. Maximum values of the deflection, stress, strain and area for the contacts defined in table I.

	FC	HC	DC	TC	CC	VC	VC43	VC32
δ_{max} (μm)	2.314	1.208	0.699	0.427	0.567	0.668	0.439	0.403
σ_{max} (MPa)	1,267	1,233	707	480	1,029	457	432	430
ε_{max} ($\times 10^{-3}$)	0.684	0.626	0.512	0.432	0.411	0.344	0.428	0.444
$A_{contact}$ (μm^2)	10,800	5,400	4,858	3,253	4,401	3,733	2,794	2,547

ACKNOWLEDGEMENTS

This technical effort was performed in support of the National Energy Technology Laboratory's on-going research in high temperature flow control hardware for advanced power systems under the RDS contract DE-AC26-04NT41817. This work was also supported in part by NSF RII contract EPS 0554328 for which WV EPSCoR and WVU Research Corp matched funds. V.R.P. was supported in part by a grant from the West Virginia Graduate Fellowship in Science, Technology, Engineering, and Math (STEM) program and the WVNano Bridge Award. V.R.P. would also like to thank A. Kabulski, R. D. Farrell, S. Kuchibhatla, and K. R. Kasarla for their helpful discussions and support.

REFERENCES

1. Q. Q. Zhang, S. J. Gross, S. Tadigadapa, T. N. Jackson, F. T. Djuth, and S. Trolier-McKinstry, *Sensors and Actuators A*, **105**, 1 (2003).
2. O. J. Aldraihem and A. A. Khdeir, *Composite Structures*, **60**, 2 (2003).
3. ANSYS® Academic Research, Release 11.0, Help System, Elements Reference, ANSYS, Inc. (2007).
4. J. G. Gualtieri, J. A. Kosinski, and A. Ballato, *IEEE Transactions on Ultrasonics, Ferroelectrics, and Frequency Control*, **41**, 1 (1994).

Mater. Res. Soc. Symp. Proc. Vol. 1129 © 2009 Materials Research Society   1129-V14-05

In-Plane Poisson's Ratio Measurement Method for Thin Film Materials by On-Chip Bending Test with Optical Interference Image Analysis

Mitsuhiro Tanaka, Takahiro Namazu, and Shozo Inoue
University of Hyogo, 2167 Shosha, Himeji, Hyogo 671-2201, JAPAN

ABSTRACT

This paper describes a simple evaluation method for in-plane Poisson's ratio of thin film materials. We designed an on-chip bending test chip that produces bending of film specimen via torsion bars, when applying normal load to loading levers. Using micro machining technologies, the test chip has been fabricated from SOI wafer. The profile of bent specimen obtained by using optical interference method enabled us to measure in-plane Poisson's ratio of the material. Average measurement value of in-plane Poisson's ratio was 0.079, which deviates by 10% from the anisotropic finite element analysis of the test chip. The measured value includes approximately 23% error as compared to ideal value for (001)[110] single-crystal silicon (SCS). This is probably caused by strong anisotropy in the material.

INTRODUCTION

By virtue of rapid growth in microeletrcomechanical systems (MEMS), primary importance is to accurately measure the mechanical properties of MEMS materials, and to integrate the measured properties into the design of MEMS devices. During the last decades, for material evaluation, tensile test is being often employed to evaluate Young's modulus, yield stress, and tensile strength of films used as structural materials in MEMS [1]. However, when dealing with a minute material, measuring a strain becomes a quite challenging task in that the strain arising from external loads to a specimen is sure to be small by minimization of specimen size. Poisson's ratio of a film material is, therefore, hard to be evaluated experimentally.

In this paper, novel Poisson's ratio evaluation technique is presented. This technique does not need complicated measurement technique during the test. Poisson's ratio of a miniature specimen can be easily evaluated by measuring interference fringes only. We have evaluated the Poisson's ratio of SCS, and its temperature dependency is discussed.

EXPERIMENTAL DETAIL

Working principle

When simultaneous bending loads are applied to both end of an isotropic beam-shaped specimen with a square/rectangular cross section, anticlastic deformation occurs to the specimen. The specimen deforms to saddle-like shape as illustrated in Fig. 1. The profile line of the saddle-like surface is a family of hyperbola. The angle, θ, between an asymptote of hyperbola and transversal to the principle bending axis is related to Poisson's ratio, v, as follows [2]:

$$\sqrt{v} = \frac{X}{Y} = \tan\theta \qquad (1)$$

where X and Y are the absolute values of coordinates. If the angle, θ, is able to be measured by the experiment, Poisson's ratio can be determined directly from the equation (1) .

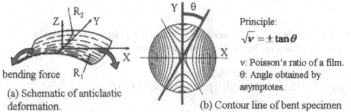

Principle:

$$\sqrt{v} = \pm \tan\theta$$

v: Poisson's ratio of a film.
θ: Angle obtained by asymptotes.

bending force R_1

(a) Schematic of anticlastic deformation.

(b) Contour line of bent specimen

Figure 1. Profile of a square/rectangular cross sectional beam bent by external force.

On-chip bending test system

In order to apply a bending load to a film specimen, we have newly designed the bending test structure that is integrated on a SCS chip. A schematic representation of the test chip is shown in Fig. 2. The test chip is composed of frame, torsion bars, loading levers, and a film specimen. When normal load is applied to the loading levers, torsion bars are twisted. Then loading levers turn around on the center of the torsion bar; consequently bending is produced in a film specimen.

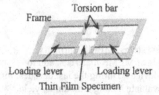

Torsion bar
Frame

Loading lever   Loading lever

Thin Film Specimen

Figure 2. Schematic of an on-chip bending test specimen.

On-chip bending test specimen was designed using finite element analysis (FEA). FEA model's specimen configuration is listed in Table 1, along with FEA conditions. The specimen configuration was determined by using Searle's analysis, which is provided an expression for differentiating between beam and plate with Searle parameter, β, as follows [3]:

$$\beta = \frac{b^2}{R_x t} \qquad (2)$$

where b, t, and R_x are specimen width, specimen thickness, and principal bending radius, respectively. Fig. 3 shows FEA results of the test chip. In Figs. 3(a) and (b), Searle's parameter increases with increasing the width-to-thickness ratio and principal bending radius. As shown in Fig. 3(c) showing relationship between measured Poisson's ratio and applied load to the levers, all the specimens takes the respective constant values throughout the applied loads. However, the specimen with high Searle's parameter has a difference of 30% at the maximum in Poisson's ratio measurement, as compared with input Poisson's ratio for FEA. Considering the accuracy of Poisson's ratio measurement, we determined the specimen sizes of type AI, AII, BI, and BII.

Anisotropic FEA was conducted to examine the measurement error of Poisson's ratio since the principle is based on an isotropic material. The contour lines of a bent specimen surface obtained by isotropic and anisotropic analyses of the test chip are depicted in Fig. 4. Difference in the measured Poisson's ratio between isotropic and anisotropic analyses is found to be about 10%. This indicates that 10% error may occur in the experiment for an anisotropic material. We emphasize the simplicity of the Poisson's ratio measurement, and apply equation (1) to measurement of Poisson's ratio for an anisotropic material.

Table 1. Specimen configuration and analysis conditions.

Specimen group	Specimen Configuration			Material property		Amount of applied load to loading levers.
	Width w [μm]	Length l [μm]	Thickness t [μm]	Elastic modulus E [GPa]	Poisson's ratio v	δ [μm]
Type	AI 500	2000	10	168.9	0.064	50
	AII 1000					
	AIII 1500				0.300	
	BI 1000					
	BII 2000	4000			0.499	
	BIII 3000					

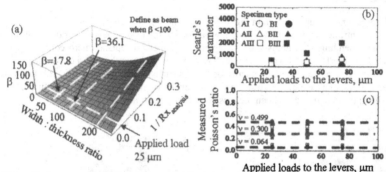

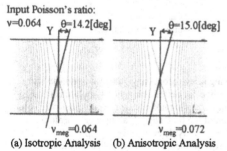

Applied load 25 μm

Figure 3. Determination of specimen configuration using Searle's analysis. These figures indicate (a) Searle's parameter as a function of the specimen's width-to-thickness ratio, and the inverse of applied bending radius, (b) relation between Searle's parameter, β, and applied load to the loading levers, and (c) Poisson's ratio measurement results obtained by FEA.

Input Poisson's ratio:
v=0.064 θ=14.2[deg]   θ=15.0[deg]

v_meg=0.064   v_meg=0.072
(a) Isotropic Analysis (b) Anisotropic Analysis

Figure 4. Comparison in contour line between isotropic and anisotropic analyses.

271

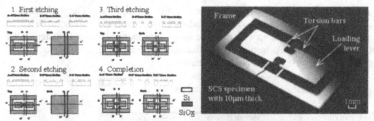

Figure 5. Fabrication process chart along with a picture of the specimen.

Fig. 5 shows a fabrication process chart and the produced on-chip bending specimen. The SCS bending specimen was fabricated through conventional photolithography and wet etching processes.

Experimental setup

Experimental setup is shown in Fig. 6. Two loading jigs whose position can be individually controlled by x-y-z stage apply normal load to the loading levers. For observation of contour line in bending, we employed optical interference method. Optical microscope equipped with Michelson interferometer was used and photographs were taken through a charge coupled device (CCD) camera attached to the microscope. A specimen holder with a cartridge heater was employed for temperature elevation test. To prevent any thermal damages to the objective lens, a thermal shield jig was placed between the lens and the specimen holder.

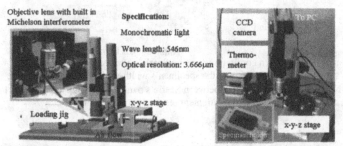

Figure 6. Experimental setup of bending test for Poisson's ratio measurement.

Optical image analysis

In this work, numerical calculation using image processing has been performed to determine the asymptotes of hyperbola. Flow chart for determining the asymptote is shown in Fig. 7. The chart can be separated into 3 major parts; (1) making a gray scale picture, (2) blob analysis, and (3) getting geometric information of the blob edge. Making a gray scale picture enabled us to shorten a calculation speed. To accurately determine the asymptote, blob analysis was conducted. Then the closure line of blob was defined, and numerical calculation was performed at each line to derive the points which the asymptote was laid on. The angle, θ, was finally obtained by the slope of the line that was drawn from intersection points.

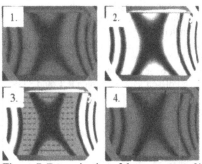

1. Original Picture

2. Make a monochrome picture based on original picture.

3. Blob analysis was conducted, then line scan was performed to obtain geometrical information of a hyperbola .

4. Draw an asymptotes of hyperbola.

Figure 7. Determination of the asymptote of hyperbolic interference fringe pattern.

EXPERIMENTAL RESULT

Interference fringe pattern

Fig. 8 depicts the typical interference pattern of the bent specimen. This pattern was obtained using the specimen of Type BII when the displacement of 50 μm was applied to the loading levers. The interference fringes formed to a hyperbolic pattern, similar to a schematic in Fig. 1. This indicates that a thin film specimen could deform to a saddle-like shape by the application of the bending force. The slope of the asymptotes in this pattern is 15.6 degree. By the equation (1), Poisson's ratio is found to be 0.078, 8.6% difference from the anisotropic analysis of SCS.

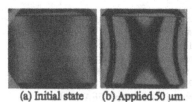

(a) Initial state   (b) Applied 50 μm.

Figure 8. Typical interference fringe pattern of SCS thin film specimen.

Evaluation of Poisson's ratio of SCS at R.T. and 323K

Fig. 9 shows relations of the angle of fringe and Poisson's ratio with applied displacement of the loading levers at R.T. and 323K. The square plots represent the angle that is obtained by the asymptotes and the transversal to the principal stress direction, and the circle plots represent calculated Poisson's ratio from the angles. The open and closed plots represent obtained data at R.T. and 323K, respectively. At R.T., when the displacement of loading levers was 40 μm, the angle measured by the asymptotes of hyperbola was 15.9 degrees. This angle was stayed constant over a displacement range of 0 to 80μm. However, measurement values tend to slightly increase above 80μm. This is probably related to overloading of principal bending radius to the specimen as discussed in FEA of the test chip. In the range of less than 80μm, the average of these measured angles is 15.5 degrees, resulting in Poisson's ratio of 0.077. This agrees with anisotropic analysis result of the test chip within 10%. Therefore, the proposed technique is useful for measuring the Poisson's ratio of a thin film material.

At 323K, a tendency of experimental result was similar to that at R.T. Within a range of 80μm, the measured angle showed a constant value of 15.4 degrees on average. The angle slightly increases with an increase of loading displacement over 80μm. Poisson's ratio at 323K was 0.076 in the range of less than 80μm. Difference in Poisson's ratio between R.T. and 323K is only 1%; therefore temperature dependency in Poisson's ratio of SCS was not found in temperatures ranging from R.T. to 323K.

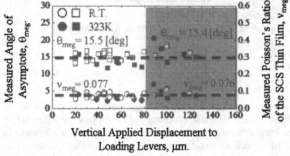

Vertical Applied Displacement to
Loading Levers, μm.

Figure 9. Experimental results obtained at R.T. and 323K.

CONCLUSIONS

We described "On-Chip Bending Test" which is a new method for simply evaluating Poisson's ratio of a film material. We fabricated the test chip on the basis of FEA results. From the experiment, we observed the saddle-like shaped deformation on film specimen by obtained hyperbolic interference fringe patterns. The angle between an asymptote of hyperbola and transversal to the principal bending axis was measured at R.T. and 323K. Experimental results showed that in-plane Poisson's ratios of SCS were 0.077 and 0.076 for R.T. and 323K, respectively. These results had an error of 10% compared with anisotropic FEA results.

REFERENCES

1. William N. Sharpe, Jr., Kmili M. Jackson, George Coles, Matthew A. Eby, and Richard L. Edwards, *Mechanical Properties of Structural Films*, pp.229-247 (2001).
2. C.G. Foster, *Exp. Mech.*16, 8, pp.311-315 (1976).
3. S. K. Kaldor and I. C. Noyan, *Appl. Phys. Lett.* 80, 13, pp.2284-2286 (2002).

Mater. Res. Soc. Symp. Proc. Vol. 1129 © 2009 Materials Research Society   1129-V14-01

RF Microwave Switches Based on Reversible Metal-Semiconductor Transition Properties of VO$_2$ Thin Films: An Attractive Way to Realise Simple RF Microelectronic Devices

Frédéric Dumas-Bouchiat[1,2,3], Corinne Champeaux[1], Alain Catherinot[1], Julien Givernaud[2], Aurelian Crûnteanu[2] and Pierre Blondy[2],
[1]SPCTS UMR 6638, Université de Limoges / CNRS, France,
[2]Xlim UMR 6172, Université de Limoges / CNRS, France,
[3]Institut Néel UPR 2940, Université de Grenoble I / CNRS, France.

ABSTRACT

Microwave switches in both shunt and series configurations have been developed using the semiconductor to metal (SC-M) transition of vanadium dioxide (VO$_2$) thin films deposited by *in situ* pulsed laser deposition on C-plane sapphire and SiO$_2$/Si substrates. The influence of geometrical parameters such as the length of the switch is shown. The VO$_2$-based switches exhibit up to 30-40 dB average isolation of the radio-frequency (RF) signal over a very wide frequency band (500 MHz-35 GHz) with weak insertion losses, when thermally activated. Furthermore, they can be electrically activated. Finally, these VO$_2$-based switches are integrated in the fabrication of innovative tunable band-stop filters which consist of a transmission line coupled with four U-shaped resonators and operate in the 9-11 GHz frequency range. Its tunability is demonstrated using electrical activation of each VO$_2$-based switch.

INTRODUCTION

The development of advanced communication systems for defense or space applications (antennas, multi-standard communication systems etc.) requires reconfigurable RF-microwave and millimeter-wave circuits. In order to address these demands with high performance devices, the integration of active electronic components such as semiconductor-based diodes or transistors or RF MEMS-based solutions [1,2] is investigated. The performance of such devices is often limited, for example, by the high consumption of the semiconductor components or by the weak reliability of the MEMS switches. These devices demand not only an improvement of the properties of conventional electronic materials, but also the introduction of new materials with new functionalities.

One potential candidate is a thermochromic vanadium dioxide [3] that undergoes an abrupt reversible semi-conductor to metal (SC-M) transition around 341K. At this temperature the crystalline structure changes from the low-temperature monoclinic to the high-temperature tetragonal phase. During the transition, the resistivity [4] of the material and especially its optical transmission in the near-infrared domain [5] decrease abruptly. The VO$_2$ SC-M transition can also be induced in an electrical [6,7] or optical [8,9] way. In these cases, the material response time is very fast and can achieve values as low as 1 ps. Because of its controllable properties, VO$_2$ is an interesting material for applications in a wide variety of devices such as microbolometers, light modulators and optical or electrical switches [10-12]

In this paper we describe the properties of vanadium dioxide thin films deposited by in situ pulsed laser deposition and their integration in RF shunt and series switches and in a more complex component, a four pole band-stop filter operating in the range 11-13 GHz which can be thermally or electrically activated.

EXPERIMENTAL SET UP – STRUCTURAL AND OPTICAL CHARACTERIZATION

Vanadium dioxide thin films are deposited by reactive pulsed laser deposition [13,14] using a KrF laser (wavelength 248nm, pulse duration 25ns, repetition rate 10Hz) focused on a high purity grade (99.95%) vanadium metal target. The ablation takes place in an ultrahigh vacuum deposition chamber under an oxygen atmosphere. According to the numerous phases [15] in the vanadium oxide phase diagram, from V_4O to V_2O_5, the deposition of pure VO_2 is obtained with a fluence, i.e. energy per irradiated unit surface area, of 3 J/cm², oxygen pressure of 2.2×10^{-2} mbar and substrate temperature of 500°C and without any post-treatment on C-plane sapphire and SiO_2(1μm)/Si substrates.

The film morphology consists of compact quasispherical crystallites with a root mean square roughness of about 10 nm on the two substrates as shown by the atomic force microscopy (AFM) image in figure 1. The apparent non-dependence of the substrate nature suggests that the growth mechanism of the film seems to be imposed mainly by the laser/target interaction.

X-Ray Diffraction (θ, 2θ) patterns of VO_2/Al_2O_3(C) are characterised by two peaks near 40.2° and 86.8°, corresponding respectively to the (020) and (040) peaks of the monoclinic VO_2 phase (fig. 2). On SiO_2/Si substrates, the larger lattice mismatch between substrate and film leads to the growth of (011) planes of an orthorhombic VO_2 structure such as already observed [16].

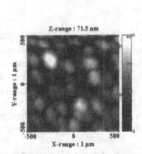

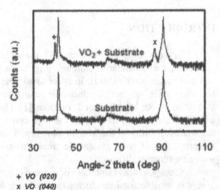

+ VO (020)
x VO (040)

Figure 1: Typical AFM image obtained on VO_2 films (thickness 75 nm) deposited onto silicon or sapphire substrates.

Figure 2: (θ, 2θ) X-ray scan for a 100 nm VO_2 thin film deposited on C-plane sapphire substrate.

Optical transmission spectra are measured as a function of the activation temperature using a Varian Carry 5000 UV Visible NIR spectrophotometer equipped with a sample heater. The transmittance decreases by a factor of about 4 during the transition from the semiconducting to the metallic phase. These curves present a particular point where the transmittance is the same for all the temperatures. This isosbestic point is situated at a wavelength of 850 nm, confirmed by other authors and investigations [17].

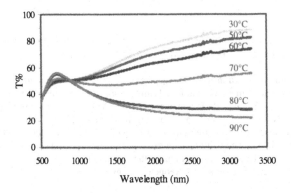

Figure 3: Transmission spectra of a 50 nm VO₂ thin film deposited on a sapphire substrate

ELECTRICAL CHARACTERIZATION AND MICROWAVE SWITCHING

The SC-M transition of the VO_2 thin films has been investigated during a heating-cooling loop in the 20-100 °C temperature range using a four-probe system showing typical hysteresis cycles. Films on sapphire substrates exhibit a change in resistivity between the SC and the metallic states of the order of 10^3 and a hysteresis width of the order of 5°C whereas for the films deposited on SiO_2/Si, the contrast is only 10^2 and the hysteresis width 10°C because of the difference in crystalline structure and film quality [12].

To study the switching properties of the VO_2 films, microwave switches (in both shunt and series configurations) were designed and fabricated on both types of substrates using micro fabrication process (wet etching, lift off). A microwave co-planar waveguide (CPW) is realised by thermal evaporation of gold (thickness 200nm). A VO_2 layer (thickness 200nm) is deposited by PLD all over the device and patterned. Finally the CPW metal lines are thickened up to ~800 nm in order to minimize the propagation losses of the signal. As a result, the CPW is "charged" by a VO_2 thin film structure. In shunt configuration, the VO_2 film covers the gap between the signal line (Fig. 4a) like a "bridge" with a width, hereafter labeled W. In series configuration, both parts of the central signal line are linked together through a VO_2 thin film (Fig. 4b) with a length labeled L.

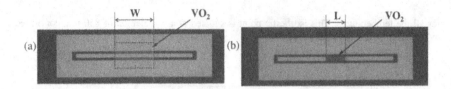

Figure 4: Microwave switch in parallel configuration (a) (length of VO_2: W=1mm) and (b) series configuration (length of VO_2: L=500μm)

The evolution of the microwave transmission S_{21} parameter is reported in figure 5 for a shunt switch on sapphire substrate with a VO_2 width of 1mm (fig. 4a) for different temperatures in the frequency range 500MHz – 35 GHz. At low temperatures, the VO_2 line is in the semiconducting state and introduces only very low losses on the propagation signal ($|S_{21}| < 3dB$ at 300K). An increase of temperature leads to a decrease of the resistivity of VO_2, and thus induces an increase of the losses of the RF signal, eventually leading to a short circuit to the ground ($|S_{21}| > 30dB$ at 400K). This result shows the possibility to monitor the attenuation of the signal from a few dB to 30-40 dB with the temperature over a large frequency domain.

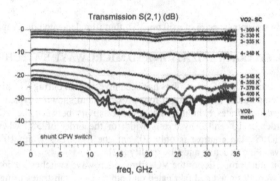

Figure 5: Thermal monitoring of the transmission S_{21} parameter of a microwave parallel switch (width of VO_2: W=1mm).

The influence of the width of the VO_2 shunt (W) is studied for devices on sapphire and on SiO_2/Si (Fig. 6). The insertion losses are about 1-2 dB at 300K when VO_2 is in SC state and is independent of the length of the VO_2 shunt. The average signal attenuation measured at 400K where VO_2 is in the metallic state increases from about 8dB to 25dB for SiO_2/Si devices when W increases from 250μm to 1mm. These values are weaker than those for sapphire devices which change from about 27dB to 40dB, due to the difference of the VO_2 film structure and of the resitivity jump. These experimental results indicate that the width of the VO_2 shunt can be adapted in order to obtain a chosen attenuation value.

From these results, we consider that the tunability of the attenuation can be reached by using several cascaded VO_2 –shunts on the signal line as shown in figure 7i. The comparison of the transmission parameter of a CPW line "charged" by a 150μm-width VO₂ bridge and by three cascaded 150μm-wide VO₂ "bridges" on a sapphire substrate is given in figure 7.

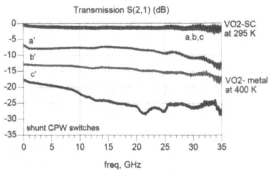

Figure 6: S_{21} parameter of VO₂ shunt switches on an SiO₂/Si substrate for different lengths of VO₂ : (a) 250μm, (b) 500μm and (c) 1mm. (a,b,c) measured at 300K with SC VO₂ and (a',b',c') measured at 400K with metal VO₂.

The attenuation level of the switch with one bridge is between 20 and 30 dB and as expected, the attenuation obtained with the switch with three bridges is higher, between 30 and 50 dB, depending on the frequency.

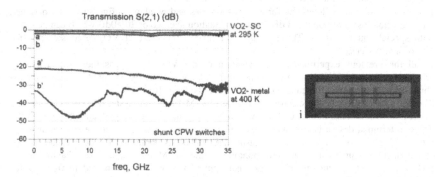

Figure 7: S_{21} parameter of VO₂ shunt switches on a sapphire substrate: (a) width of VO₂: W=150μm, (b) width of VO₂: 3x150μm as given in (i). (a,b) measured at 300K with SC VO₂ and (a',b') measured at 400K with metal VO₂

Similar switching investigations have been undertaken on switches in series configuration (fig. 4b). In this case, when the VO₂ line is semiconducting, i.e. for temperature lower than 341K, the CPW is electrically discontinuous and the signal is strongly attenuated. The signal is transmitted with weak insertion losses when the VO₂ line is metallic, above the transition temperature. Figure 8 shows the influences of the substrate nature and of the length L of the VO$_2$ line on the transmission parameter in the frequency range 500MHz-35GHz.

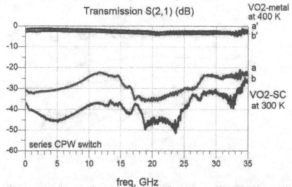

Figure 8: S$_{21}$ parameter of VO$_2$ serial switches on a sapphire substrate for different lengths L of VO$_2$: (a) 100µm, (b) 250µm. (a,b) measured at 300K with SC VO$_2$ and (a',b') measured at 400K with metal VO$_2$.

The insertion losses at 400K are about 3 dB for switches on sapphire. The signal attenuation at 300K increases with L (average value of about 30 dB for L=100µm and 40dB for L=250µm) because of the increase of the resistance of the VO$_2$ line.

In summary, similar switching performances for series and parallel configurations are obtained with insertion losses of about 2-3 dB and an attenuation average of 20-40dB which can be tuned with special shunt devices. These results underline the interest of VO$_2$-based switches for microwave devices.

In all the previous experiments, the transition of the VO$_2$ material is thermally activated. Obviously its electrical activation would be preferred, particularly for the tunable cascaded VO$_2$-shunt devices. Moreover the transition is expected to be faster [6].

Therefore to investigate the electrically induced SC-M transition, a special series device was fabricated with 200 nm-thick VO$_2$ films with a VO$_2$ line of 40 µm in length and 95µm in width. The two-terminal device is introduced in an electrical circuit with a cc voltage source, an ampmeter and a resistor R$_S$. When an increasing DC voltage is applied to the device, the current slowly increases up to a jump corresponding to the SC-M transition. The behavior of the measured current as function of the external applied voltage V$_{ap}$ is reported in figure 9 for different values of R$_S$.

At a given threshold of the applied voltage (about 12.6V for R$_S$=200Ω, 16.3V for R$_S$ = 600Ω, and 20V for R$_S$ = 1000Ω), the current increases abruptly in the circuit, due to the abrupt change of the resistivity of the VO$_2$ layer on passing from the SC to the metallic state. The device remains in the activated state as long as the voltage is maintained in the circuit. The presence of two different positive slopes before and after the jump in current underlines the SC-M transition. The ratio of the slopes is higher than about 300, in good agreement with the contrast in the resistivity measured during thermal heating for these specific two-terminal devices.

When the applied voltage is reduced, the VO$_2$ film goes back to the SC state with an hysteresistic behavior.

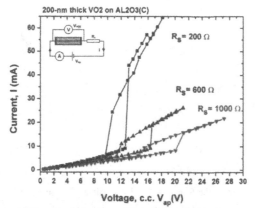

200-nm thick VO2 on AL2O3(C)

$R_s = 200 \, \Omega$

$R_s = 600 \, \Omega$

$R_s = 1000 \, \Omega$

Voltage, c.c. V_{ap}(V)

Figure 9: Current (I) – Applied voltage (V_{ap}) characteristic of a two-terminal VO$_2$ device: electrically -induced SC-M transition.

The estimation of the switching time for these switches reported in [12] is as low as some hundreds of nano-seconds, which is better than the current MEMS-based switches and not so far from semiconductor capabilities. Further optimization of the switch and its electrical activation system is in progress in order to improve the time response of these devices.

DESIGN AND CHARACTERIZATION OF A BAND-STOP FILTER TUNED BY THE VO$_2$ SC-M TRANSITION

The band-stop filter operating in the 9-13 GHz frequency range consists of a 50 Ω signal line coupled to four U-shaped resonators, in microstrip configuration. The extremities of each U are linked by a VO$_2$ film. The activation of the VO$_2$-switches induces a change in the frequency resonance of the resonator leading to a rejection band in the signal transmission spectrum. The experimental validation of the filter, according to the simulated characteristics was reported previously [18] using simultaneous thermal activation of the four resonators. Since the individual heating of each VO$_2$ switch is difficult, requiring the integration of individual micro-heaters which may disturb the signal propagation, we developed a filter using electrical activation of VO$_2$ switches, as shown in figure 10. The geometrical dimensions of the filters are optimized to obtain a similar filter response in the same frequency domain. Each U-shaped resonator is closed by a VO$_2$ rectangular switch of size 200µm x 150µm, accompanied by its electrical activation pads.

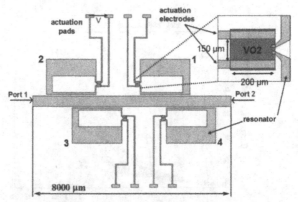

Figure 10: Design of the four pole band stop filter tuned by electrical activation of VO_2-based resonators.

The simulated response of the filter is shown in Fig. 11. When VO_2 films are in the SC state (at 300K), the switches are not activated and the filter presents a four-pole band rejection with attenuation up to 25dB around 10-11GHz. When all the switches are activated (VO_2 in the metallic state) the rejection band is shifted away from the investigated frequency domain (5-15GHz) and the propagating signal is slightly attenuated.

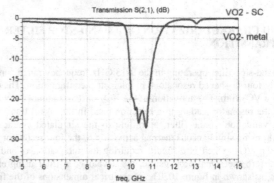

Figure 11: Simulated S_{21} transmission parameter of the four pole band stop filter when the VO_2 resonators are "open" (VO_2 SC) and "closed" (VO_2 metal).

Figures 12 and 13 present the measured filter transmission for thermal and electrical activation, respectively, of all the resonators. The results are in good agreement with the simulation concerning the transmission contrast and the position of the rejection band, considering the deviation from the theoretical values of the material constants taken into account in the simulation and of the size of the switches. Evidently the measured responses are the same at room temperature (VO_2 switches in insulating state). Thermal or electrical activation of the filter

(Fig.12 and 13, VO$_2$-metal) leads to the elimination of the rejection band from the studied frequency range. The relatively high insertion losses of about 4dB are due to the losses of the connecting cables and connectors and also to an impedance miss-match between the connectors and the ends of the transmission line of the filter. The parasitic band between 10 and 12 GHz (Fig. 13) for the electrical activation of all four resonators most probably reflects the influence of the actuation pads and of the wire bonding process during the housing of the filter.

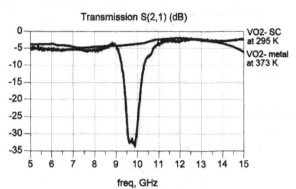

Figure 12: Measured responses of the four pole band stop filter, at room temperature (curve VO$_2$-SC) and in the thermally activated state (curve VO$_2$-metal).

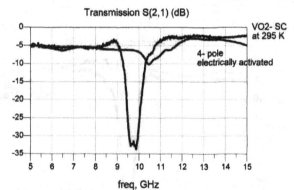

Figure 13: Measured responses of the four pole band stop filter at room temperature (curve VO$_2$-SC) and in the electrically activated state (curve 4-pole electrically activated).

The tunability of the filter can be demonstrated by electrical activation of selected resonators. When the VO$_2$-switch of two resonators, e.g. resonators 1 and 4 (Fig. 11), becomes metallic, the rejection band of the filter has a central frequency of about 9.8GHz and a full width at half maximum (FWHM) of 0.3 GHz, as shown in figure 14. The electrical activation of two other resonators, e.g. resonators 2 and 3 in Fig. 11, leads to a displacement of the central frequency of the filter to lower frequencies (Fig. 15) at 9.55 GHz (FWHM = 0.5GHz). Note that the parasitic

283

influence also appears when electrical actuation of individual resonators was further substracted from the measured responses presented below.

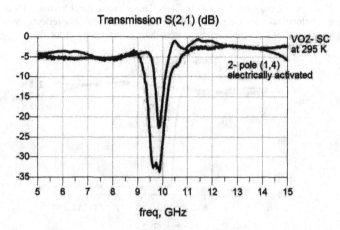

Figure 14: Measured response of the four pole band stop filter, when resonators 1 and 4 (Fig. 11) are simultaneously electrically activated compared to the measured response without activation (curve VO2-SC).

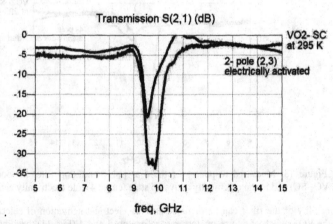

Figure 15: Measured responses of the four pole band stop filter, when two resonators (2,3) (Fig. 11) are electrically activated compared to the measured response without activation (curve VO2-SC).

284

CONCLUSIONS

Vanadium dioxide thin films were deposited by in situ pulsed laser deposition. Their microstructure and crystalline orientation, which depend on the substrate (i.e. monoclinic on sapphire and orthorhombic on SiO_2/Si), control their physical properties: resistivity, optical transmission,... These films are integrated in millimeter-wave coplanar switching devices in both configurations (shunt, series). The thermal commutation of switches leads to an attenuation of 30 dB of the signal propagation over a very wide band frequency 500MHz-35 GHz and can be tuned by choosing appropriate geometrical parameters. The VO_2-based switches were used to design a four pole band stop filter. Electrical activation of individual switches allows the tuning of the filter in the frequency range 9-11 GHz.

In conclusion, VO_2-based switches are promising candidates for realizing efficient and simple microwave switches as well as more complex devices for telecommunication applications.

ACKNOWLEDGMENTS

The authors would like to acknowledge the financial support from "Agence Nationale de la Recherche" (ANR, France) through the research grant "Admos-VO2" no. JC07_190648.

REFERENCES

1. D. M. Pozar, *Microwave Engineering – 3rd ed.*, J. Wiley & Sons, (2005).
2. G. M. Rebeiz, *RF MEMS Theory, Design, and Technology*, New Jersey: J. Wiley & Sons, (2003).
3. F. Morin, Phys. Rev. Lett. 3, 34 (1959).
4. A. Zylbersztejn, N.F. Mott, Phys. Rev. B, 11[11], 4383 (1975)
5. H.W. Verleur, A. S. Barker Jr., C.N. Berglund, Phys. Rev. 172 [3], 788 (1968).
6. G. Stefanovich, A. Pergament, D. Stefanovich, J. Phys.: Condens. Matter 12, 8837 (2000).
7. C. Chen, R. Wang, L. Shang, C. Guo, Appl. Phys. Lett. 93, 171101 (2008).
8. A. Cavalleri, C. Tóth, C.W. Siders, J.A. Squier, F. Ráksi, P. Forget, J.C. Kieffer, Phys. Rev. Lett. 87(23), 237401 (2001).
9. T. Ben-Messaoud, G. Landry, J.P. Gariépy, B. Ramamoorthy, P.V. Ashrit, A. Haché, Opt. Commun., doi:10.1016/j.optcom.2008.09.027.
10. L.A.Luz de Almeida, G.S. Deep, A.M.Nogueira Lima, IEEE Trans. Instrument. Measure. 50[4], 1020 (2001).
11. S. Chen, H. Ma, X. Yi, H. Wang, X. Tao, M. Chen, X. Li, C. Ke, Infrared Physics 1 Technology 45, 239 (2004).
12. F. Dumas-Bouchiat, C. Champeaux, A. Catherinot, A. Crunteanu and P. Blondy, Appl. Phys. Lett. 91, 223505 (2007).
13. D B Chrisey and G K Hubler, *Pulsed Laser Deposition of Thin Films* New York: Wiley, (1994).

14. R. Eason, *Pulsed laser deposition of thin films; Applications-led growth of functional materials*, Wiley Interscience, (2007).

15. C H Griffith and H K Eastwood, J. Appl. Phys.45, 2201 (1974).

16. D. Youn, J. Lee, B. Chae, H. Kim, S. Maeng, and K. Kang, J. Appl. Phys.95, 1407 (2004).

17. M. M. Qazilbash, M. Brehm, Byung-Gyu Chae,P.-C. Ho, G. O. Andreev, Bong-Jun Kim, Sun Jin Yun, A. V. Balatsky, M. B. Maple, F. Keilmann, Hyun-Tak Kim,3 D. N. Basov, Science 318, 1750 (2007).

18. J. Givernaud, C. Champeaux, A. Catherinot, A. Pothier, P. Blondy, A. Crunteanu, IEEE MTT-S, IMS 2008, Atlanta, WEP1D-02.

Mater. Res. Soc. Symp. Proc. Vol. 1129 © 2009 Materials Research Society 1129-V11-24

Microstructure and Service Life of Silver Copper Oxide Contact Materials After Reactive Synthesis Fabrication and Severe Plastic Deformation

Zhou Xiao-Long *[1,2], Cao Jian-chun [3], Chen Jing-chao [1,2], Zhang Kun-Hua [4], Du Yan[1,2], Sun Jia-ling [4]

[1]Key Laboratory Of Advanced Materials of Yunnan Province, Kunming University of Science and Technology, Kunming 650093, P. R. CHINA
[2]Key Laboratory of Precious and Nonferrous Metal Advanced Materials (Kunming University of Science and Technology), Ministry of Education, Kunming 650093, P. R. CHINA
[3]Faculty of Materials and Metallurgy Engineering, Kunming University of Science and Technology, Kunming 650093, P. R. CHINA
[4]Kunming Institute of Precious Metal, Kunming 650221, P. R. CHINA

Abstract: A novel reactive synthesis was developed to fabricate silver copper oxide contact materials in this paper. The x-ray diffraction (XRD) and scanning electron microscopy(SEM) were adopted to analyze phase and characterize microstructure. The results showed the CuO particles were agglomerate and fine, which is good for a better AgCuO composites microstructure dispersion in the process. Specially ringed microstructure was formed, in which those fine CuO particles were distributed around Ag matrix particles which is larger in size,. The severe plastic deformation was utilized to homogenize AgCuO composites. SEM analysis results showed microstructure of severe plastic deformation AgCuO composites was uniform and CuO particles were well dispersed. Furthermore, the service life of the silver copper oxide composites as contact materials was tested, which was compared with that of other Ag matrix contact materials. The results showed that the service life of prepared AgCuO contact materials was close to that of $AgSnO_2$ contact materials fabricated by reactive synthesis and contain 8% SnO_2 and was approximately two times of that of AgCdO and AgCuO contact materials fabricated by powder metallurgy. Therefore, the AgCuO composites fabricated by reactive synthesis can substitute for AgCdO composites as contact materials.
Keywords: AgCuO, Contact materials, Reactive synthesis, Severe plastic deformation

1.Introduction

In the electrical contact materials, the silver metal oxide (AgMeO) composites are used extensively to the lower-electrical equipment owing to resistance against contact erosion, contact sticking and better electrical conductivity [1], high melting temperature and hardness. The silver cadmium oxide (AgCdO) composites are representative of the AgMeO contact materials and used to electrical switching contact or relays in the electrical equipment. But The AgCdO composites have health hazards in using. On behalf of environmental concerns the European Union is going to restrict the use of AgCdO contact materials in electrical and electronic consumer applications in 2006[2]. So, the alternative another AgMeO was studied in the world [1-8]. The AgCuO composites are one of the alternative AgMeO electrical contact materials [6-8].

Reactive synthesis [6-9] is a new technique to synthesis AgMeO contact materials, which can be defined as a solidification reaction according to the chemical reactions in sintering process:
$$AgMe + Ag_2O \rightarrow 3Ag + MeO$$
Comparing with traditional fabrication technique which have powder metallurgy and internal oxidation to fabricate AgMeO composites, the reactive synthesis have following characters: firstly, the reinforcement-matrix interfaces are produced by the reaction, which contributes to improve the wettability for the interface and increase the bonding strength of interface between reinforcement and matrix. Secondly, the reinforcement particles produced by reactive synthesis method are finer in size. The metal oxidation particles are circularity distributed and around the

matrix particles. Thirdly, since it is a one-step process for reaction and sinter, the fabrication cost of the reactive synthesis is lower.

The severe plastic deformation (SPD) methods were rapidly developed in recent years [10-15] to refine microstructure and fabricate bulk nano-structured materials. It bases on the fact that heavy deformations can result in significant refinement of microstructure in metal materials at low temperature and the Hall–Petch relationship specifies that the yield stress of a material varies inversely with the square-root of the grain size [11]. Moreover, the SPD provides an opportunity to homogenize the microstructure of metal matrix composites.

In this paper, the AgCuO composites were successful fabricated by reactive synthesis. The microstructure homogenization was investigated in the different deformation condition. The service life testing had been adopted to compare with several AgMeO composites as contact materials.

2.Experiment

The raw materials included the silver powder with particles size of 35-45μm (purity >99.5%), silver copper alloy with particles size of 30-40μm (28%Cu, at the weight percent), and another reactant materials with particles size of 30-40μm (purity >99.8%). The raw materials were mixed, formed, sintered, reactive synthesized, severe plastic deformed and drawn to wires in diameter of 3mm. The CuO content is 10% in the AgCuO contact materials. The process parameters were as follows:

1)Forming: take the cold isostatic pressing forming at 120MPa for 5min.

2)Sintering and reactive synthesis: the conditions of the sinter and reactive synthesis are 650H, 4 hour and 2×10^{-3}Pa (vacuum degree) in the vacuum reactive synthesis sinter furnace.

3) Severe plastic deforming and drawing: take extrusion severe plastic deformation with the true strain of 4.0 or 8.0.

The products, i.e. the AgCuO composites, were examined by XRD (D8 ADVANCE, Bruker company of Germany) for phase determination, as well as SEM (XL30 ESEM, Philips company of Netherlands) with an EDS facility for microstructure characterization. For the service life test, the wires in 3mm were fabricated sample to test the service times in the contact frequency of 105 times/min. The test was run according to the standard ways in the laboratory on insulating material product accredited by the state administration of import and export commodity inspection of china (IMAL).

3. Results and discussion
3.1 Phase

Fig.1 shows the X-ray diffraction analysis results for AgCuO composites from reactive synthesis. The composites consist of only silver and copper oxide and the AgCuO composites have been fabricated by reactive synthesis.

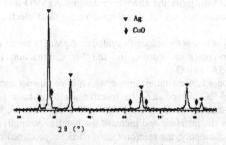

Fig.1 X-Ray diffraction pattern of reactive synthesis AgCuO

288

3.2 Microstructure

Fig.2 The sinter microstructure of reactive synthesis AgCuO composites

The sinter microstructure of reactive synthesis AgCuO contact materials is presented in Fig.2. From Fig.2(a), the reinforcement particles, i.e., CuO particles, distribute in specially ringed pattern, i.e., fine CuO particles are around Ag matrix particles (black-gray color in Fig.2) and Ag matrix particles (gray-white color in Fig.2) is larger in size. It is difficult to homogenize this microstructure. So, it is important that a special method, for example, severe plastic deformation, is used to obtain uniform microstructure and program properties of AgCuO contact materials. Fig.2(b) shows that CuO particles are fine and about several μm to 25 μm in size. As well those CuO particles agglomerate. These particles fine and agglomerate can be testified from the deformed composites showed in Fig.3 and Fig.4. These agglomerate and fine particles are very important for the process of severe plastic deformation, which can ensure the CuO particles dispersion in the processing.

Fig.3 and Fig.4 show microstructure of AgCuO composites after severe plastic deformation, in

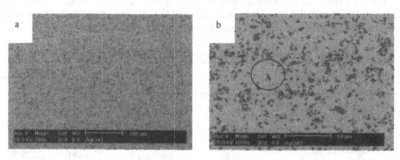

Fig. 3 The microstructure of reactive synthesis AgCuO after severe plastic deformation (true strain 4.0, 200×(a) and 1000×(b))

which the true strains are 4.0 and 8.0 in Fig.3 and in Fig.4 respectively. Comparing Fig.2(a) with Fig.3(a) or Fig.4(a), the microstructure tends to be homogenized after severe plastic deformation. In Fig.3 and Fig.4, the CuO particles are about several μm to 10 μm, which testifies the CuO particles are agglomerate after reactive synthesis in Fig.2. But the degree of homogenization can not be distinguished from Fig.3(a) and Fig.4(a) in the low magnified multiple(200×). For the sake of seeing about the degree of disperse CuO particles, the high magnified multiple microstructure are showed in Fig.3(b) and Fig.4(b) . The size of Ag matrix in Fig.4(b) is found smaller than that in Fig.3(b) when area A in Fig.3(b) is compared with area B in Fig.4(b).The degree of CuO particles dispersion in Fig.4(b) is better than that in Fig.3(b). The true strain of severe plastic deformation is the major reason degree of CuO particles dispersion. However, the CuO particles could not be perfectly completed to disperse at true strain of 8.0(Fig 4(b)). The true strain must be increased for a better dispersion of CuO particles, for example 12.0, in the process of severe plastic deformation.

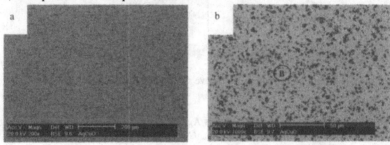

Fig. 4 The microstructure of reactive synthesis AgCuO after severe plastic deformation (true strain8.0, 200×(a) and 1000×(b))

3.3 Service life

The data in Table1 shows the service life test results of the prepared AgCuO contact materials. The electric service life of the AgCuO contact materials are 92925 times in alternating current (AC) and 108000 times in direct current (DC). As showed in the Table2, the service life of AgCuO contact materials is close to that of $AgSnO_2$ contact materials which were fabricated by reactive synthesis (RS) and contain 8% SnO_2[9] and is approximately two times that of AgCdO and AgCuO contact material by powder metallurgy(PM) [16]. Therefore, AgCuO composites fabricated by reactive synthesis can substitute for AgCdO composites as contact materials.

Table 1 Service life testing results of AgCuO contact materials

Test item		Unit	Test results	
			AC	DC
Weight loss of moving contact	1	mg	-16.4	-4.6
	3		-12.4	-8.1
	5		-18.0	-7.7
Weight loss of stillness contact	2	mg	-7.3	-17.7
	4		-8.0	-28.3
	6		-9.8	-21.3
Average weight loss of ever pair of contacts	1-2	mg	-23.7	-22.3
	3-4		-22.4	-36.4
	5-6		-27.8	-29.0
Average weight loss of there pairs of contacts		mg	-24.63	-29.63
Electric life of contacts		times	92925	108000
Average weight loss of ten thousand times		mg/ (10000 times)	-2.65	-2.74

290

| Weld or adhesive (Average) | times | 1.3 | 0.6 |

Remark: AC: Voltage 220V Current 15A Contact frequency: 105 times/min
DC: Voltage 24V Current 15A Contact frequency 105 times/min
Closing force: 100g Breaking force: 75g

Table.2 Comparison of service life test among different Ag matrix contact materials

Composites	Cycles($\times 10^4$)	
	AC(220V- 15A-105)	DC(24V- 15A-105)
AgCuO (10) (RS)	9.29	10.8
AgCuO (10) (PM)	4.23	5.86
AgSnO$_2$(8)(RS)	9.2	13.4
AgSnO$_2$(10)(RS)	11	14.1
AgSnO$_2$(10)(PM)	9.5	6.1
AgCdO(12)(PM)	6.5	5.1

Remark: The data of AgSnO$_2$ and AgCdO materials come from [9] and[16].
The data of AgCuO(10)(PM) materials come from [6].

4.Conclusions

By reactive synthesis technique, the AgCuO composites were fabricated successfully and specially ringed microstructure of the agglomerate and fine CuO particles around Ag matrix particles are formed. This microstructure can be homogenized and the agglomerate CuO particles can be dispersed by severe plastic deformation. The service life test results show that the service life of AgCuO contact materials is close to AgSnO$_2$ contact materials which were fabricated by reactive synthesis(RS) and contain 8% SnO$_2$ [9,16], and is approximately two times that of AgCdO and AgCuO contact materials fabricated by powder metallurgy(PM). AgCuO composites from reactive synthesis can substitute for AgCdO composites as contact materials.

Acknowledgements

The authors would like to acknowledge National Key Technologies Research and Development Program of China for financial support (NO.2001BA326C), acknowledge key laboratory of Advanced Materials of Yunnan Province for financial and free analysis support, acknowledge the laboratory on insulating material product accredited by the state administration of import and export commodity inspection of china (IMAL) for Service life test.

References

[1] Liu Jun, The development and application of silver metal oxide contact materials. Jiangsu wiring(in chinese). (1)(2000)29.

[2] Chen Jingchao, Sun Jialin, Zhang Kunhua, et al. Restrictive Policy of European Union in Silver / Cadmium Oxide and Development of Other Silver / Metal Oxide Electrical Contact Materials, Electrical engineering materials(in Chinese). (4)(2002)41.

[3] B. Gengenbach, K. W. Jäger, et al. Mechanism of arc erosion on silver-tin oxide contact materials, Proc. 11 ICECP, Berlin, 1982,p208.

[4] G. Zhang, Deformation and fracture analysis for AgSnO2 composite, Precious Metal (in Chinese), 20(4)(1999)1.

[5] F.Hauner, D.Jeanot,K.Mcneilly,in Proc.20th Int. Vonf. Elect. Contact, Advanced AgSnO2 Contact materials for Current Contacters, Edited by IEEE(Stockholm Sweden,2000,p19.

[6] Zhou Xiaolong,Cao Jianchun,Chen Jingchao, et al. Effects of Preparation Method on Microstructure and Electrical Contact Performance for AgCuO Composites Precious Metals(in Chinese). 3(29)(2005)25.

291

[7] Yan Xingli, Chen Jingchao, Zhou Xiaoiong, et al. Microstructure and Properties of Silver-Copper Oxide Electrical Contact Material Fabricated by Reactive Synthesis, Precious Metals(in Chinese). 4(25)(2004)22.

[8] Zhou Xiaolong, Cao Jianchun, Chen Jingchao, et al. Microstructure Com pared of AgSnO z and AgCuO Composites by Reactive Synthesis Fabrication, Electrial Engineering Materials(in Chinese).(3)(2004)7.

[9] Chen Jingchao, Sun Jialin, Zhang Kunhua, et al. Silver Tin Oxide Electrical Contact Material Fabricated by Reactive Synthesis, ICEC. Zurich. 2002, p447.

[10] Masahide Sato, Nobuhiro Tsujib, Yoritoshi Minamino, et al. Formation of nanocrystalline surface layers in various metallic materials by near surface severe plastic deformation, Science and Technology of Advanced Materials. (5)(2004)145.

[11] Cheng Xu, Minoru Furukawa, Zenji Horita, et al. Severe plastic deformation as a processing tool for developing superplastic metals, Journal of Alloys and Compounds. 378(2004)27.

[12] Kyung-Tae Park, Duck-Young Hwang, Young-Kook Lee. High strain rate superplasticity of submicrometer grained 5083 Alalloy containing scandium fabricated by severe plastic deformation, Materials Science and Engineering. A341(2003)273.

[13] S.S.M. Tavares, D. Gunderov, V. Stolyarov, et al. Phase transformation induced by severe plastic deformation in the AISI 304L stainless steel, Materials Science and Engineering. A358(2003) 32.

[14] R.S. Mishra, R.Z. Valiev, S.X. McFadden, R.K. Islamgaliev, et al. Severe plastic deformation processing and high strain rate superplasticity in an aluminum matrix composite, Scripta Materialia. 10 (40)(1999)1151.

[15] Irina G. Brodova,Denis V. Bashlykov, Alexander B. Manukhin.etal. Formation of nanostructure in rapidly solidified Al-Zr alloy by severe plastic deformation, Scripta Materialia. 44(2001)1761.

[16] Chen Jingchao, Research Report of Yunnan Province Key Technologies Research and Development Program of China, 2005.

Mater. Res. Soc. Symp. Proc. Vol. 1129 © 2009 Materials Research Society   1129-V07-10

Temperature Dependent SPR study of ZnO thin Film

Shibu Saha[1], K. Sreenivas[1] and Vinay Gupta[1,a]
[1]Department of Physics and Astrophysics, University of Delhi, Delhi-110 007, India
[a] E-mail address of corresponding author: vgupta@physics.du.ac.in

Abstract
In the present study the optical properties of a RF-sputtered Zinc Oxide(ZnO) thin film has been investigated using a simple Surface Plasmon Resonance (SPR) based on Kretschmann-Reather configuration. The SPR data is taken over a wide range of temperature from room temperature to 250°C. The dielectric constant at 6328 Å wavelength is obtained by fitting the experimentally obtained data with the theoretically generated SPR curve. The refractive index(RI) of the ZnO thin film, was found to increase from 1.95 to 1.98 with an increase in ambient temperature from room temperature to 250°C. The observed linear increase in the refractive index with temperature shows the promising application of ZnO film as an effective and reliable temperature sensor.

Introduction

Zinc oxide (ZnO) is a direct band gap semiconductor [E_g = 3.30 eV] having good electro-optic, photoconducting, piezoelectric, elasto-optic and optical wave guiding properties. ZnO normally has a hexagonal wurtzite crystal structure with a=3.25Å and c=5.12Å. It finds a variety of applications in solar cell, SAW devices, gas sensors, transparent conductors, heat mirrors, etc[1]. Studies on ZnO nanoparticles have revealed confinement effect of optical and acoustic phonons [2]. ZnO has a broad chemistry leading to many opportunities for wet chemical etching, radiation hardness, biocompatibility and low power threshold for optical pumping [3]. ZnO is a promising material for application in optical devices [4]. Hence, it is very important to have knowledge of the behavior of dielectric properties of the material in the optical frequency range. The properties of the material are known to depend upon the temperature and may affect the device performance. Though studies have been carried out on temperature dependent variation of dielectric constant of ZnO, even with different doping, but they are mainly in the lower frequency range, and the reports in the optical frequency range has been scantly available [4-7].

Surface plasmon resonance (SPR) based sensor has become an active research area and are being used in creating sensitive surface effect sensors [8]. Surface plasmon waves (SPWs) are electromagnetic waves guided along a metal dielectric interface. SPR studies are generally based on the change in the angle of the minima with any deviation in the refractive index of the thin film, which may be due to its physical characteristic or any variation in the ambient condition. It is quite surprising that this powerful technique has not been exploited for the study of dielectric behavior at optical frequencies especially as a function of temperature.

In the present study a simple SPR technique has been used to study the temperature dependant dielectric constant of ZnO thin film at optical frequency.

Theory

Charge density oscillations (i.e. surface plasmon wave, SPW) that may exist at the interface of two media with dielectric constants of opposite sign are known as Surface plasmons and are excited when the surface-parallel component of the momentum of incident transverse magnetic (TM) light wave is equal to that of SPW propagating between the metal and the dielectric surface (eqn. 1).

$$\frac{2\pi}{\lambda} n_p \sin\theta_{in} = \frac{2\pi}{\lambda}(\frac{\varepsilon_1'\varepsilon_2}{\varepsilon_1' + \varepsilon_2}) \tag{1}$$

where n_p is the refractive index of the prism, λ is the wavelength of the light used for exciting the surface plasmons, θ_{in} is the angle of incidence on the metal film and $\varepsilon_1 (= \varepsilon_1' + i\varepsilon_1'')$ and ε_2 are the dielectric constants of the metal film and prism respectively.

Light incident on a metal-layer directly cannot excite surface plasmons along the metal-dielectric interface as the conservation laws are not satisfied. Prism coupling is used to enhance the incident wave vector of the incident p-polarized light. The resonance is observed in terms of a sharp dip in the intensity of the reflected light output at the resonance angle.

Kretschmann-Raether configuration has been used where metal film deposited on a prism is kept in contact with a bulk dielectric medium, which is air in our case. The p-polarized light is incident with varying angle on the metal film through the prism, and the intensity of the reflected light is measured to know the resonance angle. The resonance angle depends on the metal film and the refractive index of the dielectric media [9].

The reflectance of a two layer system (prism-gold-air) at an incident angle θ is given by

$$R_{013} = \left| \frac{r_{01}r_{13}\exp(2ik_{z1}d_1)}{1+r_{01}r_{13}\exp(2ik_{z1}d_1)} \right|^2 \tag{2}$$

where d_1 is the thickness of the metal film and

$$r_{ik} = \left(\frac{k_{zi}}{\varepsilon_i} - \frac{k_{zk}}{\varepsilon_k}\right) \Big/ \left(\frac{k_{zi}}{\varepsilon_i} + \frac{k_{zk}}{\varepsilon_k}\right) \tag{3}$$

and,

$$k_{zi} = \frac{2\pi}{\lambda}(\varepsilon_i - \varepsilon_0 \sin^2\theta) \tag{4}$$

ε_0, ε_1 and ε_3 are the dielectric constants of glass, metal film and the bulk media respectively.

Once, a dielectric layer (say, ZnO film) is deposited on the metal film we get a three-layer system (prism-gold-ZnO-air). The Fresnel's equation gives the reflectance R_{0123} corresponding to an angle of incidence θ [10].

$$R_{0123} = \left| \frac{r_{01} + R_{123}\exp(2ik_{z1}d_1)}{1+r_{01}.R_{123}\exp(2ik_{z1}d_1)} \right|^2 \tag{5}$$

where,

$$R_{123} = \left| \frac{r_{12} + r_{23}\exp(2ik_{z2}d_2)}{1+r_{12}.r_{23}\exp(2ik_{z2}d_2)} \right|^2 \tag{6}$$

ε_2 is the dielectric constant of the dielectric layer and d_2 is the thickness of the deposited dielectric film. r_{ik} and k_{zi} are given by equation 3 and equation 4 respectively.

From, the dielectric constant the RI and the band gap are calculated.

Experiment

A thin film (~ 40 nm) of gold was deposited on the hypotenuse face of a right angled BK7 glass prism in vacuum (7×10^{-6} mbar) by thermal evaporation at a deposition rate of about 5 Å per second. The thickness of the film was controlled in-situ using quartz thickness monitor. The Au-film was annealed in air for one hour. X-Ray diffraction (XRD) studies were carried out to study the crystallographic orientation of the deposited film. The reflectance measurements were made using indigenously developed SPR setup comprising of prism table loaded on XYZθ rotation stage attached with an angle detector of accuracy 0.01°C. SPR reflectance measurements were made using a p-polarized He-Ne laser incident on the Au-film coated face of the prism fixed on the rotating stage.

ZnO film (~ 240 nm) was then deposited on the gold coated prism by RF-sputtering. A Zn-metal target (99.99% pure) was sputtered in an Argon-Oxygen atmosphere (40% Ar and 60% O_2) at a pressure of 25mTorr. The SPR reflectance measurements were made for both the two-layer and three-layer system as a function of angle of incidence.

A specially designed heating arrangement, with an accuracy of 1°C, was made and fixed on prism table for temperature dependant studies.

Results and discussions

The experimental SPR reflectance curve obtained for Au-air interface as a function of angle of incidence is shown as symbols in fig 1. A sharp reflectance dip is observed at an angle of 36.82°. The observed curve was theoretically fitted using Fresnel's equations (eqn. 2). The dielectric constant of the gold film was taken as the variable parameter and the others were kept constant. The best fit theoretical curve is shown in the fig. 1 as the solid line. The fitting parameters yielded the complex dielectric constant for the gold film to be -12.8+ i2.7 which is found to be quite close to -12.7+ i1.9 and -13.9+ i1.8 as obtained by Pokrowosky by two different methods [11]. The values obtained by other workers, tabulated by Pokrowsky, vary from -13.1+ i1.3 to -11.5+ i1.6 [11].

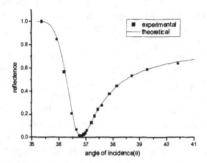

FIG. 1.SPR spectrum for prism-Au-air system

Figure 2 shows the XRD spectra of the ZnO film deposited on glass substrate. Only one peak at 2θ =34.2° is observed corresponding to (002) plane of wurtzite structure indicating the growth of preferred oriented ZnO thin film with c-axis normal to the substrate. The broadening of the XRD peak indicates the formation of naocrystalline film.

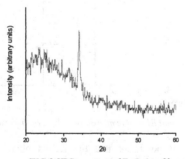

FIG.2 XRD spectra of ZnO thin film

The SPR reflectance curve of the three-layer system (prism-*gold-ZnO-air*) is recorded and shown in fig.3 as symbols. The experimental data is fitted with theory using equation 5 and taking the dielectric constant of the ZnO as the variable parameter while keeping the dielectric constant of the underneath gold film at a constant value (which was obtained from the previous fitting). The best fitted theoretical curve is shown in fig.3 as a solid line.

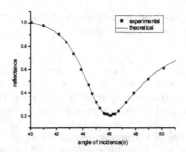

FIG.3.SPR spectrum for prism-Au-ZnO-air system

The estimated value of the room temperature dielectric constant of ZnO film obtained at λ=633nm is 3.81+i0.08, from which the refractive index was found to be 1.95 and is in close agreement to the values 1.9797 and 1.9849 reported by Mehan $et.al.$ for c-axis oriented ZnO film using optical wave guiding method[12].

The SPR curves of the same three layer system was carried out over a wide range of temperature (27°C to 250°C) and the dielectric constant at the optical frequency was estimated using theoretical fitting. The SPR minimum was found to shift continuously towards higher angle with an increase in the temperature. The variation of the real part of dielectric constant and the refractive index of the c-axis oriented ZnO film is shown in fig.4 as a function of temperature. A continuous increase in the real part of the dielectric constant ε' from 3.81 to 3.92 was observed with increase in temperature from 27°C to 250°C. The temperature dependant dielectric constant of ZnO film in the optical frequency region has not been available in the literature. However, increase in ε' with increase in temperature has been reported by various other workers for ZnO thin film in low frequency region (upto 1MHz) [4-7]. The variation in the imaginary part of the dielectric constant with temperature is also shown in fig.5 and slight increase in its value with temperature was observed.

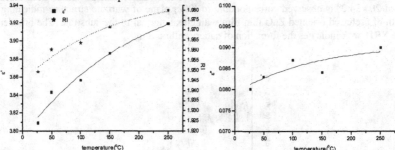

FIG. 4.Variation of ε' and RI with temperature   *FIG.5.Variation of ε'' with temperature*

It is important to point out that the results obtained in the present study were reproducible in repeated experiments with same ZnO film and also with the films prepared under different batches with same deposition parameter, which proves the prepared system (prism-*gold-ZnO-air*) promising for sensing of fine change in ambient temperature.

Conclusion

The temperature dependence of dielectric properties of ZnO thin film (λ=633nm) has been studied in the range 27°C to 250°C. The dielectric constant increased from 3.81+i0.08 to

3.92+ i0.09 with increase in temperature from 27 °C to 250°C. Owing to good reproducibility SPR based system can be used as a reliable temperature sensor.

Reference:

1. S. Das and S. Chaudhuri, Phys. Stat. Sol., **244**(2007)2657.
2. H. K. Yadav, V. Gupta, K. Sreenivas, S. P. Singh, B. Sundarkannan and R. S. Katiyar, Phys. Rev. Lett., **97**(2006)085502.
3. V. A. Coleman and C. Jagdish, "Basic properties and applications of ZnO" in Zinc Oxide Bulk, Thin Flms and Nanostructures, (Edited by . Jagdish and S. J. Pearton) Elsevier, UK, Chap.1, 2006.
4. D. M. Bagnall, Y. F. Chen, Z. Zhu, T. Yao, S. Koyama, M. Y. Shen and T. Goto, Appl. Phys. Lett., **70**(1997)2230.
5. C. K. Ghosh, K. K. Chattopadhyay and M. K. Mitra, J. App. Phys., **101**(2007)124911.
6. V. Gupta and A. Mansingh, Phys. Rev. B, **49**(1994)1989.
7. M. A. Seitz and T. O. Sokoly, J. Electrochem. Soc., **121**(1974)163.
8. R. Levy and S. Ruschin, Sens. Act. B, **124**(2007)459.
9. Rajan , Subhash Chand and B.D. Gupta, Sens. Act. B, **123**(2007)661.
10. O. S. Heavens, "Optical Properties of Thin Solid Films", Butterworths Scientific, London, Chap.4, 1955.
11. P. Pokrowsky, Appl. Opt., **30**,(1991) 3228.
12. Navina Mehan, Vinay Gupta, Kondepudy Sreenivas and Abhai Mansingh, J. App. Phys., **96**(2004)3134.

Mater. Res. Soc. Symp. Proc. Vol. 1129 © 2009 Materials Research Society   1129-V05-01

Metal Hydride Fluidic Artificial Muscle Actuation System

Alexandra Vanderhoff and Kwang J. Kim
Active Materials and Processing Laboratory, Mechanical Engineering Department, University of
Nevada, Reno, Reno, NV 89557, U.S.A.

ABSTRACT

The study determines the feasibility of a new actuation system that couples a fluidic
artificial muscle designed by Festo [1] with a metal hydride hydrogen compressor to create a
compact, lightweight, noiseless system capable of high forces and smooth actuation. An initial
model for the complete system is developed. The analysis is restricted in some aspects
concerning the complexity of the hydriding/dehydriding chemical process of the system and the
three-dimensional geometry of the reactor, but it provides a useful comparison to other actuation
devices and clearly reveals the parameters necessary for optimization of the actuation system in
future work. The system shows comparable work output and has the benefits of biological
muscle-like properties [2] for use in robotic systems. When compared to other previously
developed metal hydride actuation systems the potential for increasing the reaction kinetics and
improving the overall power output of the system is revealed. A comparison of the system to
common actuation devices, including a biological muscle, shows similar stress and strain
relations, but a lower power and frequency range due to the slow actuation time. Improving the
reaction kinetics of the system will be the first approach to enhancing the system, along with
optimization of the mass and type of metal hydride used in the reactor to produce a full actuation
stoke of the fluidic muscle while minimizing system weight.

INTRODUCTION

The actuation system of this study couples a fluidic artificial muscle that is driven by
pressure input and a metal hydride compressor designed to reversibly absorb and desorb
hydrogen gas. The actuator and metal hydride reactor create a closed system that operates by
heating the reactor containing the activated metal hydride resulting in the desorption of hydrogen
gas which pressurizes the artificial muscle causing it to contract; the reactor is then cooled and
the hydrogen gas is reabsorbed by the metal hydride causing the artificial muscle to relax. The
fluidic artificial muscle used in the system was developed by Festo [1] and is an improved design
over the common Braided Artificial Pneumatic Muscle (BPAM). The reactor design is a copper
containment unit for the metal hydride pellets that is heated and cooled by thermoelectric
elements attached to the outer walls of the unit, along with cooling fans to improve the heat
transfer rate. The system was designed with the goals of creating a lightweight (high force to
weight ratio), compact, innovative system that is durable, has smooth actuation, and noiseless
operation. The actuator was characterized following a general method used for pneumatic
muscles, and a thermal circuit model was created as a baseline for the system feasibility and the
future development of the project. The two models (of the actuator and reactor) were coupled
and the overall system performance and capabilities were examined and compared to other
common actuation devices.

EXPERIMENT

BPAM actuators consist of a closed rubber tubing (with an inlet for gas) sheathed in a braided sleeve. As pressure is input to the rubber tubing it expands radially against the braided shell. The braid angle of the shell increases and the muscle contracts in the axial direction producing the actuation displacement and force (when loaded) as shown in Figure 1(a). The Festo actuator (see Figure 1(b,c)) used in this study is an improvement to the BPAM because the braiding that controls the actuation is made of strong aramide fibers embedded in a rubber chloroprene membrane, reducing the friction caused by the rubbing of the braided sleeve in the BPAM design, and creating a more durable and longer lasting fluidic muscle, determined through testing to be safe for operation using hydrogen gas with nearly zero leakage.

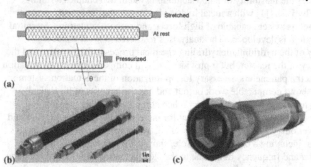

Figure 1. (a) Concept diagram of BPAM operation [3], where θ is the braid angle of the outer sleeve, (b) the Festo fluidic muscle in various sizes [1], and (c) an internal membrane view of the Festo actuator [1].

The copper reactor containment unit houses the metal hydride $Ca_{0.6}Mm_{0.4}Ni_5$ (initially in the form of pellets prior to activation) in three chambers. The pellets were activated by pressurizing the reactor with hydrogen gas, causing the metal hydride to absorb the hydrogen into the interstitial molecular spaces of the compound and expand to fill the reactor chambers (neglecting the small hole through original pellets created by a pellet spacer).

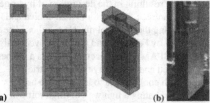

Figure 2. (a) Reactor containment unit design showing metal hydride pellets enclosed in reactor unit chambers, and (b) actual sealed containment unit.

Once activated and removed from the hydrogen source, the actuator and reactor were coupled to create a closed system. The reactor cycles of desorption and absorption were controlled by temperature input and output from thermoelectric elements and cooling fans, and the maximum allowable suspended mass was added to the actuator for testing (~30kg). The room temperature pressure equilibrium for $Ca_{0.6}Mm_{0.4}Ni_5$ is high, and with the cooling capability of the set-up, the minimum achievable system pressure (after absorption) was ~400kPa (60psig), which only allowed for partial actuator stroke (the Festo operates between 0-116psig).

(a)   (b)

Figure 3. (a) Metal hydride reactor unit with thermoelectric elements and cooling fans, and (b) Festo fluidic artificial muscle with displacement transducer and suspended mass on pulley.

THEORY

Characterization of the Festo fluidic muscle follows a previous procedure for the modeling of a BPAM actuator [3,4]. A quasi-static model was created based on the relationship of the change of volume with respect to length, which is a function of pressure and the amount and type of loading on the actuator. The model results for the actuator behavior were verified with experimental data from isotonic testing. The basic principle governing all fluidic artificial muscles is, as the membrane is pressurized internally, radial expansion occurs simultaneously with axial contraction. Mechanical power is transferred along the actuator membrane to produce unidirectional forces on a load. As an infinitesimal amount of gas (or fluid), df, is introduced into the muscle, an infinitesimal change of volume, dV, is produced doing a net amount of work given by Equation 1.

$$W_f = PdV \tag{1}$$

Within the same time span, dt, as the change of volume, there is an axial change of length, dl (less than 0 for contraction), and a force is produced, described by Equation 2.

$$W_t = -Fdl \tag{2}$$

Neglecting the work required to deform the actuator membrane (assumed negligible), it can be said that the work done by the fluid to create the change in volume is equal to the work to produce the change in length, resulting in the force capability of Equation 3.

$$F = -P\frac{dV}{dl} \tag{3}$$

In the previous equation, P is the gage pressure of the actuator and dV/dl is determined experimentally assuming a cylindrical shape for the actuator volume. This assumption was verified by comparing the cylindrical volume from diameter measurements (averaged over the actuator length) to the measured actuator volume from water displacement testing over the contraction/relaxation of the actuator for various loading conditions. The previous model [3] considered the muscle tension as a function of pressure and length. The stiffness is defined as K in Equation 4 and is proportional to the pressure, and the stiffness per unit pressure is given by Equation 5.

$$K \equiv \frac{dF}{dl} \qquad (4)$$

$$K_g \equiv \frac{dK}{dP} \qquad (5)$$

Because K_g is approximately constant over an observable range (there is a linear range of actuator behavior), the force can be written in a linearized form as shown in Equation 6, where L_{min} is the theoretical minimum length of the actuator at zero force ($F = 0$).

$$F = K_g P(L - L_{min}) \qquad (6)$$

Considering the energy stored in the actuator membrane, the final form of the force is given by Equation 7.

$$F = \begin{cases} K_g(P - P_{th})(L - L_{min}) + K_p(L - L_0) + nl(L) & \text{if} \quad P > P_{th} \\ K_p(L - L_0) + nl(L) & \text{if} \quad P < P_{th} \end{cases} \qquad (7)$$

In the above equation, P_{th} is the threshold pressure to overcome the radial elasticity of the membrane for expansion, L_0 is the rest length of the actuator, K_p is described in the previous study [3] as the linearized equivalent parallel passive elastic constant for the shearing force of the bladder material and the braided shell (should be almost negligible for the Festo actuator with embedded braiding), and $nl(L)$ is the nonlinear term for extreme lengths of the actuator. The model of this study considers the linear actuator behavior (for $P > P_{th}$) with Equation 7 rewritten as $F = K_g(P - P_a)(L - L_{min}) + F_a$, where $P_a = P_{th} - (K_p/K_g)$ and $F_a = K_p(L_{min} - L_0)$. To experimentally estimate the four unknown parameters, ten sample points consisting of pressure, length and force coordinates were taken from the isobars of an isotonic test—contraction under constant pressure—(Figure 6) and fit to the force equation through linear optimization. The results were compared to data from the suspended mass test set-up (referred to as the weight set-up in figures). The internal volume with respect to length was determined for six different loading cases on the Festo actuator (refer to Figure 4). The results were averaged and the fit with a second order polynomial curve in order to determine the dV/dl relationship.

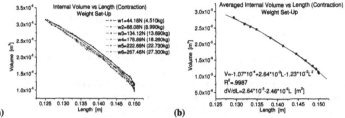

Figure 4. (a) Internal actuator volume w/ respect to length, and (b) averaged curves and fit.

The dV / dl relation with respect to length in Figure 5(a) was used to determine the stiffness, defined previously. An approximately linear range for stiffness gave a constant for stiffness per unit pressure, K_g, equal to $2.46 \times 10^{-2} m$ displayed in Figure 5(b).

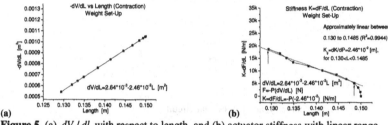

Figure 5. (a) dV / dl with respect to length, and (b) actuator stiffness with linear range.

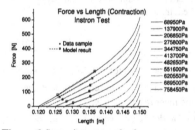

Figure 6. Isotonic test results for actuator model verification with sampled data.

The results of the linear optimization fit to Equation 7 gave an average difference error in the output force compared to sampled data force of 3.9%. The linear fit value of $K_g = 2.06 \times 10^{-2} m$ is very close to the model value of $2.46 \times 10^{-2} m$. The minimum length ($L_{min} = 10.8 \times 10^{-2} m$ with force equal to zero) from linear fit is very close to the minimum length of the Festo specification of 25% maximum contraction ($L_{min} = 0.75(L_0) = 11.3 \times 10^{-2} m$). $P_{th} = 250kPa$ is reasonable when observing from Figure 6 that the actuator does not display much linear behavior for isobars below ~275kPa. K_p was found to be almost negligible (~4kPa), as expected.

303

The thermal model of the reactor is a simplified approach, intended for use as a starting point for future work on the project and to provide a preliminary comparison of the entire system to other actuation systems when coupled with the actuator model. Due to the inherent complexity of the reactor and the change of system temperature and pressure with time, the thermal input to the reactor over finite time steps is used with the assumption that the slow desporption/absorption cycles are quasi-static and the measured instantaneous state of the system is the equilibrium state of each time step. The model is convenient and has been used with previous studies on metal hydride phenomena [5, 6], but it does not consider the complex transient nature of heat and mass transfer in the hydriding/dehydriding process. Only the desorption cycle is discussed here, as it was used to evaluate the overall system performance for the actuator contraction. The one-dimensional approach to the reactor geometry and the thermal circuit for the overall heat transfer coefficient are shown in Figure 7. The outer side walls of the reactor are assumed adiabatic (and were insulated for testing), the thin inner chamber walls are neglected as well as the small pellet holes, and a constant heat transfer coefficient is used. Half of the reactor is modeled due to symmetry.

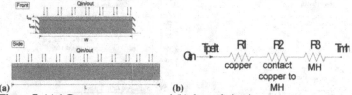

Figure 7. (a) 1-D reactor geometry, and (b) thermal circuit.

The reversible desorption process is an endothermic chemical reaction where the hydrogen stored in the metal hydride is released as a gas. The heat input is given by Equation 8.

$$\dot{Q}_{in} = UA\left(T_{peltier,de} - T_{reactor,de}\right) \qquad (8)$$

UA is the overall heat transfer conductance, $T_{peltier,de}$ is the measured temperature at the interface between the thermoelectric element and the outer reactor wall and is referred to as the Peltier temperature throughout the analysis, and $T_{reactor,de}$ is the inner reactor temperature. The resistances of the thermal circuit of Figure 7(b) are used to define the overall heat transfer conductance. R_1 is the thermal resistance of the copper wall, R_2 is the contact resistance between the copper and the metal hydride [7], and R_3 is the thermal resistance of the metal hydride [8]. Due to the lack of data concerning the contact resistance between copper and the metal hydride, the value is approximated at 10^{-3} m²K/W [7]. The conductivity and effective heat transfer coefficients are assumed constant at 300K and the conductivity of the metal hydride is assumed to be 5 W/mK [8]. $T_{peltier,de}$ was measured over three cycles of desorption and the average of the data was fit with an exponential decay curve to obtain an equation for $T_{peltier,de}$ as a function of time. The outer wall temperature was considered the driving input to the system since the system pressure was a response to heating/cooling of the reactor. The system pressure was therefore considered a function of $T_{peltier,de}$ and was graphed for the three cycles with respect to the outer wall temperature and the average data was fit with a polynomial curve. The outer reactor wall and system gage pressure data for the desorption cycle analysis is shown in Figure 8.

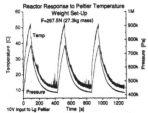

Figure 8. Outer reactor wall temperature and system gage pressure data for the desorption cycle analysis and average curve fits.

The inner reactor temperature could not be measured directly throughout the cycles and was estimated using the van't Hoff equation (Equation 9) which is based on the absolute system pressure to obtain $T_{reactor,de}$ as a function of pressure. The enthalpy and entropy of desorption are $\Delta H_{H2} = -1.276 \cdot 10^4\, kJ/kg_{H2}$ and $\Delta S_{H2} = -51.4 kJ/kg_{H2}$ [9], respectively. Figure 9 displays the model results.

$$\ln P_{eq} = \frac{\Delta H}{R_u T} - \frac{\Delta S}{R_u} \tag{9}$$

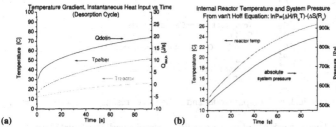

(a) (b)

Figure 9. (a) Thermal analysis results, and (b) inner reactor temperature and absolute system pressure over the average desorption cycle time of 94 seconds.

To evaluate the overall system performance the thermodynamic first and second law efficiencies of the system were considered, along with the power per kilogram of metal hydride. The first law efficiency compares the work output to the overall heat input. The second law efficiency is given as the ratio of the useful system work to the ideal work. The useful mechanical work output of the system is clearly defined by the actuator model. To determine the ideal work of the system, the isentropic work based on an energy balance over the control volume of the actuator and system plumbing is performed. These results are compared to the measured work from the complete system testing over the desorption cycle. The measured results from the complete

305

system testing are shown in Figure 10 where the total work (4.58 J) was calculated using $W = Fx$, with the force, F, defined by the suspended mass constant load on the actuator, and the actuator displacement, x, with a correction for the actuator pre-stretch.

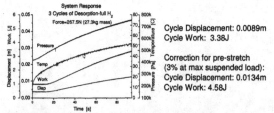

Cycle Displacement: 0.0089m
Cycle Work: 3.38J

Correction for pre-stretch
(3% at max suspended load):
Cycle Displacement: 0.0134m
Cycle Work: 4.58J

Figure 10: Measured system response for comparison with coupled model result.

The instantaneous useful mechanical work for the system is found using Equation 10, where F is the actuator equation for force (see Equation 3). The dL/dt relation was determined from the derivative of the polynomial curve fit of length with respect to time from the complete system testing (refer to Figure 10, where displacement is converted to length and fit with a third-order polynomial curve). The model results of Figure 11 show a close comparison to the measured data in terms of the useful work (4.69 J compared to 4.58 J), neglecting some system losses.

$$\dot{W}_{mech} = FdL = P_{de}(t)dV = P_{de}\frac{dV}{dt}dt = P_{de}\frac{dV}{dL}\frac{dL}{dt}dt \tag{10}$$

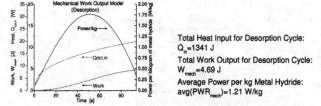

Total Heat Input for Desorption Cycle:
Q_{in}=1341 J
Total Work Output for Desorption Cycle:
W_{mech}=4.69 J
Average Power per kg Metal Hydride:
avg(PWR$_{mech}$)=1.21 W/kg

Figure 11: Useful mechanical work output model and summary of results for desorption cycle.

To determine the ideal work capability of the system the isentropic work is calculated using Equation 11. The hydrogen is assumed to behave as an ideal gas with constant specific heat and uniform control volume temperature equal to the reactor temperature with the desorption cycle considered as a reversible process.

$$\dot{W}_{IC} = C_{v,H2}\left(\gamma\dot{m}_{H2}T_{reactor,de} - \frac{M_{H2}}{R_u}\left(P\dot{V} + V\dot{P}\right)\right), \qquad m_{H2} = \frac{M_{H2}PV_{CV}}{R_u T_{reactor,de}} \tag{11}$$

In the above equation, γ is the specific heat ratio for hydrogen at 300K, R_u is the universal gas constant, and M_{H2} is the molar mass of hydrogen. The isentropic work is composed of the energy of the mass entering the control volume and the boundary work. The mass of hydrogen, m_{H2}, available to produce work in the system is increasing throughout the desorption process, and was calculated using the ideal gas equation for hydrogen. It is unknown what percentage of the total hydrogen contained in the metal hydride is desorbed during the process because desorption was halted when the system pressure reached ~800kPa (110psig)—the pressure limit for the Festo. The results of the isentropic work model are graphed in Figure 12. The system efficiencies are shown in Figure 13.

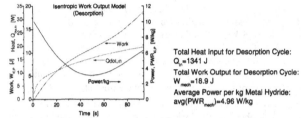

Total Heat Input for Desorption Cycle:
Q_{in}=1341 J
Total Work Output for Desorption Cycle:
W_{mech}=18.9 J
Average Power per kg Metal Hydride:
avg(PWR_{mech})=4.96 W/kg

Figure 12. Isentropic work model and summary of results for desorption cycle

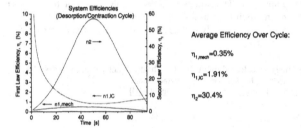

Average Efficiency Over Cycle:

$\eta_{1,mech}$=0.35%

$\eta_{1,IC}$=1.91%

η_2=30.4%

Figure 13. Thermodynamic efficiencies (1st and 2nd law) and summary of results for desorption cycle

DISCUSSION

Actuator and Reactor Modeling

The method for modeling fluidic artificial muscle actuators proved to be acceptable when applied to the Festo actuator. The graphs of Figures 4 and 5 show relationships to length that are very close to those of the BPAM that was modeled by the previous study [3]. The coefficients determined through linear optimization of the data points and fit to the model equation were all within range of expectation. The Festo actuator showed lower stiffness than the BPAM, but a higher threshold pressure to overcome the radial elasticity, which can be observed by comparing the isotonic test data of the two actuators [3]. The heat input for the cycle is well defined by the temperature gradient over the cycle time using the method covered in the theory and for the purposes of this study. The thermal model is a simplified analysis for the metal hydride reactor.

Further work is necessary to create a more thorough analysis of the complex heat transfer process.

System Performance Evaluation

Although the work output is small, the system has a high force to weight ratio. The weight of the Festo actuator is 61.9g, and the system weight (the thermoelectric elements, fans, heat sinks, and the reactor with pellets) is a total of 1.29kg. The system also requires plumbing, but this can be minimized, and can be composed of lightweight Teflon tubing. Comparing the force output capability to system weight based on pressure and loading (calculated using Equation 3) the maximum ratio is 436.2 N/kg_{sys}. Another comparison ratio, force to weight of metal hydride, has a maximum value of 13,641 N/kg_{MH}. Table I is a comparison to an existing metal hydride actuation system using metal bellows as the actuator. The load on the Festo actuator is much greater than for the bellows. The displacement for the Festo is much smaller due to the nature of the actuator (fluidic muscles have a smaller contraction ratio, $\Delta l / l_0$, than the bellows) and based on the limited stroke due to the high equilibrium pressure of $Ca_{0.6}Mm_{0.4}Ni_5$ at room temperature which prevented the absorption of hydrogen below ~60psig system pressure. The output work is comparable, but contraction time is much slower for the system of this study and greatly affects the power comparison. The parameter of desorption/absorption time needs to be minimized in future studies by improving the heat input and output to the reactor. The benefits of using the actuation system of this study are the applications to artificial limbs and biorobotic systems of the Festo fluidic muscle.

Table I. Comparison to previously developed metal bellows hydride actuation systems [10, 11].

	Festo Fluidic Muscle MH Actuator	Metal Bellows MH Actuator [10]	Metal Bellows MH Actuator [11]
Load [N]	270	98	196
Disp [m]	0.0134	0.05	0.08
Work [J]	4.6	4.9	15.7
Contraction Time [s]	94	6	35
Mass Hydride [kg]	0.04125	0.006	0.01
Mass System [kg]	1.29	0.3 (no fans)	----
Power [W]	0.049	0.875	0.45
Power/kg_MH [W/kg_MH]	1.19	146	45
Power/kg_sys [W/kg_sys]	0.038	2.92	----

Figure 14 gives a comparison of the actuation system to other common actuation devices [12]. The Festo fluidic muscle was operating under reduced stroke with the metal hydride reactor due to the minimum pressure being 60psig, so this reduces the range of strain, which has the potential of 20% at maximum pressure, as was shown in the pneumatic materials tests (refer to Figure 6). The strain capability of the actuator is shown in Figure 14(a) as the "H_2 System Potential." The reactor is capable of producing stresses up to 150MPa (at ~140°C) using $Ca_{0.6}Mm_{0.4}Ni_5$, and this is shown as the potential stress range for the system. Another comparison of actuation devices is the relation of power output per unit volume to actuation frequency. The range for the metal hydride system is between the minimum and maximum power outputs (refer to Figure 11) and the corresponding volume of the Festo actuator. The power considered over the desorption cycle, which has an average cycle time of 94s. Also displayed is the power per unit volume of metal hydride, which is the driving mechanism of the actuator and produces a large amount of power, considering the slow desorption process.

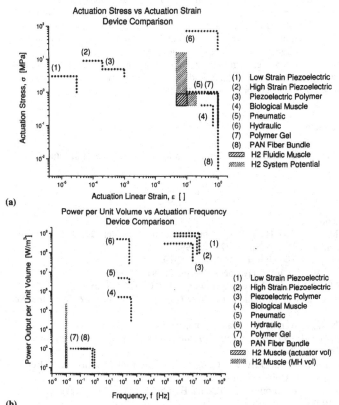

(a)

(b)

Figure 14: Comparison of common actuation devices [12] in terms of (a) actuation stress vs. linear strain, and (b) power output per unit volume vs. frequency.

CONCLUSIONS

The study of the metal hydride actuation system using a fluidic artificial muscle gave the feasibility of creating an innovative, compact, high force to weight system with smooth actuation that can be developed further to optimize performance and be more compatible with (or potentially outperform) other existing actuation devices. An initial model was used for the system performance evaluation. When comparing the actuation system of this study to other devices, the main drawback is the low frequency. Future work should target a reduction in cycle time of at least one order of magnitude, which is possible from both actuator and reactor viewpoints; the fluidic muscle actuator has been rated at 3-5Hz [13] over full actuator stroke, and the comparison with a previous study on a metal hydride actuation system using metal bellows shows the desorption cycle almost one order of magnitude faster. The system can be optimized by minimizing the amount metal hydride and maximizing the actuator stroke. This can be achieved either by using a smaller reactor with less metal hydride content, or by increasing the size of the Festo muscle (with higher force capability being the trade-off for a less-compact overall system). Another important consideration is the operating environment. Hydrides have very different desorption characteristics and the hydride used in this study produced a pressure at room temperature that was not optimal for the overall system.

ACKNOWLEDGEMENT

Special thanks to Nevada Space Grant Consortium for partial support of the study.

REFERENCES

1. Festo, www.festo.com
2. G.K. Klute and J.M Czerniecki, Hannaford, B. Int. J. Robotics Research. 21(4), 295 (2002).
3. C. Chou and B Hannaford, Report No. G93268R, 1996.
4. F. Daerden, Ph.D Thesis, Vrije Universiteit Brussel, 1999.
5. M.R. Gopal and S.S. Murthy, Int. J. Hydrogen Energy. 20, 911 (1995).
6. M.R. Gopal and S.S. Murthy, Int. J. Hydrogen Energy. 17(10), 795 (1992).
7. F.P. Incropera and D.P Dewitt, *Introduction to Heat Transfer*. (John Wiley & Sons, Inc. New York, 2002).
8. M. Lee, Master Thesis, University of Nevada, Reno, 2007.
9. K.J. Kim, *Active Materials and Processing Laboratory* (metal hydride property information) University of Nevada, Reno.
10. S. Shimizu and S. Ino, Sato, M., Odagawa, T., Izumi, T., Takahashi, M., Ifukube, T. Report No. N12 W6, Hokkaido University, 1993.
11. W. Yuichi and M. Muro, Kabutomori, T., Takeda, H., Shimizu, S., Ino, S., Ifukube, T. IEEE, 5, 148 (1997).
12. C. Kiyoung and K.J. Kim, Kim, D., Manford, C., Heo, S., Shahinpoor, M. J. Intelligent Material Systems and Structures. 17, 563 (2006).
13. <http://www.festo.com/cat/en-us_us/data/doc_enus/PDF/US/DMSP-MAS_ENUS.PDF> (fluidic muscle specifications) 2008.

Mater. Res. Soc. Symp. Proc. Vol. 1129 © 2009 Materials Research Society   1129-V14-06

Mechanical Characteristics of Al-Si-Cu Structural Films by Uniaxial Tensile Test with Elongation Measurement Image Analysis

Hiroki Fujii[1], Takahiro Namazu[1], Yasushi Tomizawa[2], and Shozo Inoue[1]

[1] Department of Mechanical and Systems Engineering, University of Hyogo, 2167 Shosha, Himeji, Hyogo 671-2201, JAPAN

[2] Mechanical Systems Laboratory, Corporate Research & Development Center, Toshiba Corporation, 1 Komukai-Toshiba-cho, Saiwai-ku, Kawasaki, Kanagawa 212-8582, JAPAN

ABSTRACT

This paper presents mechanical characteristics of sputtered Al-Si-Cu film used as a structural material in microelectromechanical systems (MEMS). Novel tensile test technique with an elongation measurement image analysis system has been developed. Uniaxial tensile tests were conducted at temperatures ranging from room temperature (R.T.) to 358 K. No specimen-size effect on Young's modulus, yield stress, and fracture strength, was found, whereas the annealing and temperature effects were clearly observed. To study mechanical degradation mechanism, cyclic loading tests were conducted under constant stress- and strain-amplitude modes. In constant stress-amplitude mode, stain amplitude increased with an increase of loading cycles due to creep-like deformation. Almost of the specimens fractured during the tests. Meanwhile, in constant strain-amplitude mode, stress amplitude gradually decreased with increasing cycles due to stress-relaxation, but no fatigue failure was observed.

INTRODUCTION

Accurate evaluation of the mechanical properties of thin film materials becomes a significant issue for optimum design toward the high-performance and reliability of microelectromechanical systems (MEMS) [1]. For characterization of MEMS materials, the tensile, bending, and bulge tests are being widely employed. The uniaxial tensile test that is the most popular test method is used for direct measurement of the mechanical properties, such as Young's modulus, Poisson's ratio, and yield strength. In the case of film specimens that are targeted, it is difficult to accurately measure the microscopic deformation of specimens during the test as compared with macroscopic specimens. Many researchers have proposed a variety of the specimen elongation measurement techniques; for example, optical interferometry [2], atomic force microscopy [3], and X-ray diffractometry [4]. Accurately evaluating MEMS materials is unavoidable for the practical use of MEMS because the outcome is reflected to improvement of the device design and durability of the devices.

The objective of this study is to develop the uniaxial tensile test technique with elongation measurement image analysis system for thin film materials. This paper focuses on evaluating Al-Si-Cu film used as a constituent material of a MEMS device. The uniaxial tensile test has been firstly conducted to study the influences of specimen size, annealing, and test temperature on Young's modulus and yield strength. Then, the fatigue test has been performed to investigate the material response to cyclic loading. The degradation mechanism of Al-Si-Cu structural film is discussed.

EXPERIMENTAL PROCEDURE

Tensile Test System

Fig. 1 shows the developed uniaxial tensile test system for film specimens. The tensile tester has been designed to become a compact size. The tester consists of piezoelectric actuator for applying tensile force, actuator case, a load cell for force measurement, a linear variable differential transformer (LVDT), specimen holders, and a heating system. The actuator case has a hinge structure with a lever to amplify the actuator's elongation. A micro heater is set into specimen holders. The tester was set under a charge coupled device (CCD) image analysis system for direct measurement of specimen elongation. During the test, the film specimen section was shot using a CCD camera to monitor two gauge marks of triangle shape on the specimen section with developed original software. The elongation was calculated directly from relative displacement of the gauge marks. At the same time, elongation of the entire specimen including a fillet section was estimated from the displacement between chucking holes by LVDT. The test system is able to conduct the fatigue test under constant stress/strain amplitude at frequency of up to 100 Hz as well as the quasi-static tensile and creep/stress-relaxation test, at temperatures ranging from RT to 423 K.

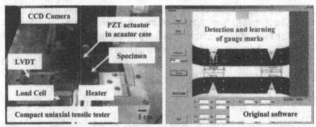

Fig. 1 The developed uniaxial tensile test system with CCD-image analysis for direct measurement of specimen elongation.

Specimen

Fig. 2 shows the produced Al-Si-Cu film specimens. The six types of specimens having different size and annealing condition were prepared on a 6-inch Si wafer, as listed in Table 1. All the specimens consist of a 2 μm-thick sputtered Al-Si-Cu film specimen section with triangle-shaped gauge marks, Si springs for supporting the thin film specimen section, chucking holes, and frame. These specimens have been fabricated through conventional micromachining technologies. To obtain force-displacement relation of Al-Si-Cu film alone, the tensile tests were performed twice a specimen. In the first test, both Al-Si-Cu film and Si spring are tensioned simultaneously. After fracture of the film, only Si spring was tensioned as the second test. By subtracting force-displacement relation of Si spring from that of the sum of Al-Si-Cu film and Si spring, the relation of Al-Si-Cu film alone was obtained [1].

Fig. 2 *Al-Si-Cu film specimens fabricated in a 6-inch wafer.*

	Annealing	Width [μm]	Length [μm]	Thickness [μm]
Type A	None	100		
Type B	623 K		600	
Type C	None	50		2
Type D	623 K			
Type E	None	100	240	
Type F	623 K			

Table 1 *Specimen dimensions for material test.*

RESULTS AND DISCUSSIONS

Uniaxial Tensile Test

Quasi-static tensile test was firstly performed to study the influences of specimen size, annealing, and test temperature on Young's modulus and yield strength. Fig. 3 shows typical stress-strain curves of non-annealed and 623 K-annealed Al-Si-Cu specimens at R.T. and 358 K. All the specimens were tensioned to failure. The stress-strain curves of all the specimens show the typical shapes for ductile materials. The effect of test temperature clearly exists. Annealed specimens exhibit a similar trend to non-annealed specimens. However, a yield stress drop with increasing temperature in the annealed specimens was smaller than that in the non-annealed specimens.

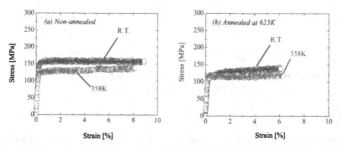

Fig. 3 *Stress-strain curves of non-annealed and 623 K-annealed Al-Si-Cu films at RT and 358 K.*

Table 2 and Fig. 4 show comparisons in Young's modulus and yield stress between non-annealed and 623 K-annealed specimens that were tested. In Fig. 4, differences in a color and

313

shape of plots indicate those of specimen-size and test temperature, respectively. No specimen-size effect on these properties was found, whereas there were the annealing and temperature effects. For non-annealed specimen, mean Young's modulus of 67.6 GPa and yield stress of 161.3 MPa were obtained at R.T. After thermal treatment at 623 K for an hour, these characteristics changed to 69.6 GPa and 125.0 MPa on average, respectively. From AFM observations, surface morphology of annealed specimen was relatively rough as compared with that of non-annealed specimen though the data are skipped here. If surface roughness is related to grain size, yield stress is thought to decrease with an increase of grain size, based on Hall-Petch law. In the tests of non-annealed specimen at 358 K, mean Young's modulus of 57.7 GPa and yield stress of 129.0 MPa were obtained. Temperature elevation provided a significant drop of the yield stress for annealed Al-Si-Cu film. On the other hand, for annealed specimen, mean Young's modulus and yield stress were found to be 59.4 GPa and 117.7 MPa, respectively, which were a little bit smaller than those at R.T. Therefore, annealing at 358 K after deposition would be a significant process to keep the mechanical characteristics constant if the film is employed under a temperature fluctuation environment.

Table 2 *Summary of the quasi-static tensile tests of Al-Si-Cu film.*

Temperature	Anneal	Specimen type	Young's modulus [GPa]		Yield stress [MPa]	
			LVDT	CCD image	LVDT	CCD image
R.T.	None	A, E	43.8	67.6	158.9	161.3
	623 K	B, F	46.3	69.6	122.9	125.0
	None	C	49.4	69.2	154.6	153.1
	624 K	D	53.5	67.3	116.8	119.4
358 K	None	A, E	39.3	57.7	128.3	129.0
	625 K	B, F	49.5	59.4	113.1	117.7

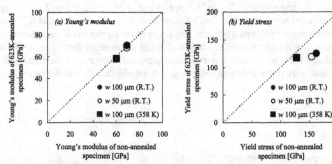

Fig. 4 *Comparisons in Young's modulus and yield stress between non-annealed and annealed Al-Si-Cu films tested at RT and 358 K.*

Fatigue Test

The mechanical degradation mechanism of Al-Si-Cu film was considered by conducting cyclic loading tests at 358K. Fig. 5 shows waveforms of stress and displacement in the tests. Stress or strain amplitude was maintained constant throughout the test. In the constant displacement amplitude condition, stress amplitude gradually decreased with increasing loading cycles. This is because stress relaxation has occurred in Al-Si-Cu film. On the other hand, in the constant stress amplitude condition, stain amplitude increases with an increase of cycles due to creep-like deformation.

Fig. 6 shows the changes of stress and strain amplitudes in cyclic loading tests. In the constant displacement amplitude condition, strain gradually decreased throughout the tests. The amount of stress drop depended on the stress amplitude at the initial state. Fatigue failure was not observed. In the constant stress amplitude condition, strain amplitude abruptly increases around 10^5 cycles, and then fatigue fracture occurred by fatigue crack nucleation and propagation. These results indicate that creep-like deformation is dominant in Al-Si-Cu film in spite of dynamic forces.

Fig. 7 shows snapshots of Al-Si-Cu film specimens during the tests under constant displacement and stress amplitude conditions. In the constant stress condition, a fatigue crack was initiated at the film specimen edge, and propagated toward the film inside until the specimen fractured. In the case of constant displacement amplitude condition, Al-Si-Cu film showed creep-like deformation, but fatigue failure did not occur.

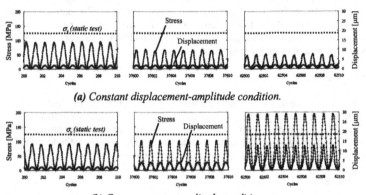

(a) Constant displacement-amplitude condition.

(b) Constant stress-amplitude condition.

Fig. 5 *Waveforms in cyclic loading tests of Al-Si-Cu films under constant displacement and stress amplitude conditions at 358 K.*

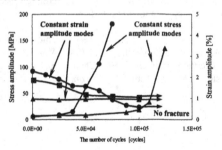

Fig. 6 *Variations in stress and strain amplitudes in cyclic loading tests at 358 K.*

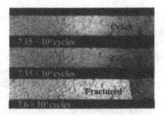

Fig. 7 *Photographs of specimen surface in cyclic loading tests.*

CONCLUSIONS

We have developed the tensile test system with elongation measurement image analysis to evaluate mechanical characteristics of sputtered Al-Si-Cu film. In the tensile tests, specimen-size effect on Young's modulus and yield stress was not found, whereas annealing and temperature effects coexisted. In the fatigue tests under constant displacement-amplitude condition, stress-amplitude gradually decreased with an increase of loading cycles and no failed specimens were obtained. In constant stress-amplitude condition, displacement-amplitude increased, and most of the specimens failed.

ACKNOWLEDGMENTS

The authors would like to thank members of Center for Semiconductor Research & Development, Toshiba Corporation Semiconductor Company and also members of Corporate Manufacturing Engineering Center, Toshiba Corporation, for sample preparation and fruitful discussion on this study. Also, the authors would like to express their gratitude to Prof. N. Araki of University of Hyogo, for extensive discussion on developing a CCD-image analysis system.

REFERENCES

1. Y. Nagai, *et al.*, "Development of Bi-Axial Tensile Tester to Investigate Yield Locus for Aluminum Film under Multi-Axial Stresses", *Proc. of MEMS 2008*, pp. 443-446 (2008).
2. J. Gaspar, *et al.*, "Comparison of improved bulge and microtensile techniques for mechanical thin film characterization – application to polysilicon", *Proc. of Transducers 2007*, pp. 575-578, (2007).
3. Y. Isono, *et al.*, "Development of AFM Tensile Test Technique for Evaluating Mechanical Properties of Sub-Micron Thick DLC Films", *J. MEMS*, Vol. 15, No. 1, pp. 169-180, (2006).
4. T. Namazu, *et al.*, "Thermomechanical Tensile Characterization of Ti-Ni Shape Memory Alloy Films for Design of MEMS Actuator", *Sensors and Actuators A*, Vol. 139, No. 1, pp. 178-186, (2007).

Mater. Res. Soc. Symp. Proc. Vol. 1129 © 2009 Materials Research Society   1129-V07-11

Optical Fiber Loop-Sensors for Structural Health Monitoring of Composites

Nguyen Q Nguyen and Nikhil Gupta
Composite Materials and Mechanics Laboratory
Mechanical and Aerospace Engineering Department
Polytechnic Institute of New York University
Brooklyn, NY 11201 U.S.A.
Phone: 718-260 3080, Fax: 718-260 3532, Email: ngupta@poly.edu

ABSTRACT

In the present work a fiber-optic loop-sensor is designed and tested for possible applications in structural health monitoring of composite materials. It is known that bending an optical fiber beyond a critical curvature leads to loss of optical power through the curved region. The optical power loss depends on the radius of curvature of the loop. The optical power can be measured by a photodetector and a change in the power due a change to the curvature can be measured. In the present research optical fiber-optic loop-sensors are developed that can exploit this concept. Single-mode optical fiber sensors having different loop radii, from 6-10 mm, are fabricated and calibrated for applied strain on the loop. The calibration is carried out using a 0.098 N load cell and a computer controlled translation stage having 50 nm step resolution. Results show that the sensors provide highly repeatable curves for loading and unloading cycles. Smaller loop radii lead to higher optical power losses, resulting in higher sensitivity. Calibration results show that such sensors can be used in structural health monitoring applications. In this approach the coating and cladding of optical fibers are maintained intact; therefore, the sensors are robust and can withstand several composites fabrication processes.

INTRODUCTION

High performance fiber- and particle-reinforced composite materials are widely used in aerospace, automobile, and marine structures. In addition, in these applications early detection of damage is desirable to enhance the safety and reliability of the structure. The detection of damage is carried out by a variety of sensors, which can either continuously monitor the structural health and the ambient conditions, or are triggered by any of the parameters that crosses the threshold. Sensors of small size and light weight are usually preferred because of the possible adverse effects of large sensors on the structural integrity of the component. Fiber-optic sensors have become an important class of sensors for structural health monitoring applications in aerospace structures and automobiles. These sensors are commonly used in monitoring stress, strain, acceleration, temperature and humidity. Several studies can be found in the published literature elaborating various fiber-optic sensing schemes, some of which are briefly discussed below.

Use of fiber-optic sensors in composite materials is advantageous because they can become an integral part of the structure and provide strengthening in addition to the sensing function. A variety of intrinsic and extrinsic fiber-optic sensors are used in structural health monitoring of composite materials. Examples of fiber-optic sensors include fiber Bragg grating (FBG) sensors [1-3], intrinsic or extrinsic Fabry-Pérot sensors (IFPI/EFPI) [4, 5], high-birefringence polarimetric sensors [6], and white-light interferometric sensors [7]. Most of these sensors are relatively expensive and require elaborate instrumentation. However, existing sensors have their own limitations. For example, FBG sensors have large size and low sensitivity, EFPI sensors do not provide a direct value of the quantity that is measured and need a demodulation, which may be cumbersome [8], and birefringence polarimetric sensor's performance is significantly affected by several factors including fluctuations in the input light source, fiber bending and detector noise [9]. The sensors based on white-light interferometry can help in avoiding some of the problems associated with the conventional interference techniques such as in FBG and IFPI/EFPI. However, applications of these techniques require substantial experience for interpretation of results and they are not sensitive enough to detect small changes in the quantity that is measured [10]. Moreover, the sensitivity of such sensors is not high enough to provide information on the crack initiation and small changes in the detection parameters. Whispering gallery mode sensors are known as one of the most sensitive class of optical sensors [11]. However, these sensors are not yet developed to the level where their applications in structural health monitoring of composite materials can be realized. Possibility of using fiber-optic sensors in composite plates and filament wound pipes has been studied, where one of the glass fibers or a fiber tow is replaced by an optical fiber [12]. These sensing schemes are successful in detecting stress, strain or temperature inside plate-like structure. Their applications in curved structures such as pipes are being developed.

A fiber-optic loop-sensor is fabricated and tested for potential structural health monitoring applications in the present study. The study parameters include loop-radius and sensor response under tensile and compressive loading conditions. The sensor is designed to be robust in construction that can be embedded in composite materials without damage.

SENSOR DESIGN

It is widely known that bending an optical fiber beyond a critical radius results in optical power losses. This concept has been applied in fiber microbend sensors where a section of the optical fiber is curved in a micrometer scale radius and attached to the specimen [13, 14]. When the lightwave propagates along the bent section of an optical fiber, local power losses occur at the bent region due to coupling between guided modes and radiation modes. The higher-order modes in a multimode optical fiber will be coupled out from the fiber at small bend radius. The optical power losses thus obtained can be calibrated with a change in the bend radius and used as the sensing principle. This phenomenon in multi-mode fibers results in almost linear change in the optical power with respect to the applied displacement. Such sensors are also used with Optical Time Domain Reflectometry (OTDR) [15, 16]. In addition, networks of microbend sensors have also been developed [17]. However, microbend sensors are not very successful in embedding in composites because of their size and problems in maintaining precise curvature in optical fibers in the harsh environments encountered during processing of composite materials.

Applications of microbend sensors can be found in the published literature in detection of stress and strain [17], cracks [18, 19], and chemical species [16]. In most applications related to stress and strain measurement these sensors are attached to the surface of the structure. Embedding such sensors inside the composite material structure is still a challenge. Some studies have also used loop sensors of multi-mode fibers [20, 21]. However, the sensitivity of these sensors is close to the linear approximation observed in [20, 21], which is almost 100 times lower than the sensitivity of our loop sensor.

In our technique instead of creating a microbend section, we create a complete loop in the fiber, as shown in Figure 1, which is easy to create, maintain, and use in sensing and these sensors can be multiplexed for use in distributed sensing applications. In addition, we use single-mode fiber to obtain very high sensitivity in the sensor system.

Use of single-mode fibers instead of multi-mode fibers causes significant differences in the physics of the optical power losses [22]. It is observed that compared to a linear approximation of optical power loss, which is generally the case in the multi-mode fibers, there are several intermediate peaks in the power vs. loop radius graphs for single-mode fibers. Most of the theories are based on the approximation that the cladding is of infinite radius and do not predict the presence of these intermediate peaks. However, the finite radius of the cladding causes internal reflections and leads to these peaks. This is an effect where the light leaked to the cladding reflects back from the outer surface of the cladding to the fiber core. The reflected light beam couples back with the remaining light in the fiber core and generates these peaks [23]. While this optical effect in single-mode fibers is known in the field of optics and communication for a long time, use of this phenomena in creating sensors are lacking. This effect can be used in creating highly sensitive force and displacement sensors.

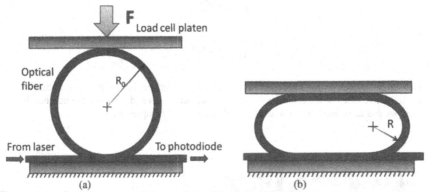

Figure 1. Configuration and working principle of the optical fiber loop-sensor.

DISCUSSION

Calibration of loop-sensor

Results of the calibration study are first discussed. These experiments have been conducted using single-mode fibers and a translation stage-load cell set up shown in Figure 2(a). Figure 2(b) shows the effect of loop radius on the power transmission through the fiber loop. It is observed that as the loop radius is decreased below a critical value the relative transmitted power ($P_{curved}/P_{straight}$) decreases. Here, P_{curved} is the optical power of the transmitted signal through the loop and $P_{straight}$ is the optical power through the straight fiber. Reducing the loop radius to smaller than 3 mm can damage the core and result in no transmission, which sets the lower limit on the sensor size using the present fibers.

In the second set of experiments a loop of 8 mm radius is loaded and unloaded and data on load, displacement, and relative transmitted optical power are collected. The relative transmitted power is defined as the ratio of output power through the loop (P_{out}) by the reference power (P_{ref}) which is that of unloaded loop. The loading is carried out upto 7500 μm displacement with a load step of 50 μm. After attaining the highest displacement, the loop is left for several hours before releasing the load in unloading cycle. The results presented in Figure 3 show that the load-displacement relationship is linear and the optical power loss is linear in certain region of displacement, where it can be used as sensor. Moreover, the loading and unloading cycles are very consistent during the deformation showing that the loop can be sustained for large deformation without losing its elasticity. Closer observation of Figure 3(a) shows intermediate peaks and dips in the optical power. These local maxima and minima appear when

$$\frac{4b\gamma R}{3\pi R_c}\left(\frac{R_c}{R}-1\right)^{3/2} = \begin{cases} 2m-1/2 \text{ for maximum} \\ 2m-3/2 \text{ for minimum} \end{cases} \quad (1)$$

where the critical radius

$$R_c = \frac{2k^2 n_2^2 b}{\gamma^2} \quad (2)$$

and

$$\gamma = \left(\beta_0 - k^2 n_2^2\right)^{1/2} \quad (3)$$

Here β_0 is complex propagation constant of the leaky fundamental mode in straight fiber, n_2 is the refractive index of the cladding, b is the cladding thickness and k is the wave number. Figure 3 shows that these sensors can be used in both tension and compression.

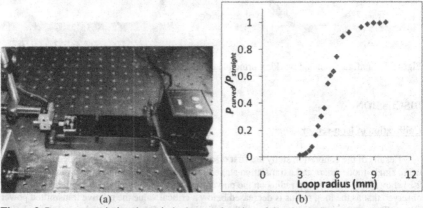

(a) (b)

Figure 2. Power transmission through the loop as function of the circular loop diameter.

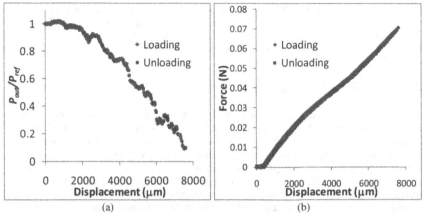

(a)   (b)

Figure 3. Loading and unloading of a loop of 8 mm radius.

Further testing is conducted on loops of 6, 7, 8 and 10 mm radius and the results are shown in Figure 4.

It can be observed that smaller loops provide higher sensitivity because the optical power decays faster in these loops. This is favorable in designing sensors because the desirable smaller sensors will provide higher sensitivity.

Intermediate peaks in the power loss curves are visible in Figure 4 also. It should be noted that with appropriate design these sensors can operate in two different measurement ranges. In a

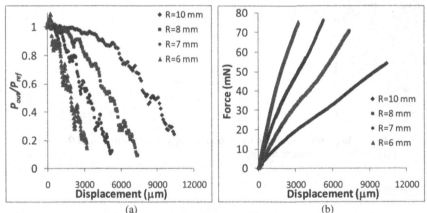

(a)   (b)

Figure 4. Displacement-optical power and load-displacement relationships for loops of four different radii.

321

region between two successive peaks their sensitivity is around 10^{-4} N for force and 10^{-5} m for displacement. Following the linear approximation in their transmitted power curve over the entire length of the displacement their sensitivity can be approximated to around 10^{-2} N for force. Increase in the resolution beyond this level while approximating the macroscopic measurement behavior leads to the possibility of false measurement due to the increased sensitivity to the intermediate peaks.

Structure health monitoring applications

Figure 5(a) shows a laminated composite specimen with a loop sensor adhesively bonded to the surface. The sensor is bonded in the center of the laminate. This specimen is tested under flexural loading conditions. The sensor is kept on the tensile side of the specimen. The loop sensor has a radius of 6.2 mm and laminate has dimensions of 26×150 mm^2. The laminate is composed of four layers of glass fabric reinforced in epoxy resin. The glass fabric layers are oriented at $\pm 45°$ angle. The span length in the flexural testing is maintained at 50 mm. Figure 5(b) presents results of relative power with respect to the displacement. The displacement, force, and transmitted optical power are collected similarly to the calibration process except that a 9.8 N load cell is attached to the translation stage in this case. The laminate is deflected with a step of 200 µm. The results show a linear trend between load and displacement and also between relative power and displacement. These trends can be correlated with each other to effectively use this sensor for measuring strain, and eventually force, in the system.

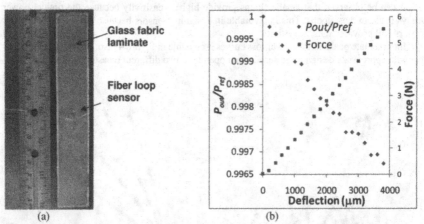

(a)   (b)

Figure 5. (a) A loop sensor bonded to the surface of a laminate composites and (b) results of sensor calibration under flexural loading.

CONCLUSIONS

An optical fiber based sensor is investigated in the present work. This sensor has a simple construction and instrumentation. The sensor is based on the principle of optical power loss from the curved section of an optical fiber. Calibration studies have shown that this sensor has sensitivity of 10^{-4} N for force and 10^{-5} m for displacement. The sensor is adhesively bonded to the surface of a glass fabric-epoxy resin laminate. The calibration studies on these laminated composites also show that these sensors maintain their level of sensitivity can be used on composite materials for structural health monitoring applications.

ACKNOWLEDGMENTS

This research work is supported by the National Science Foundation grant #CBET-0619193. Authors thank the Mechanical and Aerospace Engineering department for the support provided.

REFERENCES

[1]   W. W. Morey, in *International Conference on Optical Fiber Sensors*, Sydney, Australia **1990**, pp. 285.
[2]   S. M. Melle, K. Liu, R. Measures, *SPIE-1994* **1991**, *1954*, 255.
[3]   P. D. Foote, *SPIE-1994* **1994**, *2361*, 290.
[4]   K. S. Kim, Y. Ismailm, G. S. Springer, *Journal of Composites Materials* **1993**, *27*, 1663.
[5]   W. R. Habel, D. Hofmann, *SPIE-1994* **1994**, *2361*, 180.
[6]   S. C. Rashleigh, in *1st International Conference on Optical Fiber Sensors*, IEE, London **1983**, 210.
[7]   D. Inandi, A. Elamar, S. Vurpillot, *SPIE-1994* **1994**, *2361*, 216.
[8]   R. de Oliveira, C. A. Ramos, A. T. Marques, *Computers & Structures* **2008**, *86*, 340.
[9]   G. Liu, S. L. Chuang, *Sensors and Actuators A* **1998**, *69*, 143.
[10]   Q. Wang, K. T. V. Grattan, A. W. Palmer, *Sensors and Actuators A: Physical* **1998**, *71*, 179.
[11]   N. Q. Nguyen, N. Gupta, T. Ioppolo, M. V. Otugen, *Journal of Materials Science* **2008**, *In press*.
[12]   R. Maharsia, N. Gupta, H. D. Jerro, J. A. Peck, in *American Society for Composites 18th Annual Conference*, Vol. 125, Gainesville, FL **2003**.
[13]   J. W. Berthold, in *10th International conference on optical fibre sensors*, Vol. 11, Grassgow, UK **1994**, 182.
[14]   E. Udd, *Fiber Optic Sensors: An Introduction for Engineers and Scientists* John Wiley & Sons, Inc., **2006**.
[15]   X. Guangping, S. Leong Keey, A. Asundi, *Optics and Lasers in Engineering* **1999**, *32*, 437.
[16]   A. MacLean, C. Moran, W. Johnstone, B. Culshaw, D. Marsh, P. Parker, *Sensors and Actuators A: Physical* **2003**, *109*, 60.
[17]   F. Luo, J. Liu, N. Ma, T. F. Morse, *Sensors and Actuators A: Physical* **1999**, *75*, 41.
[18]   N. K. Pandey, B. C. Yadav, *Sensors and Actuators A: Physical* **2006**, *128*, 33.
[19]   S. Thomas Lee, B. Aneeshkumar, P. Radhakrishnan, C. P. G. Vallabhan, V. P. N. Nampoori, *Optics Communications* **2002**, *205*, 253.

[20]   G. N. Bakalidis, N. A. Georgoulas, N. J. Karafolas, C. J. Georgopoulos, *International Journal of Optoelectronics* **1993**, *8*, 187.
[21]   G. N. Bakalidis, E. Glavas, N. G. Voglis, P. Tsalides, *IEEE transactions on instrumentation and measurement* **1996**, *45*, 328.
[22]   H. Renner, *Journal of Lightwave technology* **1992**, *10*, 544.
[23]   Y. Murakami, H. Tsuchiya, *IEEE Journal of quantum electronics* **1978**, *QE-14*, 495.

Mater. Res. Soc. Symp. Proc. Vol. 1129 © 2009 Materials Research Society 1129-V11-20

Controlled Synthesis and Structure-Property Correlations in Vanadium Based Oxides

Tsung-Han Yang[1], Wei Wei [1], Chunming Jin[1], Jagdish Narayan[1]
[1] Department of Materials Science and Engineering, North Carolina State University, EB-I, Centennial Campus, Raleigh, North Carolina, 27695-7907

Abstract

Vanadium based oxides have several interesting properties with their different crystal structures and chemical compositions. V_2O_3, for example, exhibits transition from metal to insulator at a temperature of 150 k. On the other hand, VO_2 changes crystal structure from monoclinic phase (semiconductor) to tetragonal phase (metal phase) with a very sharp transition temperature close to 340 K. This property can be applied in sensor- and memory-type devices. We have successfully deposited high-quality epitaxial V_2O_3, VO_2- films on sapphire (0001) substrates by pulsed laser deposition and obtained sharp transitions without further annealing treatments. The epitaxial growth occurred via domain matching epitaxy, where integral multiples of planes matched across the film-substrate interface. We were able to control the phase and crystalline structure of the vanadium oxide by manipulating laser deposition parameters and partial pressure of oxygen. The chemical composition of vanadium oxide changed from V_2O_3 to VO_2 as the oxygen pressure increased from 10^{-5} to 10^{-1} torr. We present correlations among resistivity, stoichiometry, microstructure, and characteristics of hysteresis loop for the different phases of vanadium oxides.

* This research was sponsored by the National Science Foundation

Introduction

Vanadium oxide (VO_2) has an extremely sharp and fast phase transformation when the temperature is higher than 68 °C. Above 68 °C, VO_2 exhibits transitions monoclinic phase ($P2_1/C$), which was semi-conductor characteristics, to tetragonal phase ($P4_2/mnm$), which exhibits metallic behavior[1-2]. Vanadium oxide is also a thermo-chromic material which is able to reversibly change its optical and electrical properties, because in these oxides at the beginning of the 3d series the nonbonding (t_2g) orbitals extend out far enough to overlap and form a narrow conduction band in the high temperature phase[3]. Although Vanadium oxide (VO_2) has such attractive properties, the stoichiometry of vanadium oxide is difficult to control because of its narrow stability range in different phases, where the ratio of oxygen and vanadium varies from 1.5 to 2.5 such as V_2O_3, V_3O_5, V_4O_7, V_5O_9, VO_2, and V_2O_5[4]. Even the exact chemical composition is hard to control, the unique characteristics of VO_2, such as the large current change in electrical (resistivity) and optical (reflectivity and dielectric constants) properties around 68 °C, have been of considerable interest for device applications in optical storages, optical switching devices, and electro-chromatic materials.

There are extensive studies on the characteristics of Semiconductor to Metal Transition (SMT) in VO_2. The characteristics, sharpness (ΔT) and amplitude (ΔA) of the SMT transition, and the width of the thermal hysteresis (ΔH), for the semiconductor to metal transition are found to be a strong function of chemical composition and microstructure, specifically defect content and nature of grain boundaries. For example, poly-crystal film has the larger ΔT and ΔH with smaller ΔA because of the high defect contents and high angle grain boundary in film[5].

Experiment

While VO_2 is widely studied and surveyed, the relationship between its electrical property and microstructure is still not established. Therefore, to investigate the relationships among chemical stoichiometry, microstructures , and SMT transition characteristics of VO_2, we successfully deposited vanadium oxide (VO_2) film on sapphire (0001) under different oxygen pressure from 10^{-3} to 10^{-1} torr by KrF Excimer laser, which was operated at 248 nm with 25 ns pulse duration and 10 Hz repetition to ablate target material. The Chemical composition of VO_2 film was examined with the X-ray photoelectron spectroscopy (XPS) in a MAC 2 using Mg/Al target. The structure evolution of samples was mainly characterized with X-ray diffraction (XRD) θ-2θ scan using copper Kα wavelength (λ=1.54 Å). The microstructures of VO_2 film deposited on sapphire (0001) were also determined with a high intensity and analytical transmission electron microscope (JEM-2000Fx, JOEL) and the high-resolution image with

(JEM-2010Fx, JEOL. Based on the electrical resistivity hysteresis loop, the SMT characteristics of VO_2 films, which were deposited on sapphire (0001) at 600 °C under the different oxygen pressure, were also analyzed by HP4155B using Van Der Pauw method that in the temperature range of 25 °C to 100 °C.

Results and Discussion

Figure 1 shows the XRD patterns of vanadium-based oxide films which were grown on sapphire (0001) under the different oxygen pressure from 10^{-5} and 10^{-1} torr. For the oxygen pressure of 10^{-5} and 10^{-4} torr, the vanadium oxide films showed two peaks at 38.53° and 82.69° which were identified as the (006) plane and (0012) plane of rhomobohedral phase of V_2O_3. As the oxygen partial pressure was increased from 10^{-3} torr, 10^{-2} torr to 10^{-1} torr, the two peaks shifted to 39.74° and 85.77° which were indexed as the (002) plane and (004) planes respectively of VO_2 monoclinic structure. The lattice parameters of VO_2 film, which was deposited on sapphire (0001) under pressure of 10^{-2} torr, were a=5.751±0.01Å, b=4.537±0.01Å, c=5.382±0.01Å, and β=122.6°±0.1° based on the XRD results[8-10]. Moreover, compared with the VO_2 films, the V_2O_3 film under oxygen pressure 10^{-5} torr had larger average grain size determined from Scherzer equation[6-7]. The microstructure of VO_2 films under oxygen pressure 10^{-3} torr were textured polycrystalline with small grain size and it became epitaxy under the oxygen pressure 10^{-2} and 10^{-1} torr. Therefore, by increasing the oxygen pressure from 10^{-3} torr to 10^{-2} and 10^{-1} torr, the defects can be eliminated effectively for the VO_2 films. To obtain details on microstructure and the nature of epitaxy on sapphire (0001) substrate with oxygen pressure of 10^{-2} torr, we have taken X-section TEM image and diffraction patterns for analysis.

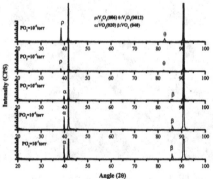

Fig. 1 The X-ray diffraction pattern of the as-deposit vanadium oxide film with different oxygen partial pressure from 10^{-5} to 10^{-1} torr.

Fig. 2(a) was the cross-section TEM image of VO_2 film deposited on sapphire (0001) under oxygen pressure 10^{-2} torr and thickness of the VO_2 film was estimated to be 105 nm. Fig. 2(b) was the Dark-field image from the same area of Fig. 2(a). The VO_2 film grown on sapphire (0001) under oxygen pressure 10^{-2} torr was the single-crystal growth because there was no obvious contrast in Fig 2(b). In addition, the structure and orientation could be determined by SAD patterns of the VO_2 film, substrate, and interface between the film and substrate, as shown in Fig. 2(c), Fig. 2(d) and Fig. 2(e), respectively. Fig. 2(c) showed the zone of VO_2 film was only in [010] while Fig. 2(d) showed the zone of sapphire (0001) was in $[2\bar{1}\bar{1}0]$ direction. In figure 2(e), the zones of the VO_2 film and substrate were identified as [010] and $[2\bar{1}\bar{1}0]$ directions, respectively. The single-phase VO_2 film on sapphire (0001) was grown epitaxially with [002] normal to substrate in oxygen atmosphere under partial pressure of 10^{-2} torr at 600 °C. The epitaxial relationship of the single-phase VO_2 film related to sapphire (0001) was as follows: $[002]_f$ I $[0006]_s$ (plane normal) and $[010]_f$ I $[2\bar{1}\bar{1}0]_s$ (in plane). Fig. 2(f) showed the high-Resolution TEM image from the film-substrate interface with VO_2 film in [010] zone and sapphire (0001) in $[2\bar{1}\bar{1}0]$ direction and the (002) d-spacing of vanadium oxide film was determined to be 2.27 Å[5].

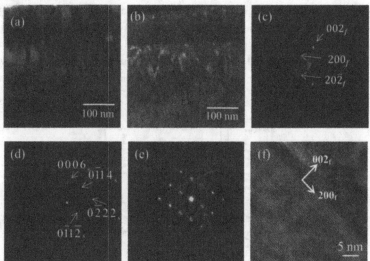

Fig. 2.(a)Cross-Section TEM image for oxygen pressure 10^{-2} torr showing 105 nmVO$_2$ deposited on C-plane Sapphire. Fig. 2(b) is the dark-field image of Fig. 2 (a). The SAD patterns: (c) in the [010] film zone only, (d) in the $[2\bar{1}\bar{1}0]$ sapphire substrate zone only,

and (e) film-substrate interface in the [010] film zone and [2$\bar{1}\bar{1}$0] sapphire substrate zone. Fig. 2(f) High-Resolution TEM image is for the film-substrate interface with the [010] film zone and [21$\bar{1}$0] sapphire substrate zone.

Further, the Bright-Field cross-section image of VO_2 film deposited on sapphire (0001) under oxygen pressure of 10^{-3} torr was shown in Fig. 3 (a) and the dark-field image was shown in Fig. 3(b). The columnar grains of VO_2 have been observed in Fig. 3 (b) and the average grain size of VO_2 film was estimated to be 40 nm. In Fig. 3(c), which is the selected-area-diffraction (SAD) pattern from the VO_2 film and substrate (sapphire (0001)) interface in Fig. 3(a), showed the VO_2 film was on [010] zone direction and sapphire was on [2$\bar{1}\bar{1}$0] zone direction. From dark-field image (Fig. 3(b)) and SAD image (Fig. 3(c)), the VO_2 film was the textured ploy-crystal film on [002] direction.

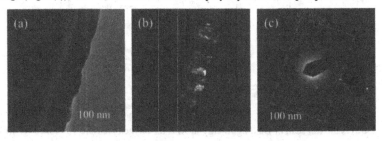

Fig. 3 (a) shows the plane-view cross-section image for VO_2 film deposited on sapphire (0001) in oxygen environment with partial pressure of 10^{-3} torr and Fig. 3(b) is the dark-field image of the area in Fig. 3(a). Fig. 3(c) is the selected-area-diffraction pattern of the VO_2 film-sapphire (0001) interface in Fig. 3 (a) and shows the [010] film zone and [2$\bar{1}\bar{1}$0] sapphire substrate zone.

Fig. 4(a) showed the full scan of Binding Energy in X-ray Photoelectron Spectroscopy (XPS) for VO_2 film grown on sapphire (0001) at 600 °C under oxygen pressure of 10^{-1} torr. The oxygen (1s), vanadium (2p), and carbon (1s) peaks were identified respectively at 532, 518, 285 eV in the full scan of XP spectrum. There was no significant characteristic peak of impurities shown in full scan of XP spectrum. Since valence charge distributions are related to chemical binding energy shift, the VO_2 and V_2O_3 films, which grown under different pressures, were scanned from 510 to 535 eV with a small step, 0.1eV, and the results were shown in Fig. 4(b). It displayed O (1s) and V (2p) core level XP spectra of V_2O_3 and VO_2 films which were deposited under various oxygen pressures ranging from 10^{-5} torr to 10^{-1} torr. The 2p core level XP spectra show two peaks of $2p_{3/2}$ and $2p_{1/2}$ due to the spin-orbital splitting. The higher binding peak of $2p_{1/2}$ peak was broader than the $2p_{3/2}$ peak because of Coster-Kronig Auger transitions[11-12]. The different valence charges of vanadium in VO_2 films under different oxygen partial pressure were evaluated by the XPS data. The percentage of V^{+5} remained approximated constant, while the percentage of V^{+4} increased and that of V^{+3} decreased with increasing oxygen pressure. To obtain detailed composition of VO_2 film, we determined the ratio of vanadium to oxygen in the VO_2 films which under 10^{-3}, 10^{-2} and 10^{-1} torr were determined to be 1.9, 2.2 and 2.3, respectively.

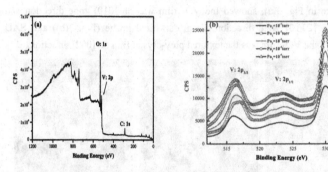

Fig. 4(a) is the full scan of Binding Energy in XPS for oxygen pressure 10^{-1} torr. Fig. 4(b) displays O(1s) and V(2p) core level XP spectra of V_2O_3 and VO_2 films deposited at various oxygen pressure.

Fig. 5 showed the resistivity as a function of temperature for different oxygen pressure, 10^{-3}, 10^{-2}, and 10^{-1} torr. The temperature of phase transformation from monoclinic to tetragonal phase of VO_2 film was obtained by differentiating the electrical hysteresis loop (resistance versus temperature) that was measured during heating up and cooling down the substrates. The transition temperatures for heating-up and cooling-down under oxygen partial pressure of 10^{-1} and 10^{-2} torr for were 71 °C and 66 °C, respectively, which were quite sharp as expected for single-crystal films with some offsets. On the other hand, 10^{-3} torr films shows a broad transition with small amplitude having a low ratio of electrical resistance [13-14]. The sharpness (ΔT) of SMT hysteresis loop was 20.1, 8.5, and 9.5 K; the width of thermal hysteresis (ΔH) was 3.24, 5.0, 5.0 K under 10^{-3}, 10^{-2}, and 10^{-1} torr; the amplitude (ΔA) of SMT characteristics determined by the ratio of the electrical resistance at 35 °C and 95 °C was 3.75×10^{1}, 1.47×10^{4}, and 2.85×10^{4} for 10^{-3}, 10^{-2}, and 10^{-1} torr, respectively. These parameters of SMT transition for the different oxygen atmosphere are summarized in table I.

Table I shows the parameters, sharpness (ΔT), the amplitude of SMT transition (ΔA), and width of the electrical hysteresis (ΔH), of the VO_2 films grown in different oxygen pressure.

	ΔT	ΔA	ΔH
$Po_2=10^{-3}$ torr	20.1 °C	3.75×10^1	3.24 °C
$Po_2=10^{-2}$ torr	8.5 °C	1.47×10^4	5.00 °C
$Po_2=10^{-1}$ torr	9.5 °C	2.85×10^4	5.00°C

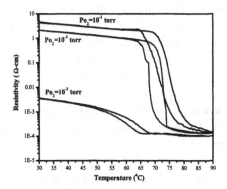

Fig. 5 The resistivity with the increase temperature for the different oxygen pressure, 10^{-3}, 10^{-2}, and 10^{-1} torr.

In our previous paper, a phase transition model correlating characteristics transition with defects and microstructures of VO_2 films was proposed. The sharpness (ΔT) of transition was determined by the overall defect content per unit volume ($\Delta T=C_t \rho_d$, where C_t is a constant and ρ_d is overall defect density, including point defects, clusters, impurities, dislocation, grain boundaries). The transition width of thermal hysteresis (ΔH) was related to $2\gamma/(r_c \Delta S_0)$. Thus, the ΔH increased with decreasing r_c and ΔS_0, indicating width of thermal hysteresis increases as the grain size decreases, as reported by Lopez's group[15-16]. The amplitude (ΔA) of SMT hysteresis loop was also toward to depend on the defect content, where it varied inversely with the defect content. Vanadium oxide (VO_2) films deposited in oxygen partial pressure of 10^{-3} torr have high density of disordered regions. It indicated that this film contained higher defect content per unit volume, so it should have small amplitude and less sharpness.

Based on this model for single-crystal film, there should be a sharp transition (ΔT) with a large amplitude (ΔA) and small width of electrical hysteresis (ΔH). The epitaxial vanadium oxide films on sapphire (0001) at 600 °C under oxygen pressure of 10^{-2} and 10^{-1} torr showed a quite small ΔH, sharp transition and large change of amplitude during SMT transition. To obtain details on defect content, which was strongly related to the sharpness (ΔT) during SMT transition, the chemical composition deviation was estimated by XPS data as follows: the ratio of vanadium to oxygen in the VO_2 films under oxygen pressures 10^{-3}, 10^{-2} and 10^{-1} torr was 1.9, 2.2 and 2.3, respectively. Although the microstructure of VO_2 film grown under 10^{-1} and 10^{-2} torr oxygen pressure were approximately the same, the small deviation in stoichiometry of VO_2 film deposited in 10^{-1} torr oxygen pressure induced higher dangling bonds and stress in film, indicating the higher defect content per unit volume in VO_2 film. This result can reasonably explain why the VO_2 film on sapphire (0001) at 600 °C under 10^{-2} torr had the sharper transition loop than for films at 10^{-1} torr[5, 15-16]. The 10^{-3} torr films contained disordered regions associated with oxygen deficiency in a crystalline matrix, which may lead to a broad transition with small hysteresis loop. Farther, atomic-resolution STEM-Z and EELS studies are being continued to unravel the role of oxygen deficiency related defects on transition characteristics.

Conclusions

We have grown successfully the epitaxial (006)-vanadium (III) oxide (V_2O_3) and (002)-vanadium (IV) oxide (VO_2) on sapphire (0001) by controlling the oxygen pressure from 10^{-5} to 10^{-2} torr. The microstructures of vanadium oxide (VO_2) films varied from textured polycrystalline to epitaxial crystal as altering the oxygen pressure varied from 10^{-3}, to 10^{-2}, and to 10^{-1} torr. The oxygen stoichiometry plays an important role, which affects the defect content and disordered regions in the films. The oxygen deficiency in $VO_{1.9}$ films deposited at 10^{-3} torr showed low resistivity and amplitude with negligible hysteresis loop. All these results strongly supported our previous model relating sharpness, amplitude and width of hysteresis loop with microstructural characteristics, defect contents, textured nature of grain boundaries and defect density of VO_2 film.

References

1. J. C. Rakotoniaina, R. Mokrani-Tamellin, J. R. Gavarri, G. Vacquier, A. Casalot, G. Calvarin, J. Solid State Chem., **103** (1993) 81

2. Joyeeta Nag and R F Haglund Jr, Journal of Physics: Condensed Mater, **20** (2008) 264016

3. F. J. Morin, Physical Review Letters, **3** (1959) 34

4. Rakel Lindström, Vincent Maurice, Henri Groult, Laurent Perrigaud, Sandrine Zanna, Camille Cohen, and Philippe Marcus, Electrochimica Acta, **51** (2006) 5001

5. J. Narayan and V. M. Bhosle, Journal of Applied Physics, **100** (2006) 103524

6. C. Grygiel, Ch. Simon, B. Mercey, W. Prellier, R. FrésardAPL, and P. Limelette, Applied Physics Letter, **91** (2007) 262103

7. D. B. McWhan, A. Menth, J. P. Remeika, W. F. Brinkman, and T. M. Rice, Physical Review B, **7** (1973) 1920

8. John B. Goodengugh, Physical Review, 117, (1960), 1442

9. R. Heckingbottom and J. W. Linnett, Nature, 194, (1962), 678

10. S. Westman, Acta Chim. Scand. 15, (1961), 217

11. Y.L. Wang, X.K. Chen, M.C. Li, R. Wang, G. Wu, J.P. Yang, W.H. Han, S.Z. Cao, L.C. Zhao, Surface and Coatings Technology **201** (2007) 5344

12. E. Antonides, E. C. Janse, and G. A. Sawatzky, Physical Review B, **15** (1977) 4596

13. P. Jin, K. Yoshimura, and S. Tanemura, Journal of Vacuum Science and Technology A, **15** (1997) 1113

14. D. H. Kim and H. S. Kwok, Applied Physics Letter, **65** (1994) 3188

15. Sihai Chen, Hong Ma, Jun Dai, and Xinjian Yi, Applied Physics Letter, **90** (2007) 101117

16. R. Lopez, T. E. Haynes, L. A. Boatner, L. C. Feldman, and R. F. Haglund, Jr., Physical Review B, **25** (2002) 224113

Mater. Res. Soc. Symp. Proc. Vol. 1129 © 2009 Materials Research Society   1129-V04-23

Properties of Intermetallic Compound Joint Made from Evaporated Ag/Cu/Sn Films.

T. Takahashi[1], S. Komatsu[2] and T. Kono[1]
[1]Corporate Research & Development Center, Toshiba Corporation,
1, Komukai-Toshiba-cho, Saiwai-ku, Kawasaki, 212-8582, Japan
[2]Toshiba Research Consulting Corporation,
1, Komukai-Toshiba-cho, Saiwai-ku, Kawasaki, 212-8582, Japan

ABSTRACT

Pb-based solders that contain a great amount of toxic Pb are still used as the high-melting-temperature-type solders to join power semiconductor chips to lead frames. For replacing the Pb-based solder, joining technique using the evaporated Ag/Cu/Sn films were developed. In this study, properties of intermetallic compound (IMC) joint made from the evaporated Ag/Cu/Sn films were investigated. The IMC joint consisting of Cu_6Sn_5, $(Ag,Cu)_3Sn$ and Cu_3Sn layers was formed under heat treatment at 573 K for 30s. The IMC joint showed sufficient strength at 543 K to hold Si chip on Cu substrate. The high-temperature strength kept even after aging at 423 K for 3.6 Ms, although some voids were formed in the microstructure of IMC joint aged. The results of nanoindentation test indicate that excessive growth of Cu_3Sn would be undesirable for joint reliability.

INTRODUCTION

Demand for Pb-free solders has globally increased in the semiconductor and electronics industries owing to the RoHS directive [1]. Sn-Ag-Cu solders have been applied as Pb-free solders to replace eutectic Sn-37 mass%Pb solder. On the other hand, Pb-based solders with Pb content of more than 85 mass% are still used as high-melting-temperature-type solders for joining power semiconductor chips on lead frames [2]. The Pb-based solders have a high-temperature performance that keeps sufficient strength to hold the semiconductor chips on lead frames at the reflow temperature. In addition, they have excellent properties of stress-relaxation characteristics, thermal fatigue resistance and cost [3, 4]. However, in view of the toxic properties of Pb, development of high-melting-temperature-type Pb-free solders is required.

The purpose of this study is to evaluate joints made from evaporated Ag, Cu and Sn films. Since IMCs formed by the reaction of Ag, Cu and Sn films have high melting points, the IMC joints are expected to have the high-temperature performance equivalent to the Pb-based solders or more. Then, properties of the IMC joints are investigated and discussed.

EXPERIMENT

Preparation of joint specimen

Evaporated films of Sn/Cu/Sn/Ag/Sn were deposited on Si wafer with sputtered Ti/Ni/Au films as joining material by electron-beam evaporation. The evaporated deposition was performed at deposition rate of about 3 nm/s under the degree of vacuum level of 10^{-4} Pa order. The sputtered Ti/Ni/Au films were applied in order to increase adhesion force between the Si wafer and the evaporated films. After deposition, the Si wafer with the evaporated films was cut to dimensions of 2.5 mm x 3.0 mm to use as Si chips. The chip size was used as typical size of power semiconductor chips. Figure 1 schematically indicates joining method for joint specimens. The joint specimens were produced to join the Si chips with the evaporated Sn/Cu/Sn/Ag/Sn films onto Cu substrates. The Cu substrate had dimensions of 10 mm x 10 mm x 0.3 mm. The joining was performed on a hot plate in N_2 atmosphere. Heating conditions were set at 573 K for 30 s, while the Si chips were pressed at about 0.1 MPa. The temperatures were measured at the surface of substrate with thermocouples. After joining, the joint specimens were air-cooled. In addition, reference specimens were prepared by joining the Si chip to Cu substrate with Pb-based solder (Pb-2 mass%Ag-3 mass%Sn) weighing 10 mg. The dimensions of Si chip and Cu substrate were the same as the above dimensions. Heating condition was set at 593 K for 10 s. Furthermore, the joint specimens were heated at 423 K for 1.8 Ms and 3.6 Ms in order to investigate the influence of aging on the joint reliability.

Table I. Initial thickness of each layer

Element	Sputtered film			Evaporated film				
	Ti	Ni	Au	Sn	Cu	Sn	Ag	Sn
Thickness (μm)	0.05	0.2	0.05	0.4	0.2	0.8	2.1	1.5

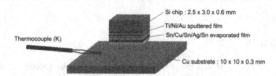

Figure 1. Schematic image of joint specimen.

Evaluation of IMC joints

The cross-sectional analysis was carried out to investigate microstructures of joint made from evaporated Sn/Cu/Sn/Ag/Sn films. The microstructures of joint were observed with transmission electron microscope (TEM) and scanning electron microscope (SEM). The phase structures of IMCs formed in joint were identified by selected-area electron diffraction (SAED) and energy-dispersive x-ray analysis (EDX) in TEM.

Shear strength of joint was measured with die shear tester at a shear speed of 0.1 mm/s. The distance between the shear tool and the Cu substrate was 10 μm. Also, the shear test was carried out at elevated temperatures of 503 K, 523 K and 543 K. Elastic modulus and hardness of each IMC were measured by by nanoindentation test. The test was performed at strain rate of 0.05/s and the maximum indentation depth of 200 nm. Each property was determined from indentation load-depth curve.

DISCUSSION

Cross-sectional analysis

The joint between Si chip and Cu substrate was analyzed with TEM. Figure 2 shows a cross-sectional TEM micrograph and SAED pattern obtained from each layer. Although some voids were observed in the TEM micrograph, they would have formed during sample making process. The IMC joint had three layers that were two thick layers with a few μm and a thin layer with about one μm. The results of SAED identified the layers as the following IMCs: the upper layer was Cu_6Sn_5, the middle layer was Ag_3Sn, and the lower layer was Cu_3Sn. However, about 4.5 at% Cu was detected in the middle layer by EDX analysis as seen in Table II. Since Ag in Ag_3Sn can be partially replaced with Cu [5], the middle layer would be $(Ag,Cu)_3Sn$. In addition, Cu_6Sn_5 and $(Ag,Cu)_3Sn$ layers were formed by the reaction among the evaporated films, and Cu_3Sn layer was formed by reacting molten Sn in evaporated film with Cu diffused from the substrate. Since melting points of Cu_6Sn_5, Ag_3Sn and Cu_3Sn are 688 K, 753 K and 949 K respectively [6], the IMC joint is expected to continue to be solid at 523 K that is often used for the reflow temperature, and to have sufficient high-temperature strength to hold Si chips on substrates at the reflow process.

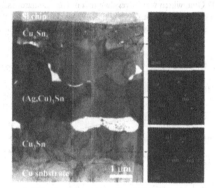

Figure 2. TEM image of IMC joint soldered at 573 K for 30 s, corresponding SAED patterns with the indices of the diffraction spots.

Table II. Composition of each layer

Element (at%)	Ag	Cu	Sn
Upper layer	0.3	57.0	42.7
Middle layer	73.6	4.5	21.9
Lower layer	0.6	24.6	74.8

Mechanical properties

The high-temperature strength was measured by die shear test. Figure 3 shows the shear strength of IMC joint and Pb-based solder joint. Shear strength of IMC joint kept higher strength than that of Pb-based solder joint until 543 K, and the strength at 543 K showed 21.3 MPa. The result indicated that the IMC joint had high-temperature strength required for high-melting-temperature-type solder. Therefore, joint with good high-temperature performance could be formed under joining condition at 573 K for 30 s by using the evaporated film.

On the other hand, excessive increase of shear strength might result in several degradation of the mechanical performance owing to the brittleness of IMCs. Then, mechanical properties of each IMC were investigated by nanoindentation test. Figure 4 shows indentation load-depth curves of $(Ag,Cu)_3Sn$, Cu_3Sn obtained from the nanoindentation test. From the curves, hardness and elastic modulus were estimated [7, 8]. Figure 5 shows the relationships between the hardness and the elastic modulus of each IMC, including measurement data of Ag, Cu and Si obtained from the same test as references. This figure gave a generous correlation with the hardness and the elastic modulus. The hardness increased as the elastic modulus increased. Moreover, the result cleared that characteristic of $(Ag,Cu)_3Sn$ and Cu_3Sn were different. The elastic modulus of Cu_3Sn was 165 GPa, which was higher than those of Si and Cu. In contrast, the elastic modulus of $(Ag,Cu)_3Sn$ was 97 GPa, which was lower than that of Cu, and hardness was less than half that of Cu_3Sn. As a result, since the hardness and the elastic modulus of Cu_3Sn were much higher than those of $(Ag,Cu)_3Sn$, an excessive growth of Cu_3Sn would be undesirable for joint reliability.

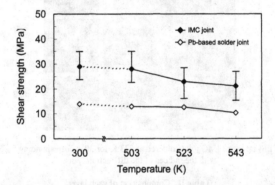

Figure 3. Influence of test temperature on shear strength of IMC joint and Pb-based solder joint.

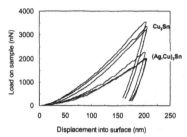
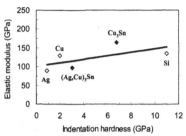

Figure 4. Indentation curves of Cu₃Sn and (Ag,Cu)₃Sn layer.

Figure 5. Comparison of hardness and elastic modulus for joint component material.

Influence of aging on joint reliability

An accelerated aging test at 423 K was performed to investigate influence of aging on the joint reliability. Figure 6 shows cross-sectional SEM micrographs of IMC joint before and after aging for 1.8 Ms and 3.6 Ms. For microstructure after aging, small-sized Cu₃Sn grains after aging for 1.8 Ms, and large-sized Cu₃Sn grains after aging for 3.6 Ms were formed in (Ag,Cu)₃Sn layer. Some voids were also formed in Cu₃Sn layers under both aging conditions. The Cu₃Sn grains would be formed by a reaction between Cu diffused from Cu₃Sn layer and Sn in (Ag,Cu)₃Sn layer. The growth of Cu₃Sn grains indicated that Cu₃Sn continued to react with Sn in (Ag,Cu)₃Sn layer as Cu was supplied from Cu₃Sn layer during aging at 423 K. The voids in Cu₃Sn layer might be formed owing to the difference of diffusion coefficient. After aging for 3.6 Ms, Cu₆Sn₅ as well as Cu₃Sn formed in (Ag,Cu)₃Sn layer. The Cu₆Sn₅ phase would be formed by the reaction between Cu₃Sn layer and Sn in (Ag,Cu)₃Sn layer.

Influence of aging on mechanical property was investigated by measuring shear strength of the IMC joint. Figure 7 shows the results of shear strength measured at 300 K and 543 K. Although both the shear strength measured at 300 K and 543 K gradually decreased according to increasing aging time, they kept more than 15 MPa until aging for 3.6 Ms. Consequently, the IMC joint had high heat-resistance at 423 K, although the voids were formed in IMC joint

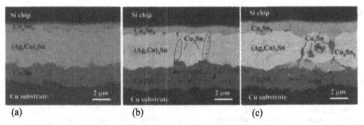

(a)   (b)   (c)

Figure 6. Influence of aging at 423 K on microstructure of IMC joint: (a) before aging test, (b) aged for 1.8 Ms, and (c) aged for 3.6 Ms.

339

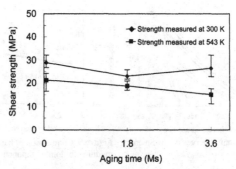

Figure 7. Influence of aging at 423 K on shear strength of IMC joint.

CONCLUSIONS

Joining technique using the Ag/Cu/Sn evaporated films was developed for replacing the high-melting-temperature-type Pb-based solder. IMC joint with higher melting point than the reflow temperature was formed in joint part between Si chip and Cu substrate under heat treatment at 573 K for 30 s. The IMC joint was composed of three IMC layers that were Cu_6Sn_5 with thickness of about 1 μm, and $(Ag,Cu)_3Sn$ and Cu_3Sn with thickness of a few μm. The shear strength of IMC joint at 543 K showed 21.3 MPa, which was higher than that of Pb-based solder joint. This result proved the IMC joint satisfied high-temperature strength required for the high-melting-temperature-type solder. The accelerated aging test at 423 K cleared that the strength of IMC joint could be sufficiently maintained until aged for 3.6 Ms, although some voids were formed in the microstructure of IMC joint. Property of each IMC layer was investigated by nanoindentation test. The test indicated that excessive growth of Cu_3Sn would be undesirable for joint reliability because hardness and elastic modulus of Cu_3Sn were much higher than those of $(Ag,Cu)_3Sn$.

REFERENCES

1. Directive 2002/95/EC of the European Parliament and of the Council on the Restriction of the Use of Certain Hazardous Substances in Electrical and Electronic Equipment.
2. L. N. Ramanathan, J. W. Jang, J. K. Lin, and D. R. Frear, *J. Electron. Mater.* **34**, 1357 (2005).
3. D. Pan, I. Dutta, S. G. Jadhav, G. F. Raiser, and S. Ma, *J. Electr. Mater.* **34**, 1040 (2005).
4. A. Scandurra, A. Porto, and O. Puglisi, *Appl. Surf. Sci.* **89**, 1 (1995).
5. S. Dunford, S. Canumalla and P. Viswanadham, 54th Electronic Components and Technology Conference, Las Vedas, 726(2004).
6. T. B. Massalski, J. L. Murray, L. H. Bennett and H. Baker, Binary Alloy Phase Diagrams, American Society for Metals, (1986).
7. J. Hey: Mechanical Testing by Indentation, Course Notes, (1997), Nano Instruments, Inc.
8. W. C. Oliver, B and M. Pharr, *J. Mater. Res.*, **7**, 1564 (1992).

Mater. Res. Soc. Symp. Proc. Vol. 1129 © 2009 Materials Research Society 1129-V04-19

Conformal Passive Sensors for Wireless Structural Health Monitoring

Sharavanan Balasubramaniam[1], Jung-Rae Park[1], Tarisha Mistry[2], Niwat Angkawisittpan[2],
Alkim Akyurtlu[2], Tenneti Rao[2] and Ramaswamy Nagarajan[1]
[1]Department of Plastics Engineering, University of Massachusetts, Lowell, MA 01854.
[2]Department of Electrical & Computer Engineering, University of Massachusetts, Lowell,
MA 01854.

ABSTRACT

The application of conductive inks for the low-cost fabrication of wireless sensors is described. Binder-free silver inks have been formulated, which can form pure metallic patterns, leading to devices with high quality factors. The conductive formulations have been screen-printed on flexible polyester films and thermally converted to silver traces at low temperatures to yield resonant inductor-interdigital capacitor elements. The radio-frequency (RF) response of the printed inductor-capacitor (LC) resonators is measured using a vector network analyzer. Structural damage such as a crack in the substrate disrupts the resonant characteristics of the sensing elements, resulting in the loss of the characteristic response of the LC circuit. The application of these screen-printable sensors for the detection of cracks is demonstrated.

INTRODUCTION

Low-cost fabrication of sensors for wireless detection of structural defects like cracks is currently a challenge. Sensors for wireless structural health monitoring are fabricated using complex or expensive methods like photolithography [1] and laser micromachining techniques [2]. The commercial success of these sensors depends on both performance and ease of manufacture. We have developed low-temperature curing conductive silver inks for screen-printing resonant LC circuits on flexible plastic substrates. Silver inks based on flaky powder and nanoparticles [3] are very popular in the area of printed electronics due to their high electrical conductivity and environmental stability. The metallo-organic decomposition (MOD) technology is well known for printing precursor inks by ink jetting [4-7] or screen-printing [8-9]. In this method, the weak bonds holding the metal atom and the organic moieties can be thermally broken, paving the way for low-temperature conversion to metal. Thus, it has been used as a simple and convenient approach to the fabrication of metallic patterns for a variety of applications such as flexible circuits, interconnections in devices and radio-frequency identification (RFID) readers.

Screen-printing has been used for the deposition of the formulated silver inks because of its compelling economic advantages and the ability to transfer a pattern of materials on any substrate at a high rate. Also, the thickness of the materials deposited by screen printing after sintering or firing is of the order of 10 microns. Thus the method lends itself very well to highly conductive circuit elements. The conformal sensor reported here employs inductor-interdigital capacitor elements printed on a flexible polyester film using conductive silver inks. The LC resonators are activated by a radio-frequency signal and can be interrogated wirelessly using a single loop antenna.

EXPERIMENTAL

Silver acetylacetonate was used as the organic silver precursor for the conductive inks. The binder-free silver inks were prepared from silver flakes (<10 μm, Aldrich), silver acetylacetonate (Aldrich) and α-terpineol. A mixture of silver flakes and silver acetylacetonate (5 wt% in the mixture) was ground into a homogeneous powder and mixed with α-terpineol solvent in a high-speed vortex for 30 minutes. Uniform pastes containing ca. 80% solids were obtained by bead-milling for one hour.

The thermal decomposition behaviour of the silver inks was studied by thermogravimetric analysis (TGA, TA Q50) in air at a heating rate of 10^0C/min. The derivative TGA curve was used to identify the temperature of maximum weight loss during the decomposition of silver acetylacetonate.

A 305-mesh monofilament polyester screen was used to print the conductive silver patterns. Printed patterns were cured in an oven at 140^0C for 10 minutes. Line patterns of dimensions 30 mm×3 mm were printed for resistance measurements. The D.C. resistance was measured as a function of cure time. Resonant inductor-interdigital capacitor circuits (Figure 1) were printed on polyester films (50 μm thickness). The geometric variables of the LC sensor design were as follows: W_L (width of the spiral inductor lines) = W_C (width of capacitor electrodes) = G_L (gap between the turns) = 1 mm; G_C (gap between electrodes) = 2 mm; N_L (number of turns) = 4.5; N_C (number of electrodes) = 6; D_{in} (inner diameter) = 25 mm; L_C (length of electrodes) = 15 mm. The radio-frequency response of the printed LC resonators was wirelessly interrogated using a single loop antenna connected to an HP8753C vector network analyzer (VNA).

Figure 1. Design of resonant LC circuit

Miniature cracks were generated in the printed plastic films to disrupt the sensing elements and detect any change in the RF response.

RESULTS AND DISCUSSION

The choice of the organic metal precursor is critical in the development of conductive MOD inks for printing electronic circuit elements. The precursor compound typically consists of organic moieties bonded to the metal through a heteroatom like oxygen, sulfur or nitrogen. This bond is a weak link which provides for easy thermal decomposition to the metal or its oxide. Figure 2 shows the TGA curve of silver acetylacetonate, used as the organic silver precursor in our conductive inks. This compound has a relatively high percentage of silver (~52%) and starts decomposing at 85^{0}C, with the maximum decomposition rate at 145^{0}C (Figure 3).

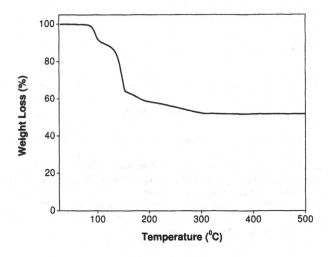

Figure 2. Thermal Decomposition of Silver Acetylacetonate

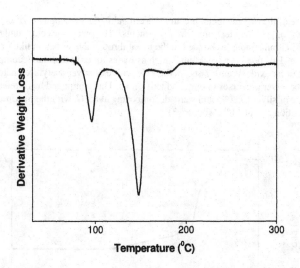

Figure 3. Derivative TGA curve of Silver Acetylacetonate

Figure 4 shows the weight loss of the prepared silver ink containing ca. 80% solids, compared with two commercially available conductive silver inks. Comparison of TGA curves of these inks with some of the commercially available silver inks clearly indicates that these inks can be thermally converted to metallic silver at relatively lower temperatures, around 140^0C. Commercial ink formulations contain resinous binders that decompose at substantially higher temperatures.

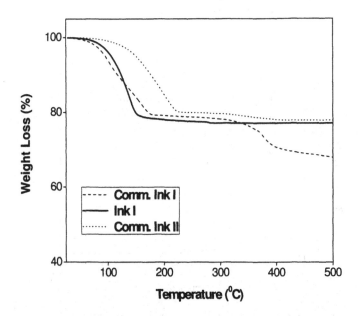

Figure 4. Thermogravimetric curves of conductive silver formulations

The D.C resistance of these silver inks during the process of thermal conversion to metallic silver line patterns compared with the commercial inks is shown in Figure 5. For line patterns of similar line dimensions (width, length and thickness) printed on polyester films and subsequently cured at 140^0C, the resistance measured by the digital multimeter was plotted as a function of the cure time. The inks described in this work showed a resistance drop of several orders of magnitude in less than 60 seconds and cured to low values of resistance.

Resonant LC patterns were screen-printed on polyester films using a monofilament polyester screen with a mesh count of 305 per inch and a rubber squeegee. The printed patterns were thermally cured at 140^0C for 10 minutes. When the LC sensor enters into the interrogation zone of the loop antenna connected to the network analyzer, a sharp resonant dip occurs at the resonance frequency (f_0) of the LC circuit. The resonance frequency is given by the equation: $f_0 = 1 / (2\pi \sqrt{(LC)})$. The characteristic dip is monitored by sweeping the reflected signal in the frequency range of 90-150 MHz. The resonant dip for the designed LC pattern occurred at 107 MHz. In principle, the electromagnetic signal generated in the VNA and radiated by the loop antenna is received by the planar inductor coils by electromagnetic coupling. An electromagnetic force (emf) is induced in the inductor coils, which in turn generate a back emf in the radiating antenna. At resonance, the magnitude of the reflected voltage is minimum, which appears as a sharp dip in the S_{11} versus frequency plot. The presence of a crack in the substrate disrupts the sensing elements resulting in a loss of the resonance peak (Figure 7).

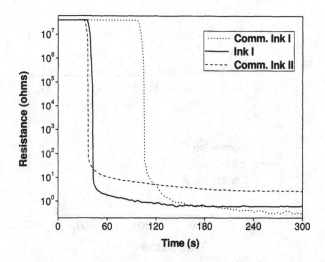

Figure 5. Cure characteristics of conductive silver formulations

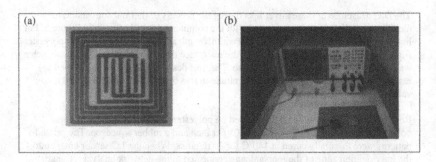

Figure 6. (a) LC resonator screen-printed on a polyester film (b) experimental set-up for the wireless interrogation of the flexible LC sensor

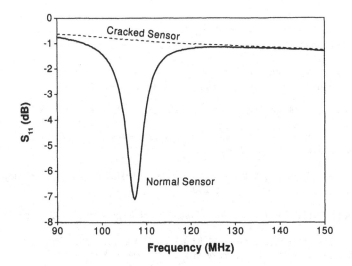

Figure 7. RF response of the LC sensor interrogated with a network analyzer

CONCLUSION

In summary, we have developed binder-free conductive inks based on silver flakes and silver acetylacetonate dispersed in terpineol solvent, which can be screen-printed on a variety of substrates. The low-temperature curing characteristics of the inks were confirmed by thermogravimetric analysis. These inks are therefore suitable for use with temperature-sensitive polymeric substrates. Inductor-interdigital capacitor circuits were screen-printed using these inks on flexible polyester films and cured at 140°C for 10 minutes. These conformal LC sensors can be wirelessly interrogated using a single loop antenna connected to a network analyzer. We have also demonstrated that these patterned could be printed on other substrates such as ceramic and can be used to monitor the structural integrity of these substrates. Structural damage to the surface (formation of cracks) containing printed RF sensors causes disruption of the circuit. Upon interrogation using an RF source, the characteristic resonance peak of the sensors disappears, thus indicating the presence of a crack in the sensing elements. These printed sensors can thus be used in wireless health monitoring, especially for the detection of structural defects like cracks.

REFERENCES

[1] Y. Jia, K. Sun, F. Agosto, M. T. Quinones, *Meas. Sci. Technol.*, **17**, 2869 (2006).

[2] J. Li, J. P. Longtin, S. Tankiewicz, A. Gouldstone, S. Sampath, *Sensors and Actuators A*, **133**, 1 (2007).

[3] B. F. Conaghan, P. H. Kydd, D. L. Richard, "High conductivity inks with low minimum curing temperatures", *U. S. Patent* 766288 (2004).

[4] J. Mei, "Formulation and Processing of conductive inks for inkjet printing of electrical components," *PhD Thesis*, University of Pittsburgh (2004).

[5] J. Perelaer, B-J de Gans, U. S. Schubert, *Adv. Mater.*, **18**, 2101-2104 (2006).

[6] F. Teng, R. W. Vest, *IEEE Transactions: components Hybrids and Manufacturing Technology*, **12**(4), 545-549 (1987).

[7] C. J. Curtis, T. Rivkin, A. Miedaner, J. Alleman, J. Perkins, L. Smith, D. S. Ginley, *Mat. Res. Soc. Symp. Proc., Vol. 730,* (2002).

[8] C-A Lu, P. Lin, H-C. Lin, S-F. Wang, *Jap. J. Appl. Phys.*, **46**(1), 251–255 (2007).

[9] M. Nakamoto, M.Yamamoto, Y. M. Kashiwagi, H. Kakiuchi, T. Tsujimoto, Y.Yoshida, *Proc. of 15th Microelectronics Symp.*, 241-244 (2005).

Mater. Res. Soc. Symp. Proc. Vol. 1129 © 2009 Materials Research Society

Microstructure Characterization of thin structures after deformation

Sabine Weiß, Tim Schnauber, Alfons Fischer
Institute of Product Engineering, Materials Science and Engineering,
University Essen-Duisburg, D-47057 Duisburg, Germany

ABSTRACT

Microscale components are used for a variety of applications. Miniaturized machines, tools for minimal invasive surgery and coronary stents are only some examples in the multidisciplinary field of medicine and metallurgy. In either case, the dimension of these types of hardware is far below the size of conventional test specimens. However, construction of those micromechanical components is still based on conventional macroscopic test data. But in crystalline samples a size effect of mechanical properties can be observed if the grain size approaches the dimension of the cross section. Experimental results revealed that there are large differences in the mechanical behavior of the material between single- and polycrystalline test specimens [1-3]. Corresponding computer simulations demonstrated that the properties of these so called oligocrystals can neither be considered as single- nor as polycrystalline behavior [4, 5]. Nevertheless, there is only few data available, which is based on investigations with small metallic samples. The deformation behavior of thin wires has been investigated with focus on the development of microstructure and microtexture during deformation. Typical size effects like differences in the inner and outer part of the structure were observed. Grain rotations which lead to orientation changes within the grains occur. Sometimes the deformation of a whole structure is concentrated on just one or only very few grains. Further investigations will demonstrate the connection between wire experiments and microscale components in use.

INTRODUCTION

During the deformation process of microscale components certain characteristics occur, which differ much from known deformation structures of polycrystalline materials. Depending on the material higher or lower strength was obtained for tension of wires compared to bulk specimens. However, for wires always a lower ductility was determined. Due to the small dimension of the wires the material often has an oligocrystalline structure (only a few grains distributed over the cross section and most of the grains are in contact to the materials surface). Deformation of these structures leads to size effects [1-3]. With regard to the importance of the orientation parameter the Electron BackScatter Diffraction (EBSD) technique has been used to compare the crystallographic orientation of the grains after several different deformation states. The present study starts with a comparison of the mechanical properties of conventional test specimens and thin wires made of 316L-type austenitic steel. After stepwise tensile deformation microstructure and microtexture development of the wires were determined and related to the differences in mechanical behavior. Some wires were investigated after multiaxial bending. Microstructure and microtexture were analyzed as well. Further investigations focused on balloon dilated stents. The influence of such deformation on the microstructure of the stents was investigated and compared to the results of the wire experiments. The present study will give a more comprehensive understanding of the structure property relationship in such thin structures with focus on the deformation behaviour of coronary stents.

EXPERIMENT

For comparison of the mechanical properties of bulk material and wires standard 316L type CrNiMo-steel specimens (e.g. 1.4441) as well as commercially available 316L type wires (ø 100 µm) were deformed in axial tensile tests. The resulting stress strain curves revealed that the wires of the same material do not only show a lower strength but also a much lower ductility (only one fourth) than the bulk material.

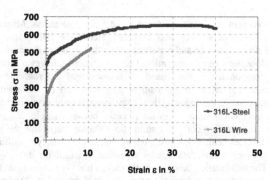

Figure 1: Typical tensile curves of bulk specimens and wires of 316L type austenitic steel

To analyze the deformation behavior of the wires (ø 100 µm, 200 µm and 300 µm) stepwise tensile tests inside a scanning electron microscope were chosen. A tensile stage (Raith GmbH, Dortmund, Germany) installed inside a scanning electron microscope DSM 940A (Zeiss, Oberkochen, Germany) was used for tensile tests. For bending deformation the wires were deformed around a guide. Because of the small radii no orientation determination of a representative area is possible directly on the surfaces. Therefore the wires were ground carefully prior to deformation to reach a small flat surface area for orientation determination. Commercially available coronary artery stents of the same material were deformed by means of balloon dilation. All specimens were electrochemically polished prior to deformation but without additional preparation after deformation. Microstructural characterization of the samples was carried out by means of scanning electron microscopy as well as single grain orientation determination EBSD in order to reveal microtextural alterations during deformation. A scanning electron microscope Gemini 1530 (Zeiss, Oberkochen, Germany) with EBSD system Crystal (Oxford Instruments, Wiesbaden, Germany) and calculation software Channel 5 from HKL-Technology (Hobro, Denmark) was used. An accelerating voltage of 20 kV and a working distance of 24 mm were applied for obtaining single grain orientations as well as orientation maps.

DISCUSSION

Tensile Test: Microstructure and microtexture of thin 316L wires (ø 200 μm) after stepwise tensile deformation were investigated. Figure 2 shows representative micrographs and orientation mappings after 5 %, 10 % and 15 % deformation by tensile stresses. After 5 % deformation the grains in such an oligocrystalline structure become visible on the former electrochemically polished surface. Increasing deformation is combined with increasing structuring of the sample surface. Also grain rotation takes place. Twins within the center of the micrographs rotate towards 45° to the stress axis. Surface roughening due to rotation of grains out of the projection plane becomes obvious.

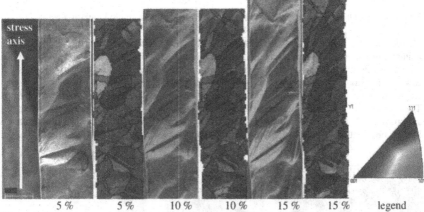

5 %   5 %   10 %   10 %   15 %   15 %   legend

Figure 2: Microstructure and orientation mappings of 316L wire after 5%, 10% and 15% tension

The orientation data were represented as mappings using the inverse pole figure colouring parallel to the longitudinal axis (y-axis) of the wire. Each orientation was represented by a special colour. The microtexture could exactly be related to the microstructure appearance. Most of the grain boundaries visible in the orientation mapping also occur as traces in the micrograph. As available from the legend most of the grains were oriented in <111> direction with respect to the wire axis. This type of texture is a typical drawing texture and despite of recrystallization and annealing after drawing this texture does not completely disappear. In this example with lots of similar oriented <111> grains only grain rotation but nearly no orientation changes could be observed. Tensile tests of single crystals identified the <111> orientation as a very "hard" and stable orientation [6]. Six independent slip systems can be active during deformation of a <111> oriented crystal. Therefore such a crystal can stand high stresses but only shows a relatively low ductility. The present results also indicate that some observations made with single crystals are likely to be valid for oligocrystalline structures as well.

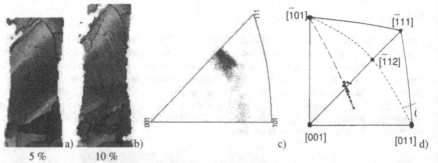

5 %   10 %

Figure 3: a, b) Orientation mapping c) inverse pole figure d) deformation path of a single crystal

The mappings in figure 3 a) and 3 b) show the development of a large grain (filling up the whole diameter of the wire) with an orientation inside the stereographic standard projection (single slip oriented). The comparison of the mappings after 5 % and 10 % gross deformation reveals that this is concentrated within only one grain. While the whole wire was elongated only 5 %, this single grain was elongated more than 15 %. Other parts of the wire remained nearly undeformed. This is combined with an orientation change of the grain. The planes of the only active slip system rotated along the longitude of the Wulff net towards the metastable <112> orientation. Similar observations were obtained in tensile tests of single crystals for "weak" orientations too [6].

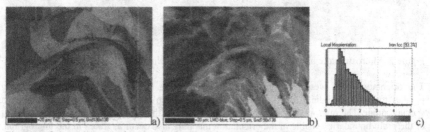

Figure 4: Orientation mapping, local misorientation mapping and legend of a bent 316L wire

Bending Test: Microstructure changes due to bending like extrusion of grains or grain rotation were also combined with changes of the crystallographic orientation. In figure 4 the orientation mapping (4 a) and the local misorientation mapping (4 b) were presented. To obtain a local misorientation mapping all orientation differences in a 7*7 grid around one measuring point were determined. Each orientation difference is color coded (4 c). Large orientation differences across the grain were available from color changes inside the grains in the orientation mapping. Internal stresses become obvious from the local misorientation mapping (4 a). Local misorientations up to 4° along the small distance of only 0.5 µm occur in the strongly bent parts on the outer radius of the wire (4 b).

352

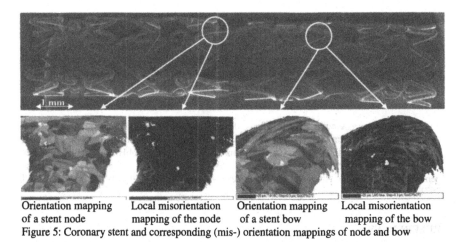

| Orientation mapping | Local misorientation | Orientation mapping | Local misorientation |
| of a stent node | mapping of the node | of a stent bow | mapping of the bow |

Figure 5: Coronary stent and corresponding (mis-) orientation mappings of node and bow

Stent Test: In figure 5 a coronary artery stent was presented. In the image circles highlighted the areas where orientation mappings were measured. Due to balloon dilation inhomogeneous deformation like a combination of tension, bending and torsion occurred. The magnitude of deformation varies within the different regions of the stent. A concentration of large orientation differences was found in the inner and outer region of the stent bow. Also available from the orientation mapping of the bow is a preferred deformation of large medium oriented grains in actually lower loaded parts of the bow. Much less deformation was found in nodes than in bows. The local misorientations can be considered as an indicator for internal stresses. Therefore the local misorientation mappings allow for describing the deformation situation in the different parts of the stent. Regions of maximum stress can be visualized. Based on current investigations the plastic deformation of stent structures follows partly the laws of polycrystalline deformation. We found typical accumulations of internal stresses in regions of the stents as expected. But the amount of stress was much lower than calculated for polycrystalline conditions [4]. As a possible reason for this behavior the ability of grain rotation and the orientation dependent extreme ductility of single grains in oligocrystalline structures could be identified.

CONCLUSIONS

Resulting deformations of oligocrystalline structures were analyzed by means of scanning-electron microscopy as well as local orientation analysis using electron backscatter diffraction. During the deformation process certain characteristics occur, which differ much from known deformation structures of polycrystalline materials. Typical size effects like differences in the inner and outer part of the structure can be observed. Grain rotation which lead to orientation changes within the grains occur. Sometimes the deformation of a whole structure is concentrated on just one or only a few grains. Even though the gross ductility of the entire microstructure decreases this is balanced by an extreme ductility of single grains. The present study helps to give a more comprehensive understanding of the structure property relationship in such thin structures. For this EBSD data can be used to describe the deformation behavior in inhomogeneously deformed oligocrystalline structures. The measurements can help in developing failure criteria for existing and future stent designs by using the orientation information (and probably grain shape) as input-data for simulations. The deformation model to be evolved can afterwards be tested and verified by means of laboratory experiments.

ACKNOWLEDGMENTS

The authors would like to thank Prof. Dr. Uta Klement (Chalmers University Gothenburg, Sweden) for cooperation and Muhammed Umar Farooq for assistance in orientation determination. Thanks to Abbot Vascular Instruments Deutschland GmbH, Rangendingen for the stents. In addition we are in debt to the MWF-NRW and the "Deutsche Forschungs-gemeinschaft" (DFG) for financial support under contracts Fi451/9-1, 9-2, WE 2671/1-3 and WE 2671/2-1.

REFERENCES

1. 1 Li Y., Laird C; "Cyclic response and dislocation structures of AISI 316L stainless steel. Part 1: single crystals fatigued at intermediate strain amplitude." Material Science and Engineering A 186 (1994), p. 65.

2. Li Y., Laird C; "Cyclic response and dislocation structures of AISI 316L stainless steel. Part 2: polycrystals fatigued at intermediate strain amplitude." Material Science and Engineering A 186 (1994), p. 87.

3. Dehm G., Motz C., Scheu C., Clemens H., Mayrhofer P.H., Mitterer C.; "Mechanical Size Effects in Miniaturized and Bulk Materials." Advanced Engineering Materials 8 (2006) No. 11, p. 1033-1045.

4. Stolpmann J., Brauer H., Stracke H.J., Erbel R., Fischer A.: "Practicability and Limitations of Finite Element Simulation of the Dilatation Behavior of Coronary Stents." Mat.-wiss. u. Werkstofftechnik 34: (2003) p. 736-745.

5. Murphy B.P., Savage P., McHugh P.E., Quinn D.F. "The stress-strain behavior of coronary stent struts is size dependent." Annals of Biomedical Engineering 31 (2003), p. 686-691.

6. Gottstein G.: „Physikalische Grundlagen der Materialkunde." Berlin u.a., Springer Verlag (1998)

AUTHOR INDEX

SUBJECT INDEX

Printed in the United States
By Bookmasters